Communications
in Computer and Information Science　　1133

Commenced Publication in 2007
Founding and Former Series Editors:
Phoebe Chen, Alfredo Cuzzocrea, Xiaoyong Du, Orhun Kara, Ting Liu,
Krishna M. Sivalingam, Dominik Ślęzak, Takashi Washio, Xiaokang Yang,
and Junsong Yuan

More information about this series at http://www.springer.com/series/7899

Cheikh Thiecoumba Gueye ·
Edoardo Persichetti · Pierre-Louis Cayrel ·
Johannes Buchmann (Eds.)

Algebra, Codes and Cryptology

First International Conference, A2C 2019
in honor of Prof. Mamadou Sanghare
Dakar, Senegal, December 5–7, 2019
Proceedings

 Springer

Editors
Cheikh Thiecoumba Gueye
Cheikh Anta Diop University
Dakar, Senegal

Pierre-Louis Cayrel
Jean Monnet University
Saint-Etienne, France

Edoardo Persichetti (iD)
Florida Atlantic University
Boca Raton, FL, USA

Johannes Buchmann
Technical University of Darmstadt
Darmstadt, Germany

ISSN 1865-0929   ISSN 1865-0937 (electronic)
Communications in Computer and Information Science
ISBN 978-3-030-36236-2   ISBN 978-3-030-36237-9 (eBook)
https://doi.org/10.1007/978-3-030-36237-9

This Springer imprint is published by the registered company Springer Nature Switzerland AG
The registered company address is: Gewerbestrasse 11, 6330 Cham, Switzerland

Preface

This volume contains the papers accepted for presentation at the First International Conference on Algebra, Codes and Cryptology (A2C 2019), in honor of Prof. Mamadou Sangharé from UCAD, Dakar, Senegal. A2C 2019 covered various topics including non-associative algebra, non-commutative algebras, coding theory, cryptology, and information security.

The first aim of this conference was to pay homage to Prof. Mamadou Sangharé for his valuable contribution in teaching and disseminating knowledge in algebra in Senegal since 1988.

The second aim of the conference was to provide an international forum for researchers from academia and practitioners from industry, from all over the world to discuss all forms of cryptology, coding theory, and information security.

The initiative of organizing A2C was started by the UCAD team and performed by an active team of researchers from all around the globe. The conference was organized in collaboration with the International Association for Cryptologic Research (IACR) and the proceedings were published in Springer's CCIS series.

The A2C Program Committee consisted of 41 members. There were 35 papers submitted to the conference. Each paper was assigned to at least three members of the Program Committee and refereed anonymously. The review process was challenging and the Program Committee was aided by reports from nine external reviewers. After this period, 14 papers were accepted. Authors then had the opportunity to update their papers. The present proceedings include all of the revised papers. We are indebted to the members of the Program Committee and the external reviewers for their diligent work.

The conference was honored by the presence of the invited speakers: L'Moufadal Ben Yakoub, Consuelo Martinez Lopez, Sylvain Guilley, and Claude Carlet. They gave talks on various topics in cryptology, algebra, and coding theory.

Finally, we heartily thank all the Local Organizing Committee members, headed by Dr. Leila Mesmoudi, all sponsors, particularly the Commission Nationale de Cryptologie, the CEA-MITIC, and the Foundation Alexander von Humboldt, and everyone who contributed to the success of this conference. We would like to thank Dr. Ousmane Ndiaye for his precious help in accelerating the schedule for writing the proceedings. We are also thankful to Springer for their help in producing the proceedings.

November 2018

Cheikh Thiecoumba Gueye
Pierre-Louis Cayrel
Johannes Buchmann
Edoardo Persichetti

Organization

General Chair

Gueye Cheikh Thiécoumba   Cheikh Anta Diop University, Senegal

Program Committee Chairs

Gueye Cheikh Thiécoumba	Cheikh Anta Diop University, Senegal
Buchmann Johannes	Technische Universität Darmstadt, Germany
Persichetti Edoardo	Florida Atlantic University, USA
Cayrel Pierre-Louis	Université Jean Monnet, France

Steering Committee

Barry Mamadou	Cheikh Anta Diop University, Senegal
Ben Maaouia M.	Gaston Berger University, Senegal
Demba Toure Sidy	Cheikh Anta Diop University, Senegal
Deme Cherif	Alioune Diop University, Mbambey, Senegal
Diankha Oumar	Cheikh Anta Diop University, Senegal
Diouf Ismaila	Cheikh Anta Diop University, Senegal
Diouf Alassane	Cheikh Anta Diop University, Senegal
Fall Mouhamed M.	Chair in Mathematics and Its Applications at AIMS-Senegal, Senegal
Gueye Anta Niane	Cheikh Anta Diop University, Senegal
Mesmoudi Laila (Chair)	Cheikh Anta Diop University, Senegal
Ndao Babacar Alassane	Commission Nationale de Cryptologie, PR, Senegal
Ndiaye Ousmane	Cheikh Anta Diop University, Senegal
Sokhna Moustapha	Cheikh Anta Diop University, Senegal
Sow Demba	Cheikh Anta Diop University, Senegal
Sow Djiby	Cheikh Anta Diop University, Senegal
Tall Amadou (Co-chair)	Cheikh Anta Diop University, Senegal

Program Committee

Azizi Abdel Malick	Mohammed Premier University, Morocco
Baldi Marco	Università Politecnica delle Marche, Italy
Barreto Paulo	University of Washington, USA
Bellini Emanuele	DarkMatter LLC, UAE
Berger Thierry	Université de Limoges, France
Boudi Nadia	Mohammed V University of Rabat, Morocco

Boulagouaz Mhammed	King Khalid University, Saudi Arabia
Buchmann Johannes	Technische Universität Darmstadt, Germany
Carlet Claude	Université Paris 8 and Université Paris 13, France
Chou Tung	University of Osaka, Japan
Cheng Chen-Mou	National Taiwan University, Taiwan
El Hajji Said	Mohammed V University of Rabat, Morocco
El Kaoutit Laiachi	Universidad de Granada, Spain
El Mamoun Souidi	Mohammed V University of Rabat, Morocco
El Mrabet Nadia	EMSE, France
El Yacoubi Nouzha	Mohammed V University of Rabat, Morocco
Futorny Vyacheslav	University of São Paulo, Brazil
Goodearl Ken	California University Santa Barbara, USA
Gorla Elisa	Université de Neuchâtel, Switzerland
Grosso Vincent	Université Jean Monnet, France
Guenda Kenza	University of Science and Technology of Algiers, Algeria
Gueron Shay	University of Haifa, Israel
Guilley Sylvain	Télécom ParisTech, France
Hirose Shoichi	University of Fukui, Japan
Loidreau Pierre	Université de Rennes, France
Lopez Ramos Juan Antonio	Universidad de Almería, Spain
Marcolla Chiara	DarkMatter LLC, UAE
Marquez Irene	Universidad de La Laguna, Spain
Martinez Lopez Consuelo	Universidad de Oviedo, Spain
Martinez Moro Edgar	Universidad de Valladolid, Spain
Maux Pierrick	Université Catholique de Louvain, Belgium
Mercedes Siles Molina	Universidad de Malaga, Spain
Misoczki Rafal	Intel Labs, USA
Nitaj Abderrahmane	Université de Caen Normandie, France
Otmani Ayoub	Université de Caen Normandie, France
Sangare Daouda	Université Nangui Abrogoua, Ivory Coast
Santini Paolo	Università Politecnica delle Marche, Italy
Shestakov Ivan	University of São Paulo, Brazil
Sol Patrick	Télécom ParisTech, France
Tour Saliou	Université Internationale de Grand-Bassam, Ivory Coast
Zelmanov Efim	UC San Diego, USA

Additional Reviewers

Alessandro Barenghi
El Yacoubi Nouzha
Florian Luca
Jose Antonio Alvarez Bermejo
Massimo Caboara

Matteo Bonini
Stefan Erickson
Simona Samardjiska
Violetta Weger

Biography

Professor Mamadou Sangharé was born in 1951 in Tatene Bambara, a village in Thiès. He has studied in a Coranic school in Medina, a district of Dakar, Senegal.

In the early 70's, he went ot Kadi Ibn Aarabi High School in Morocco, to study AlFiqa, the science of Islam.

He obtained in 1978 his baccalaureate in Experimental Sciences. Subsequently, he was enrolled in the Mathematics department of the Faculty of Sciences of the University Mohamed V. of Rabat. Mamadou Sangharé was among the best students at the University Mohamed V. In 1983 he obtained his Bachelor degree in Pure Mathematics then his Diploma of Advanced Studies in Mathematics. In 1986, Mamadou Sangharé defended his thesis with honors "very honorable with congratulations from the jury". In 1986 Sangharé was recruited as an assistant professor at the Faculty of Sciences of Rabat.

In 1988, he decided to return to his home country as a lecturer-researcher (assistant) at the Faculty of Science of the University Cheikh Anta Diop of Dakar.

In 1998, Professor Mamadou Sangharé created a research group at the Faculty of Sciences and Techniques of UCAD, named "Laboratoire d'Algebre, de cryptologie, de Géometrie Algébrique et Applications (LACGAA)".

Professor Mamadou Sangharé has advised more than 20 Ph.Ds. He has published more than fifty papers to international journals and seven books. Mamadou Sangharé is an outstanding Professor, emeritius. He is a founding member of one of the most prestigious institute of sciences in Senegal: AIMS-Senegal, where he was president for 4 years. In 2014, he received the highest state distinction: Chevalier de l'ordre national du lion.

Among the administrative position he has occupied:

- President of the Senegalese Mathematical Society
- Member of the National Academy of Sciences and Technics of Senegal
- Chair of the Innovation and Research Commission of the African Mathematical Union since 2013
- President and founding member of the African Institute of Mathematical Sciences (AIMS)-Senegal for the period 2010–2015
- Director of the Doctoral School of Mathematics and Computer Science of the Diop University Cheikh Anta of Dakar For the period 2008–2014
- Director of the LACGAA research group since 1998
- Focal point of the UMMISCO-UCAD, the Mathematical Modeling and Computational Complex Systems Unit (IRD-UCAD Project) since 2006
- President of the National Commission for Mathematics Olympiads, since 2009
- Representative in Senegal since 2007 of the International Commission for Mathematics Education (ICMI) of the International Mathematical Union (IMU)
- Head of the Department of Mathematics and Computer Science of the FST-UCAD for several years
- Coordinator of the African Virtual University (AVU) in Senegal from 1996 to 2000
- Director General of Higher Education for the period 2015–2017

- Inspector General of Education and Training since 1998 (IGEF)
- Director of the IREMPT of UCAD, a Research Institute for Mathematics, Physics and Technology Education, from 1992 to 1999

Contents

Non-associative and Non-commutative Algebra

Hopficity of Modules and Rings (Survey)

L'Moufadal Ben Yakoub[✉]

Faculté des Sciences, Université Abdelmalek Essaâdi, Tétouan, Maroc
benyakoub@hotmail.com

Abstract. Under the impulse of an elementary result that characterizes the finite dimensional vector spaces (a linear application is injective if, and only if it is surjective) and partial results which are already put in place on the commutative groups by R.A. Beaumont (1945), P. Hill and C. Megibben (1966) and P. Crawly (1968). Then, for finitely generated modules over commutative rings by J. Strooker (1966), and independently by W.V. Vasconcelos (1969–1970). Finally, towards the end of the sixties, for noetherian and artinian modules by P. Ribenboim.

In the beginning of eighties, A. Kaidi and M. Sangharé introduced the concept of modules satisfying the properties (I), (S) and (F). We say that an A-module M satisfies the property (I) (resp., (S)), if each injective (resp., surjective) endomorphism of M is an automorphism of M, and we say that M satisfies the property (F), if for each endomorphism f of M there exists an integer $n \geq 1$ such that $M = Im(f^n) \oplus Ker(f^n)$. In 1986, V. A. Hiremath introduced the concept of Hopfian modules to designate modules satisfying the property (S).

A bit later, K. Varadarajan introduced the notion of co-Hopfian modules to designate modules satisfying the property (I). Hopficity has been studied in many categories as abelian groups, rings, modules and topological spaces. In the context of the hopficity of rings and modules, K. Varadarajan studied the analogue of Hilbert's basic theorem quite extensively, that is, the transfer of Hopficity to certain polynomial extensions. He also examined various aspects of Hopkins-Levitzki's theorem related to Hopfian rings, co-Hopfian and its variants. This is a research topic, where different directions are discussed.

The subject is also of interest to several international research teams in the context of other notions related to hopficity of modules and its relationships with other classes of larger modules. In this context, we give a survey of the different notions related to the hopficity of modules, the main results of such notions and its relationships with other classes of larger modules.

Keywords: Hopfian module · Co-hopfian module · Hopfian ring · Co-hopfian ring

C. T. Gueye et al. (Eds.): A2C 2019, CCIS 1133, pp. 3–38, 2019.
https://doi.org/10.1007/978-3-030-36237-9_1

1   Generality of Modules

1.1   Notations

Let A be an unitary associative ring and U a subset of A, M a right A-module and N a nonzero subset of M, $S = End_A(M)$ and F a nonzero subset of S. We define $\mathcal{R} = \{R_a$ with $a \in A$ and $R_a(x) = xa$, for every $x \in M\}$ (If A is a commutative ring, then $\mathcal{R}$ is a subring of S). We use the following notations:

$N \leq M \Leftrightarrow N$ is a submodule of M.

$N \trianglelefteq M \Leftrightarrow N$ is a fully invariant submodule of M ($f(N) \leq N$ for every $f \in S$).

$N \ll M \Leftrightarrow N$ is superfluous in M ($M \neq N + X$, for every proper submodule X of M).

$N \leq^{\oplus} M \Leftrightarrow N$ is a direct summand of M.

$N \leq^e M \Leftrightarrow N$ is an essential submodule of M ($N \cap L = 0 \Rightarrow L = 0$, for every $L \leq M$).

$r_A(N) = \{a \in A \mid xa = 0$, for every $x \in N\}$.

$l_M(U) = \{m \in M \mid mU = 0\}$ and if $a \in A$, we denote $l_M(\{a\}) = l_M(a) = \{m \in M : ma = 0\}$.

$r_M(F) = \{m \in M \mid f(m) = 0$, for every $f \in F\}$ and $l_S(N) = \{f \in S \mid f(N) = 0\}$.

1.2   Properties (C) and (D)

Let A be an unitary associative ring and M a right A-module. M is said to satisfy the properties:

- **(C2)** $\Leftrightarrow$ (for every $N \leq M$ and $L \leq^{\oplus} M$, we have: $N \cong L \Rightarrow N \leq^{\oplus} M$) $\Leftrightarrow$ (for every $f \in S$, $Ker(f) \leq^{\oplus} M \Rightarrow Im(f) \leq^{\oplus} M$). ([30], Lemma 4.1.1 and [31], Lemma 2.1).
- **(GC2)** $\Leftrightarrow$ (for every $N \leq M$, we have: $N \cong M \Rightarrow N \leq^{\oplus} M$) $\Leftrightarrow$ ($Im(f) \leq^{\oplus} M$, for every monomorphism $f \in S$) **(C2 $\Rightarrow$ GC2)** ([5], Lemma 2.1).
- **(D2)** $\Leftrightarrow$ (for every $N \leq M$ and $L \leq^{\oplus} M$, we have: $M/N \cong L \Rightarrow N \leq^{\oplus} M$) $\Leftrightarrow$ (for every $f \in S$, $Im(f) \leq^{\oplus} M \Rightarrow Ker(f) \leq^{\oplus} M$) $\Leftrightarrow$ (for every $f, g \in S$, such that $Im(g) \leq Im(f) \leq^{\oplus} M$, there exists $h \in S$ such that $g = f \circ h$ ([44], Lemma 2.1.5).
- **(GD2)** $\Leftrightarrow$ (for every $N \leq M$, we have: $M/N \cong M \Rightarrow N \leq^{\oplus} M$) $\Leftrightarrow$ ($Ker(f) \leq^{\oplus} M$, for every epimorphism $f \in S$) **(D2 $\Rightarrow$ GD2)** ([5], Lemma 3.3).

1.3   Dedekind-Finite Module [9, 48]

A ring is said to be Dedekind-Finite if every $a, b \in A$, $ab = 1 \Rightarrow ba = 1$. A module M is Dedekind-Finite if the ring S is Dedekind-Finite. We verify that M is Dedekind-Finite if and only if M is not isomorphic to a proper factor of M ([9], p. 15 and [52], Proposition 1.25).

1.4   Endo-Regular Module [33, 49, 50]

- An element $a \in A$ is said to be **Von Neumann Regular**, if there exists $b \in A$ such that $a = aba$. Moreover if $b \in U(A)$, we say that $a \in A$ is **Unite-Regular**. A ring A is said to be a **Von Neumann Regular** (resp. **Unite-Regular**) if, every element of A is Von Neumann Regular (resp. Unite-Regular). A module M is said to be a **Endo-Regular** (resp. **Endo-Unite-Regular**) if, the ring S is Von Neumann Regular (resp. Unite-Regular). We have:

  The ring A is Endo-Regular as right A-module if and only if, A is Von Neumann Regular. If A is Unite-Regular, then A is Dedekind-Finite ([20], Lemma 2.1).

  An element $f \in S$ is Von Neumann Regular if and only if $Ker(f)$ and $Im(f)$ are direct summands of M ([56], Lemma 27). Them M is Endo-Regular if and only if, $Ker(f)$ and $Im(f)$ are direct summands of M, for every $f \in S$ ([50], Theorem 1.1 and Proposition 2.3).

- A ring A is π-**Regular** (resp. **Unite-π-Regular**) if and only if every $a \in A$, there exists $b \in A$ (resp. $b \in U(A)$) and $n \in \mathbb{N}$ such that $a^n = a^n b a^n$. A module M is **Endo-π-Regular** (resp. **Endo-Unite-π-Regular**) if and only if, the ring S is π-Regular (resp. **Unite-π-Regular**). Then M is Endo-π-Regular if and only if for every $f \in S$ there exists $n \in \mathbb{N}$ such that $Ker(f^n)$ and $Im(f^n)$ are direct summands of M ([50], Proposition 2.19 and [69], Lemma 29).

- A ring A is **Strongly-Regular** if, for every $a \in A$ there exists $b \in A$ such that $a = a^2 b$. A module M is **Endo-Strongly-Regular** if, the ring S is Strongly-Regular. M is Endo-Strongly-regular if and only if $M = Ker(f) \oplus Im(f)$, for every $f \in S$ ([50], Proposition 2.22).

1.5   Morphic Modules [11–14, 34, 56, 57]

- An endomorphism $f \in S$ is called **Morphic** if $M/Im(f) \cong Ker(f)$. A module M is **Morphic** if, every endomorphism f of M is morphic. A ring A is left (right) **Morphic** if, A considered as left (right) A-module is morphic.

- An endomorphism $f \in S$ is called π-**Morphic** if there exists $n \in \mathbb{N}$ such that $M/Im(f^n) \cong Ker(f^n)$. A module M is called π-**Morphic** if, every endomorphism f of M is π-Morphic. Any Morphic module is π-Morphique and if M is a Fitting module, then M is π-Morphic ([34], Example 2.7).

An element $f \in S$ is Unite-Regular if and only if f is Regular and morphic ([56], Lemma 27). In particular if M is Endo-Unite-Regular, then M is morphic ([56], Example 28). We deduce that M is Endo-Unite-Regular if and only if M is Endo-Regular and morphic ([56], Lemma 27).

An element $f \in S$ is Unite-π-Regular if and only if, f is π-Regular and morphic. We deduce that M is Endo-Unite- π-Regularr if and only if M is Endo-π-Regular and morphic ([34], Theorem 2.14).

1.6   Modules on a Dedekind Domain

- Let A be a principal ring and M a finitely generated free A-module. If N is a submodule of M, then N is a finitely generated free A-module and there exists a base $(e_1, ..., e_n)$ of M and nonnul elements $d_1, ..., d_r$ of A, with $0 \leq r \leq n$, such that $d_1|...|d_r$ and $(d_1 e_1, ..., d_r e_r)$ a base of N. The freedom of N is not working as soon as A is not principal. This result doesn't also mean that N has a supplementary, it doesn't mean that we can complete a base of N into a base of M. At last, it is clair in general that for every ring A, a submodule of a finitely generated free module is not necessary free and a submodule of a finitely generated free module is not necessary finitely generated.
- Let A be a principal ring and M a finitely generated A-module. There exists the nonzero elements and non invertible $d_1, ..., d_s$ of A, with $s \geq 0$, such that $d_1|...|d_s$, and an integer $r \geq 0$, such that:

$$M \simeq A^r \oplus A/ < d_1 > \oplus ... \oplus A/ < d_s > .$$

The positive intergers r and s, and the d_i depend only to M. The latters are called the invariant factors of M. We deduce that M is free if and only if M is torsion free.

- Let A be an unitary commutative domain ring and K its fractional field. A A-submodule M of K is said to be **invertible** if there exists A-submodule N of K such that $M.N = A$, where $M.N$ is the product submodule generated by the product of elements of M and N. A **fractional ideal** of A is a subset of K of the form $d^{-1}J$ where $d \in A \backslash \{0\}$ and J an ideal of A. In other words, it is a A-submodule M of K such that there exists $d \in A \backslash \{0\}$ satisfying $d.M \subseteq A$. A such fractional ideal is said to be **principal** if it is generated (as A-module) by an element, in other words, if it is in the form $d^{-1}J$ where J is a principal ideal of A.

A fractional ideal of A is not always an ideal of A. In fact, the ideals of A are exactly, among its fractional ideals, those that are included in A. We remark immediately that every invertible A-submodule of K is a fractional ideal, the set of invertible fractional ideals forms an abelian group (for the product defined previously), where the neutral element is the ring A itself. Any invertible fractional ideal is finitely generated as an A-module, in other word, it is of the form $d^{-1}J$ with a J finitely generated ideal. In particular, any invertible ideal of A is finitely generated, if a fractional ideal F is invertible then its inverse (denoted F^{-1}) is A-submodule of K constituted by the elements k such that kF is included in A. In other word, F is invertible if and only if its product by that submodule is equal to A.

- A Dedekind domain is a integral domain ring, such that every nonzero fractional ideal is invertible. The following properties are equivalent:
  (i) A is a Dedekind ring,
  (ii) Every nonzero prime ideal of A is invertible,
  (iii) Every nonzero ideal of A is invertible,

(iv) A is a domain ring and every nonzero ideal of A is a product of maximal ideals,

(v) A is a domain ring and every nonzero ideal of A is a product of prime ideals.

Moreover, if A is a Dedekind ring, the decomposition of every ideal into a product of prime ideals is unique.

Finitely Generated Modules over a Dedekind Domain

- Let A be a Dedekind ring and M a finitely generated torsion A-module, then there exists a finite family of prime ideals $\mathfrak{p}_1, \mathfrak{p}_2, ...\mathfrak{p}_r$ of A and positive integer $n_1, n_2, ..., n_r \in \mathbb{N}^*$ such that [8]:

$$M \simeq (A/\mathfrak{p}_1)^{n_1} \oplus (A/\mathfrak{p}_2)^{n_2} \oplus ... \oplus (A/\mathfrak{p}_r)^{n_r}.$$

- Let A be a Dedekind ring and M a finitely generated torsion A-module and of rang $n \geq 1$, then there exists nonzero ideal $\mathfrak{a}$ of A such that [8]:

$$M \simeq A^{n-1} \oplus \mathfrak{a}.$$

- We can extend in Dedekind rings the description of finitely generated module on a principal ring. We show that every finitely generated module over a Dedekind ring A is isomorphic to a direct sum:

$$M \simeq A^r \oplus I \oplus (A/I_1 \oplus ... \oplus A/I_s),$$

where $I, I_1, ..., I_s$ are ideals of A verifying $A \supsetneq I_1 \supsetneq ... \supsetneq I_s \supsetneq (0)$. The fact that there is one ideal in the decomposition comes from the following result: if I and J are ideals of A, we have an isomorphism of A-modules $I \oplus J \simeq A \oplus IJ$. If A is not principal, then there exist finitely generated not free torsion free module. We have also every finitely generated module is of the form:

$$M = A^r \oplus I \oplus (A/\mathfrak{p}_1^{n_1} \oplus A/\mathfrak{p}_2^{n_2} \oplus .... \oplus A/\mathfrak{p}_s^{n_s}),$$

where I is a fractional ideal of A, the $\mathfrak{p}_i$ are prime ideals, the n_i and r are positive integers.

Divisible Modules over a Dedekind Domain

- A A-module D is said to be **divisible** if, $aD = D$ for every $a \in A\backslash\{0\}$. An abelian group G is divisible if and only if, G is an injective $\mathbb{Z}$-module ([15], Corollary 3.4.1).
- Every injective module is divisible ([15], Theorem 3.4) and if A is a Dedekind domain, then we have the reverse ([15], Theorem 5.6).
- Let D be a divisible A-module and N a submodule of D, then D/N is also a divisible A-module ([15], Lemma 3.1.1).

- Let $(D_\lambda)_{\lambda \in \Lambda}$ a family of A-modules, then ([15], Theorem 3.2):

$$\prod_{\lambda \in \Lambda} D_\lambda \text{ is divisible} \Leftrightarrow \bigoplus_{\lambda \in \Lambda} D_\lambda \text{ is divisible} \Leftrightarrow D_\lambda \text{ is divisible, for every } \lambda \in \Lambda.$$

- Let A Dedekind ring and D divisible module, we have ([45], Theorem 6):
  (a) If N is a submodule of a module M and f a morphism of N into D, then f exetends to M.
  (b) If D is divisible submodule of a A-module M, then D is direct summand of M.
- Let A a Dedekind domain and D a divisible A-module. then, D is s direct sum of a torsion module and torsion free modules ([53], Remark 3.1).
- Let A Dedekind domain and K its fractional field. Then, every divisible module is the direct sum of vector space over K (direct sum of copies of K) and $E(A/\mathfrak{p})$, for certaiin nonzero prime ideals $\mathfrak{p}$ ([53], Theorem 3.2). Then, every divisible D-module M is of the form:

$$M \simeq K^{(I)} \bigoplus \left(\bigoplus_{\mathfrak{p} \in \mathcal{P}} (E(A/\mathfrak{p}))^{I_\mathfrak{p}} \right),$$

where $\mathcal{P}$ is the set of nonzero prime ideals, I and $I_\mathfrak{p}$ are sets $((E(A/\mathfrak{p}))^{I_\mathfrak{p}}$ is called p^∞ type module).

1.7 Abelian Groups

Endomorphisms of Abelian Groups. We denote $\mathbb{Z}/n\mathbb{Z}$ the group of positive integers modulo n and if p is a prime number, $\mathbb{F}_p$ is the finite field with p elements ($\mathbb{F}_p = \mathbb{Z}/p\mathbb{Z}$). We denote by $\mathbb{Z}_{(p)}$ the group of rational numbers where the denominator is prime with p and $A_p = \{\frac{a}{p^n} \mid a \in \mathbb{Z} \text{ and } n \in \mathbb{N}\}$, it is a subgroup of $\mathbb{Q}$ and has $\mathbb{Z}$ as subgroup. Then the quotient group $A_p/\mathbb{Z} = \mathbb{Z}(p^\infty) = \mathbb{Z}[\frac{1}{p}]/\mathbb{Z}$, it is the p-group of Prufer and we have: $\mathbb{Q}/\mathbb{Z} = \oplus_{p \in P} \mathbb{Z}(p^\infty)$. The ring of positive integers p-adic $\mathbb{Z}_p$ is the set of elements $(x_n) \in \prod_{n \in \mathbb{N}} \mathbb{Z}/p^n\mathbb{Z}$ such that x_{n+1} has for image x_n by the natural application of $\mathbb{Z}/p^{n+1}\mathbb{Z}$ into $\mathbb{Z}/p^n\mathbb{Z}$. $\mathbb{Z}_p$ is an integral domain ring and its fractional field is denoted $\mathbb{Q}_p = \{\frac{a}{b} \mid a \in \mathbb{Z}_p \text{ and } b \in \mathbb{Z}_p \setminus \{0\}\}$, is called field of p-adic numbers.

For every ablian group G, let us denote $E(G) = End_{\mathbb{Z}}(G)$, the endomorphism ring of G. For example, we have: $E(\mathbb{Z}) \simeq \mathbb{Z}$, $E(\mathbb{Q}) \simeq \mathbb{Q}$, $E(\mathbb{Q}^n) \simeq M_n(\mathbb{Q})$ and $E(\mathbb{Z}/n\mathbb{Z}) = \mathbb{Z}/n\mathbb{Z}$, in particular if $E(\mathbb{Z}/p\mathbb{Z}) \simeq \mathbb{F}_p$. We deduce that: $E(\oplus_{p \in P} \mathbb{Z}/p\mathbb{Z}) \simeq E(\prod_{p \in P} \mathbb{Z}/p\mathbb{Z}) \simeq \prod_{p \in P} \mathbb{F}_p$, $E(\mathbb{Z}(p^\infty)) = \mathbb{Z}_p$ and $E(\mathbb{Q}/\mathbb{Z}) \simeq \prod_{p \in P} \mathbb{Z}_p$, where $\mathbb{Z}_p$ is the ring of positive integers p-adic and $\mathcal{P}$ is the set of prime numbers. We have the isomorphisms because each composant is invariant for every endomorphism.

In general, the endomorphism ring of the group $G = \oplus_{i=1}^n G_i$ is isomorphic to the ring of square matrixes (α_{ij}), where $\alpha_{ij} \in Hom(G_i, G_j)$ ([47], Theorem 1.1). For example:

$$E(\mathbb{Z} \oplus \mathbb{Z}(p^\infty)) \simeq \begin{pmatrix} \mathbb{Z} & 0 \\ \mathbb{Z}(p^\infty) & \mathbb{Z}_p \end{pmatrix} \text{ and } E(\mathbb{Z}/p^n\mathbb{Z} \oplus \mathbb{Z}(p^\infty)) \simeq \begin{pmatrix} \mathbb{Z}/p^n\mathbb{Z} & 0 \\ \mathbb{Z}/p^n\mathbb{Z} & \mathbb{Z}_p \end{pmatrix}.$$

Properties

- The group $G = \prod_{p \in \mathcal{P}} \mathbb{Z}/p\mathbb{Z}$ is Morphic, hence Dedekind-Finite. But, $G/T(G)$ is not Dedekind-Finite, hence $G/T(G)$ is not morphic. ([11], Example 18 and [12]).
- Let $a = (\bar{1}, \bar{1}, \bar{1}, ...) \in G = \prod_{p \in \mathcal{P}} \mathbb{Z}/p\mathbb{Z}$ and $H = \mathbb{Q} \oplus P$ where $P = P(G, a) = \{g \in G \mid ng \in \langle a \rangle \text{ for certain } n \in \mathbb{N}\}$. Then, $T(H)$ and $H/T(H)$ are morphic, but, H is not morphic ([11], Example 23 and [12]).
- Any injective indecomposable abelian group is isomorphic to $\mathbb{Q}$ or $\mathbb{Z}(p^\infty)$ for a certain prime number p. ([15], Theorem 5.4). Any abelian group G has a biggest divisible subgroup denoted $D(G)$ ([15], Theorem 3.5 and [64], p. 322), meaning a divisible subgroup which contains all the others. An abelian group G for which $D(G) = \{0\}$ is said to be **réduit**. Any abelian group G has a decomposition into direct sum $G = D(G) \oplus R$, where R is a reduit group ([15], Theorem 3.5 and [64], p. 322). Any divisible abelian group is, in a unique way, the direct sum of a (finite or infinite) family of groups where each group is a Prüfer group or a group isomorphic to additive group $\mathbb{Q}$ of rational numbers ([64], p. 323). More explicitly, for every divisible abelian group G, we have : $G \simeq T(G) \oplus G/T(G)$, $G/T(G) \simeq \mathbb{Q}^{(I)}, T(G) = \oplus_p T_p(G), T_p(G) \simeq (\mathbb{Z}(p^\infty))^{(I_p)}$ where $T(G)$ is a torsion subgroup and the $T_p(G)$ its primary component. Finally: $G \simeq \oplus_{p \in \mathcal{P}} (\mathbb{Z}(p^\infty))^{(I_p)} \oplus \mathbb{Q}^{(I)}$, the cardinal of I is the dimension of the vector space over $\mathbb{Q}$ where $G/T(G)$ is the additive group and the cardinal of I_p is the dimension of a vector space over F_p, whose subgroup $\{g \in G \mid pg = 0\}$ of $T_p(G)$, is the additive subgroup. Since any subgroup of a finitely generated abelian group is itself finitely generated ([64], p. 318) and neither $\mathbb{Q}$ nor any Prüfer group is finitely generated, it results that a finitely generated divisible abelian group is necessary zero (That can be proven directly, indeed: by recurrence on the cardinal n of a finite family of generator. The statement is obvious if that cardinal is zero. If a divisible abelian group G is generated by a finite family $(x_1, ..., x_n)$, with $n \geq 1$, then $G/ < x_n >$ is generated by $n - 1$ elements, and by hypothesis of recurrence on n, $G/ < x_n >$ is zero, hence G is monogene. It can be only monogene infinitely, because $\mathbb{Z}$ is not divisible. Hence G has a finite order n, and since G is divisible, $G = nG = \{0\}$).

2   The Hopficity of Modules

2.1   Hopfian and co-Hopfian Modules

Generalities

- A module M is said **Hopfian** (resp. **co-Hopfian**) if all surjective endomorphism (resp. injective endomorphism) of M is bijective. A ring A is said left **Hopfian** (resp. left **co-Hopfian**) if A is Hopfian (resp. co-Hopfian) as a left A-module. In de similarly way, we define right Hopfian rings (resp. right co-Hopfian rings).

- All notherian module (resp. artinian module) is Hopfian (resp. co-Hopfian) ([63], p. 41).
- Let M be a Hopfian or co-Hopfian A-module, then M is Dedekind-Finite (Reciprocity failure generally) ([9], Proposition 4.2).
- Let M be a π-Morphic A-module, then M is Hopfian and co-Hopfian ([34], Theorem 2.3 and 2.4).
- Let M be a morphic A-module, then M is Hopfian and co-Hopfian [13].
- The group $G = \mathbb{Z}/2\mathbb{Z} \oplus \mathbb{Z}/4\mathbb{Z}$ is Hopfian and co-Hopfian, but it is not morphic [11].
- All direct summand of an Hopfian (resp. co-Hopfien) module is Hopfian (resp. co-Hopfian). Let $M = \bigoplus_{i \in I} M_i$ where $(M_i)_{i \in I}$ be a family of A-modules, then:
  - ⋆ If M is Hopfian (resp. co-Hopfian) then each M_i is Hopfian (resp. co-Hopfian).
  - ⋆ If for all $i \in I$, M_i is totally invariant then M is Hopfian (resp. co-Hopfian) if and only if, for all $i \in I$, M_i is Hopfian (resp. co-Hopfian).
  - ⋆ If $M^{(I)}$ is Hopfian (resp. co-Hopfian) for some set I then I is finite.
- Let $M = \bigoplus_{i \in I} T_i^{(I_i)}$ where $(T_i)_{i \in I}$ is a family of pairwise non-isomorphic simple-modules. Then, M is Hopfian (resp. co-Hopfian) if and only if for each $i \in I$, I_i is finite.
- Let $N \trianglelefteq M$, if N and M/N are Hopfians (resp. co-Hopfians, Dedekind-finites), then M is Hopfian (resp. co-Hopfian, Dedekind-finite) ([9], Corollary 4.7).

The Hopficity of Injective Modules

- A right A-module N is dit Quasi-Principally Injective or Semi-Injective if for all $f \in End_A(N)$, each morphism g from $f(N)$ to N, extended to a morphism of N.

$$
\begin{array}{c}
N \\
g \uparrow \quad \nwarrow \\
0 \longrightarrow f(N) \longrightarrow N
\end{array}
$$

- N is said Pseudo-Quasi-Principally Injective or Pseudo-Semi-Injective if, for all $f \in End_A(N)$ nonzero, all monomorphism g from $f(N)$ to N can be extended to an endomorpphism of N.

$$
\begin{array}{c}
N \\
g \uparrow \quad \nwarrow \\
0 \longrightarrow f(N) \longrightarrow N \\
\uparrow \\
0
\end{array}
$$

Quasi-Injective ⇒ Semi-Injective ⇒ Pseudo-Semi-Injective ⇒ C2 ⇒ GC2.

- M is generalized Pseudo-Quasi-Principally-Injective if for all $f \in End_A(M)$ nonzero, there is $n \in \mathbb{N}^*$ such as $f^n \neq 0$ and all monomorphism g from $f^n(M)$ to M, can be extended to an endomorphism of M.

$$
\begin{array}{c}
M \\
g \uparrow \quad \diagdown \\
0 \longrightarrow f^n(M) \longrightarrow M \\
\uparrow \\
0
\end{array}
$$

Pseudo-Semi-Injective $\Rightarrow$ Genralized
Pseudo-Quasi-Principally-Injective $\Rightarrow$ GC2
([31], Proposition 2.18; [61], Theorem 3.3 and [77], Theorem 3.7)..

- Let M be a generalized Pseudo-Quasi-Principalement-Injective right A-module. We have ([77], Theorem 3.5):

$$
M \text{ is Dedekind-Finite} \Rightarrow M \text{ is co-Hopfian.}
$$

- Let M be a generalized Pseudo-Quasi-Principally-Injective right module then ([77], Corollary 3.6):

$$
M \text{ is Hopfian} \Rightarrow M \text{ is co-Hopfian.}
$$

The Hopficity of Projective Modules

- A A-module P is Quasi-Principally-Projective or Semi-Projective (or Image-Projective, according to [14]) if, for all $f \in End_A(P)$ nonzero and for all homomorphism $g : P \longrightarrow f(P)$, there is a endomomorphism h of P such as $f \circ h = g$.

$$
\begin{array}{c}
P \\
\diagup \quad \downarrow g \\
P \longrightarrow f(P) \longrightarrow 0
\end{array}
$$

- A module P is Quasi-Pseudo-Principally-Projective or Pseudo-Semi-Projective if for all endomorphism f of P and for all epimorphism $g : P \longrightarrow f(P)$, there is a endomorphism h of P such as $f \circ h = g$.

$$
\begin{array}{c}
P \\
\diagup \quad \downarrow g \\
P \longrightarrow f(P) \longrightarrow 0 \\
\downarrow \\
0
\end{array}
$$

Quasi-Projective $\Rightarrow$ Semi-Projective $\Rightarrow$ Pseudo-Semi-Projective $\Rightarrow$ D2 $\Rightarrow$ GD2.

- A A-module P is generalized Quasi-Pseudo-Principally-Projective or generalized Pseudo-Semi-Projective if, for all endomorphism nonzero f of P, there is $n \in \mathbb{N}^*$ such as $f^n \neq 0$ and for all epimorphisme $g : P \longrightarrow f^n(P)$, there is an endomorphism h of P such as $f^n \circ h = g$.

If P is generalized Quasi-Pseudo-Principally-Projective, then P verify **GD2**. Indeed, let f be a epimorphism of P, then there is $n \geq 1$ and $h \in \mathcal{S}$ such as: $f \circ (f^{n-1} \circ h) = f^n \circ h = id_P$. So, $Ker(f) \leq^\oplus P \Rightarrow P$ satisfy **GD2**, according to ([39], Proposition 1.1.6).

- Let A be a commutative ring and M a Semi-Projective A-module. We have ([58], Theorem 2.26):

$$M \text{ is Artinian} \Rightarrow M \text{ is Noetherian.}$$

- Let M be a Quasi-Principally-Projective A-module, then ([59], Corollary 2.20):

$$M \text{ is co-Hopfian} \Rightarrow M \text{ is Hopfian.}$$

- The following assertions equivalent ([37], Theorem 2.6.1):

  (i) All projective right finitely generated A-module is co-Hopfien.
  (ii) $\mathbb{M}_n(A)$ is a co-Hopfian let ring for all $n \geq 1$.

We deuce that if A is commutative ring, then:

A is co-Hopfian $\Leftrightarrow$ all projective right finitely generated A-module is co-Hopfian.

- If A is a commutative ring, all finitely generated A-module is Hopfian ([72], Proposition 1.2).
- Let A be a commutative ring. The following assertions are equivalent [73]:
  (i) All prime ideal is maximal.
  (ii) All finitely generated A-module is co-Hopfian.
- Let A be an integral domain. Let M be a free A-module. M is co-Hopfian if and only if, M is isomorphic to A^n where A is a division ring ([41], Proposition 5.11).
- Let A be a commutative ring. The A-module A^n is co-Hopfian if and only if, A is co-Hopfian ([41], Proposition 5.13).

2.2   Generalized Hopfian Modules and Weakly-co-Hopfian Modules

- A A-module M is said **generalized Hopfian**, if for all subjective endomorphism f of M, the kernel of f is **superfluous** on M ($Ker(f) \ll M$). The ring A is said right **Generalized Hopfian** if A is **Generalized Hopfian** as left A-module. We have ([26], Corollary 1.4):

$$M \text{ is Hopfian} \Rightarrow M \text{ is Generalized Hopfian} \Rightarrow M \text{ is Dedekind-Finite.}$$

- M is said **Weakly-co-Hopfian** if all injective endomorphism f of M is essential $(Im(f) \leq^e M)$. The unitary ring A is left **Weakly-co-Hopfian** if A is **Weakly-co-Copfian** as a left A-module.
- Let M be a A-module, then ([29], Proposition 1.4):

M is co-Hopfian $\Rightarrow$ M is Weakly-co-Hopfian $\Rightarrow$ M is Dedekind-Finite.

- Let M be a A-module Quasi-Principally-Injective, we have ([16], Proposition 4.7):

M is Weakly-co-Hopfian $\Rightarrow$ M is co-Hopfian.

- Let M be a Semi-Projective right A-module, we have ([58], Proposition 2.12 and 2.13):

M is co-Hopfian $\Rightarrow$ M is Generalized Hopfian.
M is Generalized Hopfian $\Leftrightarrow$ M is Hopfian $\Leftrightarrow$ M is Dedekind-Finite.

- The following properties are equivalent ([26], Theorem 1.1):
  (i) M is Generalized Hopfian.
  (ii) M is Dedekind-Finite and $Ker(f)$ is superfluous or a direct summand of M, for all epimorphism f of M.
- The following properties are equivalent ([29], Theorem 1.1):
  (i) M is Weakly-co-Hopfian,
  (ii) M is Dedekind-Finite and $Im(f)$ is essential a proper direct summand of M, for all monomorphism f of M.
- Let M be a A-module verifying, for all endomorphism f of M, there is $n \in \mathbb{N}$ such as: $Ker(f^m) \cap Im(f^n) \ll M$ and $Ker(f^n) + Im(f^n) \leq^e M$. Then M is Weakly-co-Copfian and Generalized Hopfian ([26], Proposition 1.23).

2.3   Generalized co-Hopfian Module and Weakly-Hopfian

- A A-module M is **Generalized co-Hopfian** if all essential monomorphism f of M $(Im(f) \leq^e M)$ is an isomorphism of M. The module M is called **Weakly-Hopfian** if all superfluous epimorphism f of $M (Ker(f) \ll M)$ is an isomorphism of M. The ring A is left **Generalized co-Hopfian** (resp. **left Weakly-Hopfian**) if A is **Generalized co-Hopfian** (resp. **Weakly-co-Hopfian**) as a left A-module ([74], p. 1456):.
- If M verify the c.c.d of essential sub-modules $\Rightarrow$ M is Generalized co-Hopfian ([74], Theorem 2.4).
- If M verify the c.c.a of superfluous sub-modules $\Rightarrow$ M is Weakly-Hopfian ([74], Theorem 3.5).
- If M verify GC2, then ([5], Proposition 2.5 and [31], Lemma 2.15):

Hopfian $\Rightarrow$ Generalized Hopfian $\Rightarrow$ Dedekind-Finite $\Leftrightarrow$ Weakly-co-Hopfian $\Leftrightarrow$ co-Hopfian.

- If M verify **GD2**, then ([5], Proposition 3.6):

co-Hopfian $\Rightarrow$ Weakly-co-Hopfian $\Rightarrow$ Dedekind-Finite $\Leftrightarrow$ Generalized Hopfian $\Leftrightarrow$ Hopfian.

2.4   Strongly-Hopfian and Strongly-co-Hopfian Modules

- A A-module M is **Strongly-Hopfian** or **Generalized Noetherian**, if for all endomorphism f of M the growing sequence: $Ker(f) \subseteq Ker(f^2) \subseteq \cdots \subseteq Ker(f^n) \cdots$, is stationary. The module M is called **Strongly-co-Hopfian** or **Generalized Artinian**, if the decreasing sequence: $Im(f) \supseteq Im(f^2) \supseteq \cdots \supseteq Im(f^n) \supseteq \cdots$, is stationary, for all endomorphism f of M [23, 38]. The ring A is said **left Strongly-Hopfian** (resp. **left Strongly-co-Hopfian**) if A is Strongly-Hopfian (resp. Strongly-co-Hopfian) as a left A-module. In similar way we define right Strongly-Hopfian (resp. right Strongly-co-Hopfian) ring. The left (resp. right) Strongly-co-Hopfian rings were introduced by Kaplansky [46], Azumaya calls it Strongly π-Regular rings and prouve that for such rings A: for all $x \in A$, there exists $y \in A$, $m \geq 1$ such as $x^m = yx^{m+1}$ and $yx = xy$ then A is left Strongly-co-Hopfian if and only if A is right Strongly-co-Hopfian. All left Strongly-Hopfian (resp. Strongly-co-Hopfian) ring is Dedekind-Finite. All integral domain Strongly-co-Hopfian is a field. the jacobson radical of all Strongly-co-Hopfian is a nilideal ([37], Proposition 3.3.4 and 3.3.4, [38]). For all $x \in A$, let us consider $l(x) = l_A(x) = Ker(R_x)$ where R_x is a right endomorphism of multiplication of A, we have: A is left Strongly-Hopfian if and only if for all $x \in A$ there exists $n \geq 1$ such as $l(x^n) = l(x^{n+1})$. We deduce that integral domain is strongly-Hopfian. ([37], Proposition 3.3.2).

- Let us consider $f \in S = End_A(M)$. If there existe $n \geq 1$ such as $Im(f^n) = Im(f^{n+1})$ (resp. $Ker(f^n) = Ker(f^{n+1})$), then for $k \geq n$, $Im(f^k) = Im(f^{k+1})$ (resp. $Ker(f^k) = Ker(f^{k+1})$). We deduce that ([37], Theorem 3.2.4 and 3.2.5):

  M is Strongly-Hopfian $\Leftrightarrow$ for all $f \in S$, there exists $n \geq 1$ such as: $Im(f^n) \cap Ker(f^n) = (0)$,

  M is strongly-co-Hopfian $\Leftrightarrow$ for all $f \in S$, there exists $n \geq 1$ such that: $M = Im(f^n) + Ker(f^n)$

- Let M be a module. We have ([23], Theorem 2.2, [37], Theorem 3.2.6 and 3.2.7, [38]):

  (i) All noetherian (resp. artinian) module is Strongly-Hopfian (resp. Strongly-co-Hopfian).

  (ii) All Strongly-Hopfian (resp. Strongly-co-Hopfian) module is Hopfian (resp. co-Hopfian).

  (iii) All module of finite length is a Fitting module.

  (iv) If M is Strongly-Hopfian and Strongly-co-Hopfian $\Leftrightarrow$ M is a Fitting module.

  (v) All Fitting module is Hopfian and co-Hopfian.

  (vi) All direct summand of a Strongly-Hopfian (resp. Strongly-co-Hopfian) module is Strongly-Hopfian (resp. Strongly-co-Hopfian).

- Let N be a totally invariant sub-module of a module M. If N and M/N are Strongly-co-Hopfian (resp. Strongly-Hopfians), then M is Strongly-co-Hopfien (resp. Strongly-Hopfian) ([37], Proposition 3.4.2). There exists a finitely generated Fitting module admitting a sub-module which is no Strongly-Hopfian no Strongly-co-Hopfian and a quotient module which is not Strongly-Hopfian

([37], Remark 3.4.3). Similarly, there exists a Strongly-co-Hopfien module admitting a quotient module which is not Strongly-co-Hopfian ([37], Example 3.4.4).

- Let M be a A-module. If M is Strongly-Hopfian (resp. Strongly-co-Hopfian), then $\mathcal{S} = End_A(M)$ is right (resp. left) Strongly-Hopfian ([37], Proposition 3.4.5). The reciprocal is not true in general ([37] Remark 3.4.6). Indeed: if $M = \mathbb{Z}(p^\infty)$, then $\mathcal{S}$ is isomorphic to the ring of p-adic integers $\mathbb{Z}_p$ which is an integral domain therefore Strongly-Hopfian, but M is not Strongly-Hopfian. On the other side, if we consider $M = \mathbb{Z}$ then $\mathcal{S} \simeq \mathbb{Z}$ is Strongly-Hopfian, but M is not Strongly-co-Hopfian.
- Let M be a module, if M is Quasi-Injective (resp. Quasi-Projective) and Strongly-Hopfian (resp. Strongly-co-Hopfian), then all sub-module (resp. quotient module) of M is Strongly-Hopfian (resp. Strongly-co-Hopfian) ([37], Theorem 3.4.1).
- Let M be a module then ([3] and [37], Proposition 3.4.7):

For all $f \in \mathcal{S}$, there exists $n \in \mathbb{N}$ such as: $M = Ker(f^n) \oplus Im(f^n)$

$\Updownarrow$

$\mathcal{S}$ is Strongly-π-Regular $\Leftrightarrow M$ is a Fitting module $\Leftrightarrow M$ Strongly-Hopfian and Strongly-co-Hopfian

$\Updownarrow$

For all $f \in \mathcal{S}$, there exists $g \in \mathcal{S}$ and $n \in \mathbb{N}$ such as: $f^n = gf^n = f^n g$

- Contrary to Fitting modules, the Strongly-Hopfian (resp. Strongly-co-Hopfian) modules can not be caracterized its ring of endomorphisms. Indeed, let us consider the ring of p-adic integers $\mathbb{Z}_p$ as a $\mathbb{Z}$-module. All endomorphism of $\mathbb{Z}_p$ is in the form of $f : x \mapsto ax$ where $a \in \mathbb{Z}_p$, therefore $End_\mathbb{Z}(\mathbb{Z}_p) \simeq \mathbb{Z}_p$. Clearly, $\mathbb{Z}_p$ is not co-Hopfian, then $\mathbb{Z}_p$ is not Strongly-co-Hopfian, but $\mathbb{Z}_p$ is Strongly-Hopfian. On other side $\mathbb{Z}(p^\infty)$ is Strongly-co-Hopfian, but not Strongly-Hopfian and the both groups $\mathbb{Z}_p$ and $\mathbb{Z}(p^\infty)$ have same ont ring of endomorphisms.
- The free strongly-co-Hopfian modules are A^n where $n \in \mathbb{N}$ such that $M_n(A)$ is Strongly-π-regular. The free Strongly-Hopfian modules are A^n where $n \in \mathbb{N}$ such as $M_n(A)$ is left Strongly-Hopfian ([37], Theorem 3.4.9 and 3.4.10). Let R be a free algebra on a field K generated by $\{x_i,\ 1 \leq i \leq 8\}$, with the relation $PQ = I_2$ where $P = \begin{pmatrix} x_1 & x_3 \\ x_2 & x_4 \end{pmatrix}$ and $Q = \begin{pmatrix} x_6 & x_7 \\ x_8 & x_5 \end{pmatrix}$. Then R isleft Strongly-Hopfian, but for all $n \geq 2$, R^n is not strongly-Hopfian as left R-module ([37], Example 3.4.13).
- Let M_1 and M_2 be two modules such as $Hom(M_2, M_1) = \{0\}$, we have: ([37], Theorem 3.4.11).
  (i) If M_1 and M_2 are co-Hopfian, then $M_1 \oplus M_2$ is co-Hopfian.
  (ii) If M_1 and M_2 are Strongly-co-Hopfian, then $M_1 \oplus M_2$ is Strongly-co-Hopfian.
- Let $M = \oplus_{i \in I} T_i^{(\alpha_i)}$ be semi-simple module where $(T_i)_{i \in I}$ is a family of simple module such as $Hom(T_i, T_j) = 0$ for $i \neq j$, we have ([37], Theorem 3.4.16 and [38]):

M is Strongly-Hopfian $\Leftrightarrow M$ is Strongly-co-Hopfian $\Leftrightarrow M$ is a Fitting module $\Leftrightarrow$ there exists $n \in \mathbb{N}^*$ such as for all i, $Card(\alpha_i) \leq n$.

- Let A be a semi-simple ring, then for all A-module M, we have ([37], Corollary 3.4.17 and [38]):

  M is co-Hopfian $\Leftrightarrow M$ is artinian $\Leftrightarrow M$ is noetherian $\Leftrightarrow M$ is Hopfian.

- Let M be a Endo-Regular module. Then ([37], Theorem 3.4.14):

  M is Strongly-Hopfian $\Leftrightarrow M$ is a Fitting module $\Leftrightarrow M$ is Strongly-co-Hopfian.

- All semi-projective and Strongly-co-Hopfian module is Strongly-Hopfian. We deduce that if M is a semi-projective module Strongly-co-Hopfian, then $End(M)$ is Strongly π-regular ([37], Corollary 3.4.19 and 3.4.20, [38])
- All semi-injective and Strongly-Hopfian module is Strongly-co-Hopfian. This implies that if M is a semi-injective module and Strongly-Hopfian, Then $End(M)$ is Strongly π-regular ([37], Corollary 3.4.19 and 3.4.20, [38]). We deduce that if M a semi-projective and semi-injective module, then ([37], Corollary 3.4.21 and [38]):

  M is Strongly-co-Hopfian $\Leftrightarrow M$ is a Fitting module $\Leftrightarrow M$ is Strongly-Hopfian.

2.5   Endo-Noetherian and Endo-Artinian Modules

Definition and Properties of Endo-Artinian or Endo-Noetherian Modules

- A A-module M is called **Endo-Noetherian** (resp. **Endo-Artinian**) if, for any family $(f_i)_{i \geq 1}$ of endomorphisms of M, the sequence: $Ker(f_1) \subseteq Ker(f_2) \subseteq \cdots \subseteq Ker(f_n) \cdots$ (resp. $Im(f_1) \supseteq Im(f_2) \supseteq \cdots \supseteq Im(f_n) \supseteq \cdots$) is stationary. The ring A is said left **Endo-Noetherian** (resp. **Endo-Artinian**) if, A is **Endo-Noetherian** (resp. **Endo-Artinian**) as left A-module. The module M is **Strongly-Fitting** if, M is Endo-Noetherian and Endo-Artinian ([41], Definition 1.1).
- let M be a module, the following properties are equivalent:
  (i) M is Hopfian.
  (ii) The increasing sequence: $Ker(f_1) \subseteq Ker(f_2) \subseteq \cdots \subseteq Ker(f_n) \subseteq \cdots$, is stationary, for all family $(f_i)_{i \geq 1}$ of surjective endomorphisms of M,
  (iii) The increasing sequence: $Ker(f) \subseteq Ker(f^2) \subseteq \cdots \subseteq Ker(f^n) \cdots$, is stationary, for all epimorphism f of M,
  Proof: $(i) \Rightarrow (ii) \Rightarrow (iii)$ are obvious. To prove that $(iii) \Rightarrow (i)$, let f be a surjective endomorphism of M, there exists $n \geq 1$ such as $Ker(f^n) \cap Im(f^n) = \{0\}$, and as f is surjective we have $Ker(f^n) \cap M = Ker(f^n) = \{0\}$ this implies that f^n is bijective, indeed f is injective.

- Let M be a module, the following properties are equivalent:
  (i) M is co-Hopfian.
  (ii) The decreasing sequence: $Im(f_1) \supseteq Im(f_2) \supseteq \cdots \supseteq Im(f_n) \supseteq \cdots$, is stationary, for each family $(f_i)_{i \geq 1}$ of injective endomorphisms of M,
  (iii) The decreasing sequence: $Im(f) \supseteq Im(f^2) \supseteq \cdots \supseteq Im(f^n) \supseteq \cdots$, is stationary, for each monomorphism f of M.
  Proof: $(i) \Rightarrow (ii) \Rightarrow (iii)$ are obvious. To prove that $(iii) \Rightarrow (i)$, let f be a injective endomrphism of M, there exists $n \geq 1$ such as $Ker(f^n) + Im(f^n) = M$, and as f is injective we have $Im(f^n) = M$, this implies that f^n is bijective, indeed f surjective.
- Let A be a artinian ring having at least one non principal ideal, then there exists a Endo-Noetherian A-module which is not Noetherian. ([54], Theorem 3.9)
- Let A be a principal commutative ring verifying CCA annihilators, then ([54], Theorem 3.10)

  A is artinian $\Leftrightarrow$ (All Endo-Noetherian A-module is Noetherian).

- In general, neither a submodule nor the quotient of a Endo-Noetherian (resp. Endo-Artinian) module is Endo-Noetherian (resp. Endo-Artinian). For example ([41], Remark 1.27):
  (1) $\mathbb{Q}$ is a Endo-Artinian $\mathbb{Z}$-module, but $\mathbb{Z}$ is not Endo-Artinian.
  (2) Let R be the free ring over $\mathbb{Z}$ generated by $\{x_n, \ n \in \mathbb{N}\}$. The, R is left Endo-Noetherian but the left ideal I generated by $\{x_n, \ n \in \mathbb{N}\}$ infinite direct sum of left ideals I_n generate by $\{x_n\}$, therefore the R-module I is not Endo-Noetherian.
  (3) $\mathbb{Q}$ is a Endo-Artinian and Endo-Noetherian $\mathbb{Z}$-module but $\mathbb{Q}/\mathbb{Z} = \oplus_{p \in \mathcal{P}} \mathbb{Z}(p^\infty)$ is not Endo-Artinien and not Endo-Noetherien.
- Let N be a direct summand of M. If M is Endo-Noetherian (resp. Endo-Artinian), then N and M/N are Endo-Noetherians (resp. Endo-Artinians) ([41], Proposition 1.28).
- Let us consider $M = \oplus_{i=1}^{n} M_i$ such as $Hom(M_i, M_j) = \{0\}$, for each $1 \leq i \neq j \leq n$. Then M is Endo-Noetherian (resp. Endo-Artinian) if and only if, M_i is Endo-Noetherian (resp. Endo-Artinian), for all $1 \leq i \leq n$ ([41], Proposition 1.29).
- All Artinian (resp. Noetherian) module M is Endo-Artinian (resp. Endo-Noetherian). The $\mathbb{Z}$-module $\mathbb{Q}$ which is not Artinien and not Noetherian and as all $\mathbb{Z}$-endomorphism of $\mathbb{Q}$ is null or automorphism,we deduce that the $\mathbb{Z}$-module $\mathbb{Q}$ is Strongly-Fitting. ([41], Remark 1.5):
- If A^n is a Endo-Artinian module, then all quotient module A^n/N de A^n is a Endo-Artinian A-module ([41], Corollary 3.5).
- Let A be a ring, the following conditions are equivalent ([41], Corollary 3.6):
  (1) Ther exists $n \in \mathbb{N}^*$ such as A^n is a Endo-Artinian module.
  (2) All finitely generated A-module is Endo-Artinian.
- Let M be a Endo-Artinian (resp. Endo-Noetherian) A-module. Then M verify the CCD (resp. CCA) over direct summands ([41], Proposition 1.6).

- Let $M = \oplus_{i \in I} M_i$ be the direct sum of nonzero sub-modules. If M is Endo-Artinian or Endo-Noetherian, then I is finite ([41], Proposition 1.7).
- Let M be a semi-simple A-module. We have ([41], Proposition 1.8):

M is Endo-Noetherian $\Leftrightarrow$ M is Noetherian $\Leftrightarrow$ M is Artinian $\Leftrightarrow$ M is Endo-Artinian.

- If A is semi-simple and M is a A-module. We have ([41], Proposition 1.9 and Corollary 1.10).
  M is Dedekind-finite $\Leftrightarrow$ M is weakly-co-Hopfian $\Leftrightarrow$ M is Artinian $\Leftrightarrow$ M is Noetherian $\Leftrightarrow$ M is generalized Hopfian $\Leftrightarrow$ All homogeneous component of M is finitely generated. We deduce that, if A is semi-simple ring, then all finitely generated A-module is Strongly-Fitting.
- Let M be a nonzero Endo-Artinian or Endo-Noetherian A-module. Then M is completely decomposable (M can be decomposed to finite direct sum of non decomposable sub-modules) ([41], Corollary 1.11).
- Let M be a A-module Strongly-Fitting. Then $S = End_A(M)$ is semi-perfect and its Jacobson Radical is a nil-ideal ([41], Corollary 1.13).

Endo-Artinian or Endo-Noetherian Finitely Generated Modules

- For all ring A the following conditions are equivalent ([41], Proposition 2.1):
  (1) A is left Endo-Artinian.
  (2) All cyclic left A-module is Endo-Artinian.
  (3) All left finitely generated A-module is Endo-Artinian.
  (4) For all $n \in \mathbb{N}$, the left A-module A^n is Endo-Artinian.
- For all ring A the following conditions are equivalent ([41], Proposition 2.2):
  (1) All cyclic left A-module is Endo-Noetherian.
  (2) A verify C.C.A on left ideals of type $(I : a) = \{b \in A \mid ba \in I\}$ where I is a left ideal of A and $a \in \{b \in A \mid Ib \subseteq I\}$ (The idealisator of I).
- For all ring A, the following conditions are equivalent ([41], Proposition 2.3):
  (1) The left A-module A^n with $n \in \mathbb{N}$ is Endo-Noetherian.
  (2) All decreasing chain $ann_{A^n} M_1 \subseteq ann_{A^n} M_2 \subseteq ...$, where $M_i \in M_n(A)$ is stationary.
- For all ring A the following conditions are equivalent ([41], Corollary 2.5):
  (1) All left finitely generated A-module is Endo-Noetherian.
  (2) For all $n \in \mathbb{N}^*$ and for all sub-module N of A^n, the increasing sequence $(N : M_1) \subseteq (N : M_2) \subseteq ...$ where $M_i \in M_n(A)$ verifying $NM_i \subseteq N$ for each $i \geq 1$, is stationary.

Injective Endo-Artinian or Endo-Noetherian Modules

- If M is Pseudo-Quasi-Principally-Injective, the ([77], Theorem 2.6):

M is Endo-Noetherian $\Rightarrow$ S is left Endo-Artinian $\Rightarrow$ M is Strongly-co-Hopfien.

- If M is Quasi-Principally-Injective, then ([41], Theorem 3.2, Corollary 3.4 and [77], Corollary 2.7):

  M is Endo-Noetherian $\Leftrightarrow S$ is left Endo-Artinian $\Rightarrow M$ is Strongly-co-Hopfian.

- Let M be a module Quasi-Principally-Injective. Then ([41], Corollary 3.8):

  M is Endo-Noetherian $\Leftrightarrow M^n$ is Endo-Noetherian, for all $n \in \mathbb{N}^*$.

Projective Endo-Artinian or Endo-Noetherian Modules

- Let M be a projective A-module, Then ([41], Corollary 1.14):

  M is Strongly-Fitting $\Rightarrow M$ is finitely generated.

- If M is Quasi-Pseudo-Principally-Projective, then ([58], Theorem 2.21 and [60], Theorem 3.10):

  M is Endo-Artinian $\Rightarrow S$ is right Endo-Artinian.

- If M is Quasi-Principally-Projective, then ([41], Theorem 3.3):

  M is Endo-Artinian $\Leftrightarrow S$ is right Endo-Artinian.

- If M is Quasi-Principally-Projective and Endo-Artinian (resp. Quasi-Principally-Injective and Endo-Noetherian). Then M is a Fitting module ([41], Corollary 3.4).
- If M is Quasi-Pseudo-Principalement-Projective, then ([58], Theorem 2.21 and [60], Theorem 3.10):

  M is Endo-Artinian $\Rightarrow M$ is Strongly-Hopfian.

Endo-Notherian or Endo-Artinian Module over a Dedekind Domain

- A module M over a Dedekind Domain A is Endo-Artinian if and only if, M is isomorphic to the direct sum of: $A/\mathfrak{p}^n$, modules of type $\mathfrak{p}^\infty$ for different prime ideals $\mathfrak{p}$ and K^m where $n, m \in \mathbb{N}^*$ ([41], Proposition 5.7).
- A module M over a Dedekind Domaine A is Strongly-Fitting if and only if, M is isomorphic to the finite direct sum of: $A/\mathfrak{p}^n$ and K^m where $n, m \in \mathbb{N}^*$ and $\mathfrak{p}$ is a prime ideal ([41], Proposition 5.8).
- Let A be a Dedekind domain and M a divisible A-module. Then, M is co-Hopfian if and only if, M is isomorphic to a sum of K^n and modules of type $\mathfrak{p}^\infty$ for different prime ideals $\mathfrak{p}$. The module M is Hopfian if and only if, M is isomorphic to K^n ([41], Proposition 5.12).

- Let A be a Dedekind domain and M a Projective A-module. We have ([41], Proposition 5.9):
  M is Noetherian $\Leftrightarrow M$ is endo-Noetherian $\Leftrightarrow M$ is Strongly-Hopfian $\Leftrightarrow M$ is Hopfian $\Leftrightarrow M$ is finitely generated.
- Since any Projective abelian group is free, we deduce that a Projective abelian group G is Endo-Noetherian if and only if, G is isomorphic to $\mathbb{Z}^n$ with $n \in \mathbb{N}^*$ ([41], Remark 5.6).

2.6 Semi-Hopfian and Semi-co-Hopfian Modules (in the Sense of Aydogdu-Ozcan)

- An A-module M is said to be **Semi-Hopfian** (resp. **Semi-co-Hopfian**) if, $Ker(f)$ (resp. $Im(f)$) is a direct summand of M, for every surjective (resp. injective) endomorphism f of M. A ring A is said to be left **Semi-Hopfian** (resp. **Semi-co-Hopfian**) if, A is **Semi-Hopfian** (resp. **Semi-co-Hopfian**) as a left A-module. By analogy we define right Semi-Hopfians and Semi-co-Hopfians ring. There exists a left Semi-co-Hopfian which is not right Semi-co-Hopfian ([5], Example 2.20).
- If M satisfies the property **D2**, then M is Semi-Hopfian. We deduce that the notion of Semi-Hopfian is a generalization of the hopficity and the property **D2**. If M satisfies the property **C2**, then M is Semi-co-Hopfian. The same way the notion of Semi-co-Hopfian is a generalization of the co-hopficity and the property **C2** ([5], Lemma 2.1 and Lemma 3.3).
- If M is Quasi-Pseudo-Principally-Projective, then M satisfies the properties **D2**. Hence M is Semi-Hopfian (([58], Lemma 3.24 and [60], Corollary 3.5).
- If M is Quasi-Pseudo-Principally-Injective, then M satisfies the property **C2**. Hence M is Semi-co-Hopfian ([7], Lemma 2.2 and [77], Corollary 2.3).
- Let M be an A-module and $\mathcal{S} = End_A(M)$, we have ([5], Proposition 2.17 and Corollary 2.18):
  $\mathcal{S}$ is left Semi-co-Hopian $\Rightarrow M$ is a Semi-co-Hopian A-module. If M is free, we have: $\mathcal{S}$ is Semi-co-Hopian $\Leftrightarrow M$ is a Semi-co-Hopian A-module. In particular:

  A^n is a Semi-co-Hopian A-module $\Leftrightarrow M_n(A)$ is a left Semi-co-Hopian ring.

- Let p be a prime number and M a infinite direct sum of copies of $\mathbb{Z}/p^2\mathbb{Z}$. Then M is not Hopfian as $\mathbb{Z}$-module, but M satisfies the property $GD2$. Indeed, it is clair that M is not Hopfian and let $f : M \mapsto M$ an epimorphism, since $p^2 M = 0$, then f is also a $\mathbb{Z}/p^2\mathbb{Z}$-epimorphism. As M is a free $\mathbb{Z}/p^2\mathbb{Z}$-module, then $Ker(f) \leq^{\oplus} M$ which implies that M satisfies **GD2** ([5], Example 3.2). Remark that there exists torsion free modules that does not satisfy **GD2**. ([5], Example 3.5).
- A direct sum of modules satisfying **GD2** does not satisfy necessary **GD2**. For example, let p be a prime number $M_1 = \mathbb{Z}/p\mathbb{Z}$ and M_2 a direct sum of copies of $\mathbb{Z}/p^2\mathbb{Z}$, then M_1 is simple and M_2 satisfying **GD2**. But, $M = M_1 \oplus M_2$ does not necessary satisfy **GD2** as $\mathbb{Z}$-module. Indeed, let $f : M \mapsto M$ such

that $f(a_1 + p\mathbb{Z}, a_2 + p^2\mathbb{Z}, a_3 + p^2\mathbb{Z}, ...) = (a_2 + p\mathbb{Z}, a_3 + p^2\mathbb{Z}, ...)$. Then, f is an $\mathbb{Z}$-epimorphism and $Ker(f) = \mathbb{Z}/p^2\mathbb{Z} \oplus p\mathbb{Z}/p^2\mathbb{Z} \oplus 0 \oplus 0 \oplus ....$ is not a direct summand of M, because $p\mathbb{Z}/p^2\mathbb{Z}$ is not a direct summand of $\mathbb{Z}/p^2\mathbb{Z}$. The direct product of modules satisfying **GD2** does not necessary satisfy **GD2**. Indeed, as previously, we check that the $\mathbb{Z}$-module $M = \prod_{n \geq 1} \mathbb{Z}/p^n\mathbb{Z}$ does not satisfy **GD2** ([5], Example 3.9).

3   Commutative Case

In this section, A is an unitary commutative ring and M a right A-module.

3.1   Semi-Hopfian and Semi-co-Hopfian modules (in the Sense of Diviani-Azar and Mafi)

Let $a \in A$, we define R_a the endomorphism of M such that $R_a(m) = ma$, for every $m \in M$. M is said to be **Semi-Hopfian** (resp. **Semi-co-Hopfian**) in the sense of Diviani-Azar and Mafi if, R_a is an isomorphism for every endormorphism R_a surjective (resp. injective). The ring A is Semi-Hopfian (resp. Semi-co-Hopfian as a right A-module if and only if A is Hopfian (resp. co-Hopfian) as a right A-module.

- Let N be a nonzero A-module Hopfian (resp. co-Hofian) and $M = \oplus_{i \in \mathbb{N}} N$. then M is Semi-Hopfian (resp. Semi-co-Hopfian), but M is not Hopfian (resp. co-Hopfian). ([19], Example 2.2).

3.2   S.S.H and S.S.C.H Modules

- A A-module M is said to be **S.S.H** (resp. **S.S.C.H**), if for every $a \in A$ the chain: $l_M(a) \subset l_M(a^2) \subseteq l_M(a^3) \subseteq ...$ (resp. $Ma \supseteq Ma^2 \supseteq ...$) is stationary. The ring A is said to be **S.S.H** (resp. **S.S.C.H**) if A is **S.S.H** (resp. **S.S.C.H**) as A-module. We have the following implications where the reverses are not always true ([24], Definition 2.3 and Example 2.4):

$$\text{Strongly-Hopfian} \Rightarrow \text{S.S.H} \Rightarrow \text{Semi-Hopfian.}$$
$$\text{Strongly-co-Hopfian} \Rightarrow \text{S.S.C.H} \Rightarrow \text{Semi-co-Hopfian}$$

- Since $End_A(A) \cong A$, then ([24], Proposition 2.5 and 2.6):

A is Strongly-co-Hopfian $\Leftrightarrow$ A is S.S.C.H $\Leftrightarrow$ A is strongly π-Regular.
For every $a \in A$, there exists $n \in \mathbb{N}$ such that: $l_M(a^n) = l_M(a^{n+1})$.

$$\Updownarrow$$

M is S.S.H.

$$\Updownarrow$$

For every $a \in A$, there exists $n \in \mathbb{N}$ such that: $l_M(a^n) \cap Ma^n = 0$.
For every $a \in A$, there exists $n \in \mathbb{N}$ such that: $Ma^n = Ma^{n+1}$.

$$\Updownarrow$$

M is S.S.C.H.

$$\Updownarrow$$

For every $a \in A$, there exists $n \in \mathbb{N}$ such that: $M = l_M(a^n) + Ma^n$.

- M is said to be **Semi-Fitting** if, for every $a \in A$, there exists $n \in \mathbb{N}$ such that: $M = l_M(a^n) \oplus Ma^n$. We have ([24], Proposition 2.7):

$$M \text{ is S.S.H and S.S.C.H } \Leftrightarrow M \text{ is Semi-Fitting.}$$

- The $\mathbb{Z}$-module $\mathbb{Q}$ is Semi-Fitting, but $\mathbb{Z}$ is not S.S.C.H and $\mathbb{Q}/\mathbb{Z}$ is not S.S.H. We verify that the $\mathbb{Z}_{(p)}$ the group of rational numbers where the denominator is prime with p, is a S.S.H $\mathbb{Z}$-module, but $\mathbb{Z}_{(p)}/\mathbb{Z}$ is not S.S.H ([24], Example 2.8).

- The ring $A = \bigoplus_{n=1}^{\infty} \mathbb{Z}/p^n\mathbb{Z}$ is a (finitely generated) A-module but it is neither S.S.H nor S.S.C.H ([24], Example 4.1).

- The A-module $M = \bigoplus_{i=1}^{n} M_i$ is S.S.H (resp. S.S.C.H) if and only if, M_i is S.S.H (resp. S.S.C.H), for every $1 \leq i \leq n$ ([24], Corollary 2.10). We remark that for every $n \geq 1$, the $\mathbb{Z}$-module $\mathbb{Z}/p^n\mathbb{Z}$ is S.S.H (resp. S.S.C.H). But, the $\mathbb{Z}$-module $M = \bigoplus_{n=1}^{\infty} \mathbb{Z}/p^n\mathbb{Z}$ is neither S.S.H nor S.S.C.H. Indeed, it is clair that M is a Hopfian and co-Hopfian $\mathbb{Z}$-module and hence M is Semi-Hopfian and Semi-co-Hopfian $\mathbb{Z}$-module . On the other hand the sequences $Ann_M(p) \subseteq Ann_M(p^2) \subseteq Ann_M(p^3) \subseteq ....$ and $Mp \supseteq Mp^2 \supseteq Mp^3 \supseteq ...$ are not stationary.

- Let M a A-module and N a submodule of M. then ([24], Proposition 2.9)

$$M \text{ is S.S.H (resp. S.S.C.H)} \Rightarrow N \text{ (resp. } M/N) \text{ is S.S.H (resp. S.S.C.H)}.$$
If N and M/N are S.S.H (resp. S.S.C.H) $\Rightarrow M$ is S.S.H (resp. S.S.C.H).

- Let M be a finitely generated A-module. Then ([24], Lemma 4.2):

$$M \text{ is S.S.C.H } \Leftrightarrow M/N \text{ is Semi-co-Hopfian, for every submodule } N \text{ of } M.$$

- If A is a commutatif ring, then ([24], Theorem 4.3):

Every finitely generated A-module is semi-co-Hopfian $\Leftrightarrow$ every finitely generated A-module is S.S.C.H.
$$\Updownarrow$$
Every cyclic A-module is Strongly-co-Hopfian $\Leftrightarrow$ A is Strongly-π-Regular.
$$\Updownarrow$$
Every cyclic A-module is co-Hopfian $\Leftrightarrow$ Every cyclic A-module is S.S.C.H.

- Let A be a commutative Noetherian ring. Then ([24], Theorem 3.2 and [19], Theorem 1.1):

A is Artinian $\Leftrightarrow$ every A-module is Semi-Fitting $\Leftrightarrow$ every A-module Noetherian is S.S.C.H.
$$\Updownarrow$$
Every A-module is S.S.H $\Leftrightarrow$ Every A-module is S.S.C.H.
$$\Updownarrow$$
Every A-module is Semi-Hopfian $\Leftrightarrow$ Every A-module is Semi-co-Hopfian
$$\Updownarrow$$
Every Noetherian A-module is Semi-co-Hopfian $\Leftrightarrow$ Every Noetherian A-module is co-Hopfian.

4   Diagram

- Let A be an unitary associative ring and M a left A-module. We have the following implications where the reverses are not always true.

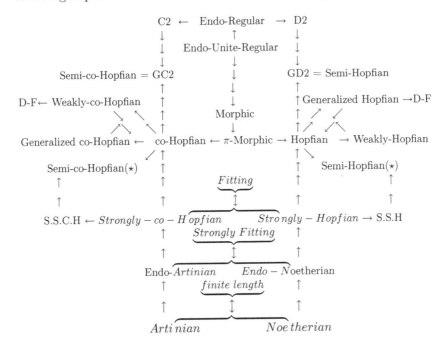

The notions: **S.S.H, S.S.C.H, Semi-Hopfian** ($\star$) **and Semi-co-Hofians** ($\star$), are defined in the case where A is commutative ring

- Every Noetherian (resp. Artinian) module verifies most properties of the hopficity (resp. co-Hopficity). Any module that doesn't satisfy the property **GD2** (resp. **GC2**) doesn't not satisfy most properties of the hopficity (resp. co-Hopficity). For example:
  (1) The $\mathbb{Z}$-module $\mathbb{Z}$ is Noetherian, hence $\mathbb{Z}$ is Endo-Noetherian, Strongly-Hopfian, Hopfian, Weakly-Hopfian, Dedekind-Finite, S.S.H, Semi-Hopfian and satisfies **GD2**. On the other hand, the $\mathbb{Z}$-module $\mathbb{Z}$ doesn't satisfy **GC2**, we deduce that $\mathbb{Z}$ is not Artinian, not Endo-Artinian, not Strongly-co-Hopfian, not co-Hopfian, not π-Morphic, not Morphic, not Endo-Unite-Regular. Finally, we verify that the $\mathbb{Z}$-module $\mathbb{Z}$ is Weakly-co-Hopfian, but it is not Generalized-co-Hopfian, not S.S.C.H, not Semi-co-Hopfian. W have also that $\mathbb{Z}$ satisfies **D2** and is not Endo-Regular, hence doesn't satisfy **C2**.
  (2) The $\mathbb{Z}$-module $\mathbb{Z}(p^\infty)$ is Artinian, hence $\mathbb{Z}(p^\infty)$ is Endo-Artinian, Strongly-co-Hopfian, co-Hopfian, Weakly-co-Hopfian, Dedekind-Finite, S.S.C.H, Semi-co-Hopfian and satisfies **GC2**. On the other hand, the $\mathbb{Z}$-module $\mathbb{Z}(p^\infty)$ is Generalized-Hopfian. But, it is not Semi-Hopfian, not Weakly-Hopfian and doesn't satisfy **D2** or **GD2**, we deduce that $\mathbb{Z}(p^\infty)$

is not Noetherian, not Endo-Noetherian, not Strongly-Hopfian, not Hopfian, not π-Morphic, not Morphic, not Endo-Unite-Regular ([5], Proposition 3.6, [24,26], Example 1.17, [38], Example 2.13 and [58], Remark 2.11 and [68]).

(3) the $\mathbb{Z}$-module $M = \mathbb{Z}(p^\infty)$ satisfies **GC2**, but $\mathcal{S} = End_{\mathbb{Z}}(M)$ which is isomorphic to the ring $\mathbb{Z}_p$ of positive integers p-adic doesn't satisfy **GC2** ([5], Proposition 2.4). We deduce that the $\mathbb{Z}$-module $M = \mathbb{Z}(p^\infty)$ is an example of module which is Artinian (resp. Endo-Artinian, Strongly-co-Hopfian, co-Hopfian) such that $\mathcal{S} = End_{\mathbb{Z}}(M)$ is not Artinian (resp. Endo-Artinian, Strongly-co-Hopfian, co-Hopfian).

(4) The quotient $\mathbb{Q}/\mathbb{Z} = \oplus_{p \in P}\mathbb{Z}(p^\infty)$ is Strongly-co-Hopfian and not Strongly-Hopfian ([38], Example 2.15 and Remark 2.16). We deduce that $\mathbb{Q}/\mathbb{Z}$ is not Quasi-Projective, hence it satisfies neither **D2** nor **GD2**, and it is not Hopfian not Endo-Regular, not Endo-Unite-Regular, not Morphic, not π-Morphic. We remark also that the module $\mathbb{Q}/\mathbb{Z}$ which is a nonzero infinite direct sum of submodules, is neither Endo-Artinian nor Endo-Noetherian ([41], Corollary 1.15 and Remark 1.27).

- Every vector space over K of infinite dimension is a module over K which is S.S.H and S.S.C.H, hence it is Semi-Hopfian and Semi-co-Hopfian, satisfies **GC2** and **GD2**. But, it is not Weakly-Hopfian, not Generalized-co-Hopfian, not Dedekind-Finite. Hence it satisfies any of the others properties of the Hopficity.

- The $\mathbb{Z}$-module $\mathbb{Q}$ is neither Artinian nor Noetherian as every $\mathbb{Z}$-endomorphism of $\mathbb{Q}$ is zero or an automorphism hence, the $\mathbb{Z}$-module $\mathbb{Q}$ is Strongly-Fitting. We deduce that the $\mathbb{Z}$-module $\mathbb{Q}$ is Hopfian (resp. co-Hopfian), Weakly-Hopfian (resp. Weakly-co-Hopfian), Generalized-Hopfian (resp. Generalized-co-Hopfian), Dedekind-Finite, S.S.H (resp. S.S.C.H), Semi-Hopfian (resp. Semi-co-Hopfian) and satisfies **GD2** (resp. **GC2**).

- The $\mathbb{Z}$-module $M = \mathbb{Q}^{(\mathbb{N})}$ is Quasi-Injective, hence satisfies **(C2)** and **(GC2)** and is not Dedekind-Finite, since $M = M \oplus \mathbb{Q}$ [5]. We deduce that M is not Artinian, not Noetherian, not Endo-Artinian, not Endo-Noetherian, not Strongly-Hopfian, not Strongly-co-Hopfian, not Hopfian, not co-Hopfian, not Generalized Hopfian, not Weakly-co-Hopfian, not π-Morphic, not Morphic.

- Let N be a nonzero module, then every infinite direct sum of copies of N is neither Hopfian nor co-Hopfian. On the contrary the $\mathbb{Z}$-module $M = \oplus_{p \in P}\mathbb{Z}/p\mathbb{Z}$, where P is the set of prime numbers, is Hopfian and co-Hopfian ([70], p. 300). More precisely, M is Strongly-Hopfian and Strongly-co-Hopfian and which is neither Endo-Artinian nor Endo-Noetherian ([23], Example 2.1 and [41], Corollary 1.15). Indeed: since $Hom_{\mathbb{Z}}(\mathbb{Z}/p\mathbb{Z}, \mathbb{Z}/q\mathbb{Z}) = \{0\}$ for every $p \neq q$ in P, we deduce that for every $f \in End_{\mathbb{Z}}(M)$, we have $f = \oplus_{p \in P}f_p$ or $f_p \in End_{\mathbb{Z}}(\mathbb{Z}/p\mathbb{Z})$, hence f_p is bijective or zero. That implies $Ker(f^n) = Ker(f^{n+1})$ and $Im(f^n) = Im(f^{n+1})$, for a certain $n \in \mathbb{N}$. We deduce that M which is not Artinian, not Noetherian, not Endo-Artinian, not Endo-Noetherian, satisfies all the other properties of the hopficity.

- Let $(p_n)_{n \geq 1}$ be a sequence of prime numbers such that $p_1 < p_2 < \cdots < p_n < \cdots$. Let $M_n = \mathbb{Z}/p_n^n\mathbb{Z}$ and consider $M = \oplus_{n \geq 1}M_n$. then, M is Hopfian and co-Hopfian, on the contrary, M is neither Strongly-co-Hopfian nor Strongly-Hopfian ([37], Proposition 3.2.8 and [38]). We deduce that M is Generalized-Hopfian (resp. Generalized-co-Hopfian), Weakly-Hopfian (resp. Weakly-co-Hopfian), satisfies **GD2** (resp. **GC2**) and Dedekind-Finite. But, M is not Noetherian, not Artinian, not Endo-Artinian, not Endo-Noetherian.

- Assume $M = \oplus_{n \geq 1} \mathbb{Z}/p^n\mathbb{Z}$. Then M is Quasi-Injective, hence M satisfies **C2**, **GC2**, Dedekin Finite, Semi-Hopfian and Semi-co-Hopfian. On the contrary M is not S.S.H, not S.S.C.H, not co-Hopfian, not Generalized-Hopfian ([5], [24], Corollary 2.10 and [26], Example 1.7). We deduce that M is not Hopfian, not Strongly-co-Hopfian, not Strongly-Hopfian, not Endo-Artinian, not Endo-Noetherian, not Noethérian, not Artinian.

- The $\mathbb{Z}$-module $M = \prod_{p \in \mathcal{P}} \mathbb{Z}/p\mathbb{Z}$ is Morphic ([11], Example 18 and [12]). We deduce that M is Hopfian (resp. co-Hopfian) Generalized-Hopfian (resp. Generalized-co-Hopfian), Weakly-Hopfian (resp. Weakly-co-Hopfian), Dedekind-Finite, satisfies **GC2** and **GD2**.

- Let p be a prime number and $M = \prod_{n=1}^{\infty} \mathbb{Z}/p^n\mathbb{Z}$, then $\mathbb{Z}$-module M satisfies neither **GC2** nor **GD2** ([5], Example 2.11 and Example 3.9). We deduce that M does not satisfy any of the Hopficity variant.

5   Abelian Groups

5.1   Hopfian and co-Hopfian Abelian Groups

- It is clear that any finitely generated infinite group G is a $\mathbb{Z}$ - noetherian module, so G is Hopfian. Since G is of the form $\mathbb{Z}^n \oplus H$ where H is a finite group and $n \in \mathbb{N}^*$ and $\mathbb{Z}$ is not co-Hopfian, it follows that finitely generated Abelian groups co-Hopfian are the finite groups.

- The Hopficity of G is deduced from that its divisible part $D(G)$ and the reduced part R such that $G = D(G) \oplus R$. Recall that an abelian group G is said **divisible** If $nG = G$, for any $n \in \mathbb{N}^*$. So, G is divisible if, and only if, [62]:

$$G \approx \mathbb{Q}^{(I)} \bigoplus \left(\bigoplus_{p \in \mathcal{P}} (\mathbb{Z}(p^\infty))^{(I_p)} \right), where \; \mathcal{P} \; is \; the \; set \; of \; prime \; numbers.$$

- Any G torsion free and non-divisible group is not co-Hopfian. Since G is not divisible, it exists $n \in \mathbb{N}^*$ and $g \in G$ such that $nx \neq g$ for any $x \in G$. Then, the multiplication by n is injective and is not surjective.

- Using the divisible group structure and the fact that $\mathbb{Z}(p^\infty)$ is not Hopfian for any prime number p, it follows that any Hopfian divisible group is of the form $\mathbb{Q}^n$ with $n \in \mathbb{N}^*$.

- Hopfian divisible groups are of the form $\mathbb{Q}^n \oplus \left(\bigoplus_{p \in \mathcal{P}} (\mathbb{Z}(p^\infty)^{n_p} \right)$, where $\mathcal{P}$ is the set of prime numbers, n and n_p are natural numbers ([37], Proposition 2.3.4).

- Abelian co-Hopfian torsion free groups are of the form $\mathbb{Q}^n$, where n is a natural number. It is deduced that Abelian co-Hopfian torsion free groups are countable ([37], Proposition 2.3.5). The group $\mathbb{Q}^n$ is Hopfian, but there is a uncountable torsion free Hopfian group [65].
- Knowing that any group is a direct sum of a divisible group and a reduced group, then the study of torsion groups is reduced to reduced p-groups to divisible p-groups. We deduce that any Hopfian p-divisible group is null, and that any co-Hopfian p-divisible group is of the form $(\mathbb{Z}(p^\infty))^n$ where $n \in \mathbb{N}^*$.
- If G is a Hopfian(or co-Hopfian) reduced p-group, then G is finite or $\aleph_0 <| G |< 2^{\aleph_0}$ [43]. As result, any infinite Hopfian(or co-Hopfian) reduced p-group is not countable (In [35], P. Hill and C. Megibben have build an Example of a co-Hopfian reduced p-group of cardinal $2^{\aleph_0}$).

5.2   Strongly-Hopfian or Strongly-co-Hopfian Abelian Groups

Strongly-Hopfian or Strongly-co-Hopfian Divisible Abelian Groups

- Commutative divisible Strongly-Hopfian rings are $\mathbb{Q}^n$ where $n \in \mathbb{N}^*$, and commutative divisible Strongly-co-Hopfian groups are of the form: $\mathbb{Q}^n \oplus \left(\bigoplus_{p \in \mathcal{P}} (\mathbb{Z}(p^\infty))^{n_p} \right)$, where $\mathcal{P}$ is the set of prime numbers and the $n_p \in \mathbb{N}$ such that $max\{n_p \,,\, p \in \mathcal{P}\} < +\infty$ ([37], Theorem 3.5.1).
- The only Hopfian divisible p-groupe is the group $\{0\}$, and the divisible Strongly-co-Hopfian p-groups are $(\mathbb{Z}(p^\infty))^n$ where $n \in \mathbb{N}$ ([37], Corollary 3.5.2).

Strongly-Hopfian or Strongly-co-Hopfian Finitely Generated Groups

- Any commutative finitely generated group is Strongly-Hopfian, and finitely generated Abelian groups Strongly-co-Hopfian are finite groups ([37], Theorem 3.5.3).
- Strongly-Hopfian p-groups are finite p-groups, and Strongly-co-Hopfian p-groups are of the form $\mathbb{Z}(p^\infty)^n \oplus K$ where K is a finite p-group and $n \in \mathbb{N}$ ([37], Theorem 3.5.5 and 3.5.6).

Strongly-Hopfian or Strongly-co-Hopfian Torsion-Free Groups

- The commutative Strongly-co-Hopfian torsion free groups, are the $\mathbb{Q}^n$ where $n \in \mathbb{N}$. All abelian torsion free group of finite rank is Strongly-Hopfian ([37], Theorem 3.5.8 and 3.5.9).
- Let's consider all of the prime numbers $p_1 < p_2 < \cdots < p_n < \cdots$. For any $n \geq 1$ set G_n the sub-group of $\mathbb{Q}$ generated by $\{\frac{1}{p_n^k}, k \geq 0\}$. Then $G = \oplus_{n \geq 1} G_n$ is a torsion free group of infinite rank and Strongly-Hopfian ([37], Theorem 3.5.11).

Strongly-Hopfian or Strongly-co-Hopfian Torsion Groups

- Strongly-Hopfian torsion groups are of the form $G = \bigoplus_{p \in \mathcal{P}} R_p$ with R_p is a finite p-group and $Sup\{l(R_p)$, $p \in \mathcal{P}\} < \infty$ ([37], Theorem 3.5.13).
- Strongly-co-Hopfian torsion groups are of the form $G = R \oplus D$ where $R = \bigoplus_{p \in \mathcal{P}} R_p$ with R_p is a finite p-group and $Sup\{l(R_p)$, $p \in \mathcal{P}\} < \infty$ and $D = \bigoplus_{p \in \mathcal{P}} \mathbb{Z}(p^\infty)^{n_p}$ with $max\{n_p$, $p \in \mathcal{P}\} < \infty$ ([37], Theorem 3.5.12).

5.3   Endo-Artinian or Endo-Noetherian Abelian Groups

- An abelian group G is Endo-Artinian if, and only if, G is isomorphic to a direct sum of finite group, $(\mathbb{Z}(p_i^\infty))^{n_i}$ and $\mathbb{Q}^m$ où $m, n_i \in \mathbb{N}^*$ and $p_i \in \mathcal{P}$ ([41], Proposition 5.1).
- An abelian group G is Strongly-Fitting if, and only if, G is isomorphic to a direct sum of finite group and $\mathbb{Q}^n$ where $n \in \mathbb{N}^*$ ([41], Proposition 5.2).
- Let G be an Endo-Artinian or Endo-Noetherian abelian group. Then, the torsion subgroup $T(G)$ is a direct summand of G ([41], Theorem 5.4). The torsion subgroup $T(G)$ of an Abelian group Strongly-Hopfian or Strongly-co-Hopfian is not, in general, a direct summand of G. For example, consider the $\mathbb{Z}$-module $M = \prod_{p \in \mathcal{P}} \mathbb{Z}/p\mathbb{Z}$ with $T(M) = \oplus_{p \in \mathcal{P}} \mathbb{Z}/p\mathbb{Z}$. The ring $End_{\mathbb{Z}}(M) \simeq \prod_{p \in \mathcal{P}} \mathbb{Z}/p\mathbb{Z}$ is Strongly-Regular. So, M is a Fitting module and $T(M)$ is not a direct summand of M ([41], Remark 5.5).

6   Hopficity of Rings

6.1   Generalities

- The study of the different variants of hopficity are analogous and similar to those of the Noetherian and Artinian modules. The main purpose of the following diagram is to highlight this analogy. Let A be a unitary associative ring, we have the following implications whose reciprocals are not generally true:

$$\text{Generalised-Hopfian} \leftarrow \text{Weakly-co-Hopfian} \rightarrow \text{Weakly-Hopfian}$$
$$\uparrow \qquad\qquad\qquad \uparrow \qquad\qquad\qquad \uparrow$$
$$\text{Generalised-co-Hopfian} \leftarrow \text{co-Hopfian} \longrightarrow \text{Hopfian} = \text{Dedekind-Finite}$$
$$\uparrow \qquad\qquad\qquad \uparrow$$
$$\text{Strongly-co-Hopfian} \longrightarrow \text{Strongly-Hopfian}$$
$$\uparrow \qquad\qquad\qquad \uparrow$$
$$\text{Endo-Artinian} -? \rightarrow \text{Endo-Noetherian}$$
$$\uparrow \qquad\qquad\qquad \uparrow$$
$$\text{Artinian} \longrightarrow \text{Noetherian}$$

- A ring A is left (resp. right) Hopfian $\Leftrightarrow A$ is Dedekind-Finite ([67], Proposition 2.2.3 and [9], Proposition 4.1). NB. it exists some left co-Hopfian rings which are not right co-Hopfian and vice versa ([67], Example A and [70], Example 1.5).

- The rings: Self-Injective, Dedekind-Finite and Regular, Strongly-π-Regular, left or right Perfect, left or right Artinian are all left or right co-Hopfian ([67], Proposition 2.2.4).

- Let $A = \begin{pmatrix} \mathbb{Z}/2\mathbb{Z} & \mathbb{Z}/2\mathbb{Z} \\ 0 & \mathbb{Z}_{(2)} \end{pmatrix}$, then A is left co-Hopfian and not left Strongly-co-Hopfian ([23], Example 2.2).

- A is left co-Hopfian $\Rightarrow A$ left Hopfian ([70], Proposition 1.10).

- A is left Artinian $\Leftrightarrow A$ is left Noetherian and A is Strongly-co-Hopfian ([37], Corollary 3.4.27).

- A is Strongly-co-Hopfian $\Rightarrow A$ is Strongly-Hopfian ([23], Theorem 2.1, [37], Proposition 3.3.5 and [38], Corollary 3.5)).

- A is Strongly-co-Hopfian $\Leftrightarrow A$ is Strongly-Hopfian and π-Regular ([37], Theorem 3.3.10).

- A is Weakly-co-Hopfian $\Rightarrow A$ is Hopfian $\Rightarrow A$ is Generalised-Hopfian ([23], Theorem 2.3 and Corollary 2.1).

- Let p be a prime number, the commutative ring $A = \prod_{n \geq 1} \mathbb{Z}/p^n\mathbb{Z}$ is Hopfian, but he is neither Strongly-Hopfian nor Strongly-co-Hopfian ([38], Remark 2.16 and [24], Example 4.1).

- If A is left principally quasi-injective, then ([61], Proposition 3.3 and [67], Proposition 2.2.4):

$$A \text{ is Dedekind-Finite } \Leftrightarrow A \text{ is right co-Hopfian.}$$

- If A is commutative, then A is Semi-Hopfian (resp. Semi-co-Hopfian) as A-module on right (in the sense of Diviani-Azar and Mafi) if, and only if, A is Hopfian (resp. co-Hopfian) as right A-module.

- Let $(A_i)_{i \in I}$ a family of rings, we have: ([5], Proposition 2.10): $A = \prod_{i \in I} A_i$ verifies **GC2** as left module $\Leftrightarrow A_i$ is a ring left co-Hopfian, for any $i \in I$.

- If 0 and 1 are the only idempotents of A, then ([5], Proposition 2.4): A verifies the property **C2** on left $\Leftrightarrow A$ is left co-Hopfian $\Leftrightarrow A$ verifies **GC2** on left.

  In particular, if A is an integral domain, the properties (i)–(iii) are equivalent to:

  (iv) A is a un division ring.

- If the set of all orthogonal idempotents of A is finite, then ([5], p. 4): A verifies **GC2** on left $\Rightarrow A$ is left co-Hopfian.

6.2 Endo-Artinian or Endo-Noetherian Rings

Generality

- A is left Endo-Noetherian if, and only if, A verifies CCA for the principal annihilators ([41], Proposition 1.16 and Example 1.17). That is to say any increasing sequence: $l(x_1) \subseteq l(x_2) \subseteq \cdots \subseteq l(x_n) \cdots$, is stationary. For example, any ring with the condition of the maximum on the left annihilators is left Endo-Noetherian, in particular, any left Goldie ring is left Endo-Noetherian.

- The notion of the left and right Endo-Artinian is not symmetrical. If A is left or right perfect, then A is a Strongly-π-Regular Ring. We deduce that if A is left or right Endo-Artinian, then A is Strongly-π-Regular.
- Let A be a left Self-Injective Ring, then ([41], Corollary 3.7):

$$A \text{ is left Endo-Noetherian } \Leftrightarrow A \text{ is right Endo-Artinian}$$

- The simplest example of ring A which is (left or right) Endo-Artinian and Endo-Noetherian, but not noetherian is the local ring $(A, \mathfrak{m})$ with $\mathfrak{m}^2 = (0)$. As, the only annihilators of A are $\{0\}, \mathfrak{m}$ and A. So A is Endo-Noetherian ([48], Exercise 24.8, [21], Example 4)
- Let $A = \{ \begin{pmatrix} a & b \\ 0 & a \end{pmatrix} \mid a \in \mathbb{Q}, \ b \in \mathbb{R} \}$. One verify $Rad(A) = \{ \begin{pmatrix} 0 & b \\ 0 & 0 \end{pmatrix} \mid b \in \mathbb{R} \}$ with $Rad(A)^2 = 0$ and $A/Rad(A) \simeq \mathbb{Q}$ is artinian, which implies A is semi-primary and hence A is a perfect ring. For any sequence $(M_i)_{i \in \mathbb{N}}$ Strictly monotonous of $\mathbb{R}$-vector subspaces of $\mathbb{Q}$, note $K_i = \{ \begin{pmatrix} 0 & m \\ 0 & 0 \end{pmatrix} \mid m \in M_i \}$, then $(K_i)_{i \in \mathbb{N}}$ is a strictly monotonous sequence of right ideals of A. It follows that A is neither Artinian nor Noetherian. On the other hand, there is no (left or right) perfect ring A which is Noetherian and which is not artinian, for if $Rad(A)$ is a Nil-ideal and A is noetherian, then A is Artinian (Hopkins-Levitzki's Theorem).
- Let A be left Endo-Artinian ring. We have ([48], Exercise 20.6):

$$A \text{ is left (resp. right) Noetherian } \Leftrightarrow A \text{ is left (resp. right) Artinian.}$$

More generally, if A is left Endo-Artinian and M a right A-module, we have ([66], Note 2.13):

$$M \text{ is Artinian } \Leftrightarrow M \text{ is Noetherian.}$$

- Let A be left Endo-Artinian, all left (or right) finitely generated A-module is co-Hopfian ([48], Exercise 23.11).
- A left Endo-Artinian $\Leftrightarrow$ any left A-module verifies DCC on the finitely generated sub-modules ([48], Exercise 23.4).

The Hopkins-Levitzki's Property

- According to Bass's theorems [6] and Jonah's one [40], we have:

$$A \text{ is left Endo-Artinian}$$
$$\Updownarrow$$
$$A \text{ verifies CCD on left principal ideals}$$
$$\Updownarrow$$
$$A \text{ is right Perfect}$$
$$\Downarrow$$
$$A \text{ verifies CCA on right principal ideals}$$

It follows that the following implications are equivalent:
(i) If A is left Endo-Artinian, then A is left Endo-Noetherian.
(ii) If A verifies the CCD to the left principal, then A verifies CCA for principal left annihilators.

To show (i) or (ii), it suffices to show that: If A verifies the CCA for the principal right ideals, then A verifies CCA for the left principal annihilators.

- For a A ring, in general we have ([20], Proposition 2.1 and Theorem 2.2):

A verifies CCD for right (resp. left) annihilators

$\Updownarrow$

A verifies CCA for left (resp. right) annihilators.

A verifies CCD for left direct summands

$\Updownarrow$

A verifies CCA for right direct summands.

- Let A be a (left and right) morphic ring we have: ([55], Theorem 35)

A is left Endo-Artinian $\Leftrightarrow$ A is Semi-Primary $\Leftrightarrow$ A is left Artinian.

- If A is right principally-Injective, the map $\Theta : \{Aa \mid a \in A\} \to \{r(a) \mid a \in A\}$ such that $\Theta(Aa) = r(a)$ is a decreasing bijection. We deduce that ([14], Lemma 6.1, [41], Example 1.18 and Corollary 3.7):

A is left Endo-Artinian $\Leftrightarrow$ A is right Endo-Noetherian.

- If A is right Generalised-Pseudo-principally-Injective and verifies CCA on right annihilators, then A is Semi-Primary, so A is Endo-Artinian ([61], Corollary 3.5).

6.3 Hilbert's Property

Hilbert's Property for Modules. Let A be an associative ring and M a A-module.

- The following conditions are equivalent ([71], Theorem 5.9):
  (i) M is a A-module Dedekind-Finite.
  (ii) $M[x]$ is a $A[x]$-module Dedekind-Finite.
  (iii) $M[x]/(x^{n+1})$ is a $A[x]/(x^{n+1})$-module Dedekind-Finite, for any $n \in \mathbb{N}$.
  (iv) $M[[x]]$ is a Dedekind-Finite $A[[x]]$-module.
- The following conditions are equivalent ([70], Theorem 2.1).
  (i) M is a Hopfian A-module.
  (ii) $M[x]$ is a Hopfian $A[x]$-module.
  (iii) $M[x]/(x^{n+1})$ is a Hopfian $A[x]/(x^{n+1})$-module, for any $n \in \mathbb{N}$.
  (iv) $M[[x]]$ is a Hopfian $A[[x]]$-module.
- The $A[x]$-module $M[x]$ it is never co-Hopfian. But, we have ([23], Theorem 3.1):
  M is a co-Hopfian A-module $\Leftrightarrow$ $M[x]/(x^{n+1})$ is a co-Hopfian $A[x]/(x^{n+1})$-module, for any $n \in \mathbb{N}$.

- We have ([5], Theorem 2.19 and 3.16, [74], Theorem 2.11 and [76], Proposition 3.7):

$M[x]$ verifies **GC2** as $A[x]$-module $\Rightarrow M$ verifies **GC2** as A-module.

$M[x]$ verifies **GD2** as $A[x]$-module $\Rightarrow M$ verifies **GD2** as A-module.

$M[x]$ is a Strongly-Hopfian $A[x]$-module $\Rightarrow M$ is a Strongly-Hopfian A-module.

$M[x]$ is a Strongly-co-Hopfian $A[x]$-module $\Rightarrow M$ is a Strongly-co-Hopfian A-module.

$M[x]$ is a Generalised-co-Hopfian $A[x]$-module $\Rightarrow M$ is a Generalised-co-Hopfian A-module.

- For any $n \in \mathbb{N}^*$, we have ([25], Theorem 2.2 and 2.4):

M is Generalised-Hopfian $\Leftrightarrow M[x]/(x^n)$ is a Generalised-Hopfian $A[x]/(x^n)$-module.

M is Weakly-co-Hopfian $\Leftrightarrow M[x]/(x^n)$ is a Weakly-co-Hopfian $A[x]/(x^n)$-module

- If A a commutative ring, The following conditions are equivalent ([24], Theorem 5.2):
  (i) M is a S.S.H A-module,
  (ii) $M[x]$ is a S.S.H $A[x]$-module,
  (iii) $M[x, x^{-1}]$ is a S.S.H $A[x, x^{-1}]$-module.
  (iv) $M[x]/(x^{n+1})$ is a S.S.H $A[x]/(x^{n+1})$-module, for any $n \in \mathbb{N}$.

Hilbert's Property for Rings

- Let A be a commutative ring and $n \in \mathbb{N}$, we have ([51], Corollary 2.6 and [76], Proposition 3.5):

A is right Strongly-Hopfian $\Leftrightarrow A[x]/(x^{n+1})$ is right Strongly-Hopfian.

- Let A be a ring with an endomorphism α and a α-derivation δ. The A-module M is said α-**compatible** if, for any $m \in M$ and $a \in A$ we have $ma = 0 \Leftrightarrow m\alpha(a) = 0$. M is said δ-**compatible** if, for any $m \in M$ and $a \in A$ we have $ma = 0 \Rightarrow m\delta(a) = 0$. If M is α-compatible and δ-compatible, one says that M is (α, δ)-**compatible**. The ring A is said (α, δ)-compatible if it is (α, δ)-compatible as right A-module ([1], Definition 2.11). Let A be a (α, δ)-compatible ring such that $A[x, \alpha, \delta]$ is reversible, we have ([51], Proposition 2.14 and Corollary 2.15):

A is right Strongly-Hopfian $\Leftrightarrow A[x, \alpha, \delta]$ is right Strongly-Hopfian.

If α is compatible and $A[x, \alpha]$ is reversible, then:

A is right Strongly-Hopfian $\Leftrightarrow A[x, \alpha]$ is right Strongly-Hopfian.

If δ is compatible and $A[x, \delta]$ is reversible, then:

A is right Strongly-Hopfian $\Leftrightarrow A[x, \delta]$ is right Strongly-Hopfian.

In particular, If A is a commutative ring, we have ([37], Theorem 3.3.6 and Corollary 3.3.8, [38], Theorem 5.1 and Corollary 5.4, [51], Corollary 2.16 and Corollary 2.20, [76], Corollary 3.8):

A is Strongly-Hopfian $\Leftrightarrow A[x]$ is Strongly-Hopfian.

- If A is commutative, then ([51], Corollary 2.20):

  A is right Strongly-Hopfian $\Rightarrow A[x, x^{-1}]$ is right Strongly-Hopfian.

- If A is Strongly-Hopfian, then $A[[x]]$ is not, generally, Strongly-Hopfian [36].

6.4   Hopficity and Extensions for Matrices

- Let A be an unitary associative ring. For $n \in \mathbb{N}^*$, we design $M_n(A)$ the square matrix ring of order n in A. We have the following implications, of which the reciprocals, are not generally true ([70], Proposition 1.12 and 7.1):
  (i) $M_n(A)$ is left Hopfian (resp. co-Hopfian) $\Rightarrow$ A is left Hopfian (resp. co-Hopfian).
  (ii) $M_n(A)$ is Hopfian (resp. co-Hopfian) as left A-module $\Rightarrow$ A is left Hopfian (resp. co-Hopfian).
- Soit A a commutative ring, for any $n \in \mathbb{N}^*$ we have ([70], Theorem 3.4):
  (i) $M_n(A)$ is Hopfian.
  (ii) A is co-Hopfian $\Rightarrow$ $M_n(A)$ is left and right co-Hopfian.
  (iii) A is co-Hopfian $\Rightarrow$ $M_n(A)$ is co-Hopfian as left and on right A-module.
- For any $n \in \mathbb{N}^*$, we have ([48], Exercise 23.11 and [76], Proposition 2.6):
  A is Strongly-Hopfian (resp. Strongly-co-Hopfian) $\Rightarrow M_n(A)$ is left and right co-Hopfian
  A is left Endo-Artinian ring $\Rightarrow$ $M_n(A)$ is left Endo-Artinian, for any $n \in \mathbb{N}^*$.
- Let A be an unitary associative ring. For $n \in \mathbb{N}^*$, we design $T_n(A)$ the ring of the upper triangular matrices of orders n in A and note $S(A, n)$ and $T(A, n)$ the sub-rings of $T_n(A)$ such that:

$$
S(A, n) = \left\{ \begin{pmatrix} a & a_{1,2} & a_{1,3} & \cdots & a_{1,n} \\ 0 & a & a_{2,3} & \cdots & a_{2,n} \\ \vdots & \vdots & \ddots & \ddots & \vdots \\ 0 & \cdots & \cdots & a & a_{n-1,n} \\ 0 & \cdots & \cdots & 0 & a \end{pmatrix} \Big/ \ a, a_{i,j} \in A \right\}
$$

$$
T(A, n) = \left\{ \begin{pmatrix} a_1 & a_2 & a_3 & \cdots & a_n \\ 0 & a_1 & a_2 & \cdots & a_{n-1} \\ 0 & 0 & a_1 & \cdots & a_{n-2} \\ \vdots & \vdots & \vdots & \ddots & \vdots \\ 0 & 0 & 0 & \cdots & a_1 \end{pmatrix} \Big/ \ a_i \in A \right\}.
$$

  we have $T(A, 2) = S(A, 2)$ and $T(A, n) \cong A[x]/(x^n)$, for any $n \in \mathbb{N}^*$.
- Let A be a Dedekind-Finite ring, generally, $M_n(A)$ is not Dedekind-Finite [17]. On the contrary, for any $n \in \mathbb{N}^*$, we have ([22], Corollary 2.10 and [9], Corollary 2.6):

  A is Dedekind-Finite $\Leftrightarrow T_n(A)$ is Dedekind-Finite.

- For any $n \in \mathbb{N}^*$, the following assertions are equivalent ([51], Corollary 2.4 and 2.6):
  - (i) A is right Strongly-Hopfian.
  - (ii) $T(A,2) = S(A,2)$ is right Strongly-Hopfian.
  - (iii) $T(A,n)$ is right Strongly-Hopfian.
  - (iv) $S(A,n)$ is right Strongly-Hopfian.
  - (v) $T_n(A)$ is right Strongly-Hopfian.
  - (vi) $A[x]/(x^n)$ is right Strongly-Hopfian.
- The following properties are equivalent ([25], Corollary 3.3 and Proposition 3.4):
  - (i) A is Generalised-Hopfian.
  - (ii) $T(A,2)$ is Generalised-Hopfian.
  - (iii) $T_n(A)$ is Generalised-Hopfian.
- $T(A,2)$ is S.S.H $\Leftrightarrow$ A is S.S.H ([24], Corollary 5.4).
- Let A and B be rings and M a (A,B)-bimodule, note the ring $T(A,B,M)$:
  $T(A,B,M) = \begin{pmatrix} A & M \\ 0 & B \end{pmatrix}$, provided with the usual matrix addition and multiplication.

  If $A = B$ we note, the ring $A \oplus M$ equipped with the usual addition and multiplication: $(a_1, m_1)(a_1, m_1) = (a_1 a_2, a_1 m_2 + m_1 a_2)$. This ring is isomorphic to the ring of the matrices:
  $T(A,M) = \left\{ \begin{pmatrix} a & m \\ 0 & a \end{pmatrix} : a \in A, \ m \in M \right\}$, equipped with the usual operations on the matrices.
- We have ([9], Corollary 2.7, [25], Proposition 3.2 and [75], Proposition 5.4):
  $T(A,B,M)$ is Dedekind-Finite $\Leftrightarrow$ A and B are Dedekind-Finite.
  $T(A,B,M)$ is Generalised-Hopfian $\Leftrightarrow$ A and B are Generalised-Hopfian .
- If for any $0 \neq a \in A$, $l(a) = 0$, then $T(A,M)$ is left Strongly-Hopfian ([76], Corollary 2.4).
- If A is Endo-Artinian, then $T(A,M)$ is left Endo-Artinian ([18], Lemma 3.3).
- A is left co-Hopfian $\Rightarrow$ A^n is co-Hopfian as left A-module ([70], Theorem 3.2).
- A^n verifies **GC2** as left A-module $\Leftrightarrow$ $M_n(A)$ verifies **GC2** as left $M_n(A)$-module ([5], Corollary 2.18).

7   Extension to Banach Spaces

7.1   Banach Space Hopfian or co-Hopfian [42]

- A Banach space X is said **Hopfian** if, any continuous linear operator surjective of X on X is bijective. It is said **co-Hopfian** if, any continuous linear operator injective of X in X, is surjective. X is said **Dedekind-Finite**, if any left invertible operator of the algebra $BL(X)$ is invertible.
- A Hopfian or Co-Hopfian Banach space is Dedekind-Finite, and any Banach space of finite dimension is Hopfian and co-Hopfian. For a Hilbert space H, we have:

  H is Hopfian $\Leftrightarrow$ H is co-Hopfian $\Leftrightarrow$ H is Dedekind-Finite $\Leftrightarrow$ H is finite dimension.

- The first example of an infinite-dimensional Hopfian (Dedekind-Finite) Banach space is given by Gowers and Maurey ([27], Theorem 17 and [28]). The question of existence of a Banach space co-Hopfian of infinite dimensional was posited in ([70]), and with very versatile methods, Avilés and Koszmider [4], constructed a $\mathcal{C}(K)$ co-Hopfian, with K infinite compact.
- Let X be a Banach space, then the following statements are equivalent:
  1. X is Hopfian
  2. X is not isomorphic to any of its own quotient spaces.
  3. X is Dedekind-Finite and the kernel of any surjective operator of $BL(X)$, is a direct summand of X.
- A Banach space X, is said **Indecomposable** if it can not be written as a direct sum of two closed subspaces of infinite dimensional and it is called **Hereditarily-decomposable** or **HI-space**, if all its closed subspaces are decomposable.
- Let X be a Banach space in which the interior of the spectrum of any operator $T \in BL(X)$ is empty $(int(\sigma(T)) = \phi)$. Then X is not isomorphic to any of its own subspaces and has none of its own quotient spaces. In particular X is Hopfian [32].
- The HI-Banach spaces are Hopfian [32]. But, there are Hopfian Banach spaces that are not HI-spaces [2].
- The strongly-Hopfian or strongly-co-Hopfian Banach spaces are finite dimensional. They exist strongly-Hopfian Banach algebras without Jacobson radical of infinite dimensional [10].
- Strongly-co-Hopfian Banach algebras are algebraic ([37], Theorem 3.3.9).

7.2   Semi-Hopfian or Semi-co-Hopfian Banach Spaces [42]

- A Banach space X is said Semi-Hopfian (resp. Semi-co-Hopfian) if, any surjective (resp. injective) operator of X in X is right invertible (resp. left invertible) in $BL(X)$.
- A Banach space X in which any operator $T \in BL(X)$, is Von Newman Regular, is finite dimensional.
- any Banach space isomorphic to an Hilbert space, is Semi-Hopfian. the Semi-Hopfian Hilbert spaces are finite dimensional.

7.3   Open Questions [42]

1. Is any co-Hopfian Banach space is Hopfian?
2. Is any Dedekind-Finite Banach space is Hopfian?
3. Characterise all Compact topology spaces K, from which the Banach space $\mathcal{C}(K)$ is Hopfian or/and co-Hopfian.
4. Characterise all Hopfian or/and co-Hopfian Banach spaces X, by properties of their Banach algebra $BL(X)$.
5. are they hereditary Hopfian or co-Hopfian Banach spaces of infinite dimensional?
6. Characterise all Semi-Hopfian or/and Semi-co-Hopfian Banach spaces

7. Is any Semi-co-Hopfian Banach space is Semi-Hopfian?
8. Study hereditable Semi-Hopfian and Semi-co-Hopfian Banach spaces (Hilbert's spaces are hereditary Hopfian).
9. Study the properties of $BL(X)$ with X Semi-Hopfian or Semi-co-Hopfian.

References

1. Alhevaz, A., Moussavi, A.: On skew-Armendariz an skew quasi-Armendariz modules. Bull. Iran. Math. Soc. **1**(38), 55–84 (2012)
2. Argyros, S.A., Lopez-Abad, J., Todorcevic, S.: A class of Banach spaces with few non-strictly singular operators. J. Funct. Anal. **22**, 306–384 (2005)
3. Armendariz, E.P., Fisher, J.N., Snider, R.L.: On injective and surjective endomorpfismes of finitely generated modules. Commun. Algebra **6**(7), 659–672 (1978)
4. Avilés, A., Koszmider, P.: A Banach space in which every injective operator is surjective. Bull. Lond. Math. Soc. **45**, 1065–1074 (2013)
5. Aydoğdu, P., Özcan, A.Ç.: Semi co-Hopfian and semi-Hopfian modules. East-West J. Math. **10**(1), 57–72 (2008)
6. Bass, H.: Finitistic dimension and a homological generalization of semi-primary rings. Trans. Am. Math. Soc. **95**, 466–488 (1960)
7. Baupradist, S., Hai, H.D., Sanh, N.V.: On pseudo-p-injectivity. Southeast Asian Bull. Math. **35**, 21–27 (2011)
8. Bouvier, L., Payan, J.J.: Modules sur certains anneaux de Dedekind. Séminaire de théorie des nombres de Grenoble **2**(3), 1–12 (1972)
9. Breaz, S., Călugăreanu, G., Schultz, P.: Modules with Dedekind-finite endomorphism rings. Mathematica **53**(76), 15–28 (2011)
10. Burgos, M., Kaidi, A., Mbekhta, M., Oudghiri, M.: The descent spectrum and perturbations. J. Oper. Theory **56**, 259–271 (2006)
11. Călugăreanu, G.: Morphic abelian groups. J. Algebra Appl. **9**(2), 185–193 (2010)
12. Călugăreanu, G.: Abelian groups with left morphic endomorphism ring. J. Algebra Appl. **17**(9), 1850176 (2018)
13. Călugăreanu, G.G., Pop, L.: Between morphic and Hopfian. An. şt. Univ. Ovidius Constanţa **21**(3), 51–66 (2013)
14. Camillo, V., Nicholson, W.K.: On rings where left principal ideals are left principal annihilators. Int. Electron. J. Algebra **17**, 199–214 (2015)
15. Campbell, R.N.: Injective modules and divisible groups. Master's thesis, University of Tennessee (2015)
16. Chaturvedi, A.K., Pandeya, B.M., Gupta, A.J.: Quasi-c-principally injective modules and self-c-principally injective rings. Southeast Asian Bull. Math. **33**, 685–702 (2009)
17. Cohn, P.M.: On n-simple rings. Algebra Univers. **53**, 301–305 (2005)
18. Diesl, A.J., Dorsey, T.J., McGovern, W.W.: A characterization of certain morphic trivial extensions. J. Algebra Appl. **10**(4), 623–642 (2011)
19. Divaani-azar, K., Mafi, A.: A new characterization of commutative artinian rings. Vietnam J. Math. **32**, 319–322 (2004)
20. Dokuchaev, M.A., Gubareni, N.M., Kirichenko, V.V.: Rings with finite decomposition of identity. Ykp. Mam. Hcyph. **63**(3), 369 (2011)
21. Faith, C.: Finitely embedded commutative rings. Proc. Am. Math. Soc. **112**(3), 657–659 (1991)

22. Foldes, S., Szigeti, J., Wyk, L.V.: Invertibility and Dedekind-Finiteness in structural matrix rings. Linear Multilinear Algebra **59**(2), 221–227 (2011)
23. Gang, Y., Zhong-kui, L.: On Hopfian and Co-Hopfian modules. Vietnam J. Math. **351**, 1–8 (2007)
24. Gang, Y., Zhong-kui, L.: On generalisations of fitting modules. Indian J. Math. Pramila Srivastava Memorial **51**(1), 85–99 (2009)
25. Gang, Y., Zhong-kui, L.: Notes on generalized Hopfian and weakly co-Hopfian modules. Commun. Algebra **38**, 3556–3566 (2010)
26. Ghorbani, A., Haghany, A.: Generalized Hopfian modules. J. Algebra **255**, 324–341 (2002)
27. Gowers, T.: A solution to Banach's hyperplane problem. Bull. London Math. Soc. **26**(6), 523–530 (1994)
28. Gowers, W.T., Maurey, B.: The unconditional basic sequence problem. J. Am. Math. Soc. **6**, 851–874 (1993)
29. Haghany, A., Vedadi, M.R.: Modules whose injective endomorphisms are essential. J. Algebra **243**, 765–779 (2001)
30. Hai, P.T.: Some Generalisations of Clas of Pseudo-Injective Modules and Related Rings. Doctor of Philosophy in Mathematics, Hue University, College of Education (2016)
31. Hai, P.T., Koşan, M.T., Quinh, T.C.: Weakly $C2$ Modules and Rings (2014)
32. Haily, A., Kaidi, A., Rodriguez Palacios, A.: Algebra descent spectrum of operators. Israel J. Math. **177**, 349–368 (2010)
33. Han, J., Lee, Y., Park, S., Sung, H.J., Yun, S.J.: On idempotents in relation with regulariy. J. Korean Math. Soc. **53**(1), 217–232 (2016)
34. Harmanci, A., Kose, H., Kurtulmaz, Y.: On π-morphic modules. Hacettepe J. Math. Stat. **42**(4), 411–418 (2013)
35. Hill, P., Megibben, C.: On primary groups with countable banc subgroups. Trans. Am. Math. Soc. **124**, 45–59 (1966)
36. Hizem, S.: Formal power series over strongly Hopfian rings (preprint)
37. Hmaimou, A.: Modules Fortement-Hopfiens et modules Fortement-co-Hopfiens. Doctorat en Sciences, Université Abdelmalek Essaâdi, Tétouan (2010)
38. Hmaimou, A., Kaidi, A., Sánchez Campos, E.: Generalized fitting modules and rings. J. Algebra **308**(1), 199–214 (2007)
39. Jasim, M., Ali, M.: Some classes of injectivity and projectivity. A thesis of the Degree of Doctor of Philosophy in Mathematics of the University of Baghdad (2006)
40. Jonah, D.: Rings with the minimumcondition for principal right ideals have the maximum condition for principal left ideals. Math. Z. **113**, 106–112 (1970)
41. Kaidi, A.: Modules with chain conditions on endoimages and endokernels (preprint)
42. Kaidi, A.: Espaces de Banach Semi-Hopfiens et Espaces de Banach Semi-co-Hopfiens. In: 3rd International Congress, Algebra, Number Theory and Applications, Oujda-Morocco, 24–27 April (2019). (preprint)
43. El Amin Mokhtar, K., Mamadou, S.: Une caracterisation des anneaux artiniens a ideaux principaux. In: Bueso, J.L., Jara, P., Torrecillas, B. (eds.) Ring Theory. LNM, vol. 1328, pp. 245–254. Springer, Heidelberg (1988). https://doi.org/10.1007/BFb0100930
44. Kaleboğaz, B.: Some generalizations of quasi-projective modules. Submitted to the Institute of Sciences of Hacettepe University as a Partial Fulfillment to the Requirements for the Award of the Degree of Doctor of Philosophy in Mathematics (2014)

45. Kaplansky, I.: Modules over Dedekind rings and valuation rings. Trans. Am. Math. Sot. **72**, 327–340 (1952)
46. Kplansky, I.: Topological representation of algebras II. Trans. Am. Math. Soc. **68**, 62–75 (1950)
47. Krylov, P.A., Mikhalev, A.V., Tuganbaev, A.A.: Endomorphism rings of abelian groups. J. Math. Sci. **110**(3), 2683–2745 (2002)
48. Lam, T.Y.: Exercises in Classical Ring Theory. Second Edition Edited by K.A. Bencs ath P.R. Halmos. Springer, Heidelberg (2003)
49. Lee, G.: Theory of Rickart modules. Dissertation Presented in Partial Fulfillment of the Requirements for the Degree Doctor of Philosophy in the Graduate School of The Ohio State University (2010)
50. Lee, G., Rizvi, S.T., Rman, C.: Modules whose endomorphism rings are Von-Neumann regular. Commun. Algebra **41**, 4066–4088 (2013)
51. Lunqun, O., Jinwang, L., Yueming, X.: Ore extensions over right strongly Hopfian rings. Bull. Malays. Math. Sci. Soc. **39**(2), 805–819 (2016)
52. Mohamed, S.H., Müller, B.J.: Continuous and Discrete Modules. London Mathematical Society, Lecture Note Series, vol. 147 (1990)
53. Mousvi, S.A., Nekooei, R.: Characterization of secondary modules over Dedekind domains. Iranian J. Sci. Technol. Trans. A Sci. **43**(3), 919–922 (2019)
54. Ndiaye, M.A., Guèye, C.T.: On commutative EKFN-ring with ascending chain condition on annihilators. Int. J. Pure Appl. Math. **86**(5), 871–881 (2013)
55. Nicholson, W.K., Sánchez Campos, E.: Rings with the dual of the isomorphism theorem. J. Algebra **271**, 391–406 (2004)
56. Nicholson, W.K., Sánchez Campos, E.: Morphic modules. Commun. Algebra **33**, 2629–2647 (2005)
57. Padashnik, F., Moussavi, A., Mousavi, H.: The ascending chain condition for principal left or right ideals of skew generalized power series rings (2016)
58. Patel, M.K., Kumar, V., Gupta, A.J.: On semi-projective modules and their endomorphism rings. Asian-Eur. J. Math. **11**(2), 1850029 (2018)
59. Patel, M.K., Pandeya, B.M., Kumar, V.: Generalization of semi-projective modules. Int. J. Comput. Appl. **83**(8), 1–6 (2013)
60. Quynh, T.C.: On pseudo semi-projective modules. Turk. J. Math. **37**, 27–36 (2013)
61. Quynh, T.C., Van Sanh, N.: On quasi pseudo-GP-injective rings and modules. Bull. Malays. Math. Sci. Soc. **37**(2), 321–332 (2014)
62. Renault, G.: Algèbre Non Commutative. Gauthier - Villars, Paris (1975)
63. Ribenboim, P.: Rings and Modules. Tracts in Math. 24. Intersciences Publ, New York (1969)
64. Rotman, J.: The Theory of Groups: An Introduction. Allyn and Bacon, Boston (1965)
65. Rowen, L.H.: Ring Theory, vol. 1. Academic Press Inc., San Diego (1988)
66. Safaeeyan, S.: Strongly duo and co-multiplication modules. J. Algebraic Syst. **3**(1), 53–64 (2016)
67. Schwiebert, R.C.: Faithful torsion modules and rings. A dissertation presented to the faculty of the College of Arts and Sciences of Ohio University (2011)
68. Thao, L.P., Sanh, N.V.: A generalization of Hopkins-Levitzki theorem. Southeast Asian Bull. Math. **37**, 591–600 (2013)
69. Ungor, B., Kurtulmaz, Y., Halicioglu, S., Harmanci, A.: Dual π-Rickart modules. Rev. Colomb. Mat. **46**(2), 167–183 (2012)
70. Varadarajan, K.: Hopfian and co-Hopfian objects. Publ. Mat. **36**, 293–317 (1992)
71. Varadarajan, K.: Analogues of IBN and related properties for modules. Acta Math. Hungarica **119**(1–2), 95–125 (2008)

72. Vasconcelos, W.V.: On finitely generated flat modules. Trans. Am. Math. Soc. **138**, 505–512 (1969)
73. Vasconcelos, W.V.: On injective endomorphisms of finitely generated modules. Proc. Am. Math. Soc. **25**, 900–901 (1970)
74. Wang, Y.: Generalizations of Hopfian and Co-Hopfian modules. Int. J. Math. Math. Sci. **9**, 1455–1560 (2005)
75. Xianneng, D.: Properties on formal triangular matrix rings and modules
76. Yan, X.F., Liu, Z.K.: Extensions of generalized fitting modules. J. Math. Res. Exposition **30**(3), 407–414 (2010)
77. Zhu, Z.: Pseudo QP-injective modules and generalized pseudo QP-injective modules. Int. Electron. J. Algebra **14**, 32–43 (2013)

Functor $(\overline{S})^{-1}()$ and Functorial Isomorphisms

Daouda Faye[1(✉)], Mohamed Ben Maaouia[2(✉)], and Mamadou Sanghare[3(✉)]

[1] Laboratory of Algebra, Cryptography and Algebraic Geometry
and Applications – LACGAA, University Cheikh Anta Diop of Dakar,
Dakar, Senegal
fayeda2@yahoo.fr
[2] Laboratory of Algebra, Codes and Cryptography Applications (ACCA), UFR SAT,
University Gaston Berger (UGB) - St. Louis Senegal, Saint Louis, Senegal
maaouiaalg@hotmail.com
[3] Doctoral School of Mathematics-Computer – UCAD-Sénégal, University Cheikh
Anta Diop of Dakar, Dakar, Senegal
mamadou.sanghare@ucad.edu.sn

Abstract. The main result of this paper is the following:
    Let B be a unitary noetherian ring, $A = Z(B)$, the center of B, S a central saturated multiplicative subset of A satisfying the left (respectively right) conditions of Ore (respectively the subset of regular elements of $A - P$ where P is a prime ideal of A), $S^{-1}A$ the ring of fractions of A in S, $(\overline{S})^{-1}B$ the ring of fractions of B in $\overline{S}$, $_BM_A$ a free $(B - A)-$ bimodule of finite type, $A - Mod_{ff}$ (respectively $S^{-1}A - Mod_{ff}$) the subcategory of $A - Mod$ (respectively $S^{-1}A - Mod$) containing the free A-modules of finite type (respectively the free $S^{-1}A$-modules of finite type); then the covariant functors:
    $Ext^n_{(\overline{S})^{-1}B}(S^{-1}M, -)$: $(\overline{S})^{-1}B - Mod_{ff} \rightarrow S^{-1}A - Mod_{ff}$ and
$Tor^{S^{-1}}_n A(S^{-1}M, -) : S^{-1}A - Mod_{ff} \rightarrow (\overline{S})^{-1}B - Mod_{ff}$ are adjoint.

Keywords: Ring · Left (right) conditions of Ore · Closed multiplicative subset · Letf A-module · Ring of fractions · Module of fractions · Categories A-Mod · Mod-A · $A - Mod_{ff}$ · $S^{-1}A - Mod_{ff}$ · Functors $S^{-1}()$ · Ext and Tor

1   Introduction

In this **paper**, unless otherwise stated, B means a unitary noetherian ring, $A = Z(B)$, the center of B, S a central saturated multiplicative subset of A satisfying the left (respectively right) conditions of Ore (respectively the subset of regular elements of $A - P$ where P is a prime ideal of A, $\overline{S}$ the saturated multiplicative subset of B generated by S satisfying the left (respectively

---

Submission 17 A2C Conference.

right) conditions of Ore, $S^{-1}A$ the ring of fractions of A with respect of S, $(\overline{S})^{-1}B$ the ring of fractions of B with respect of $\overline{S}$ and $_BM_A$ a finitely generated free $(A\text{-}B)$-bimodule, $A - Mod_{ff}$ (resp. $S^{-1}A - Mod_{ff}$) the subcategory of $A - Mod$ (respectively $S^{-1}A - Mod$) containing the finitely generated free A-modules(respectively the finitely generated free $S^{-1}A$-modules).

We are interested in the adjoint notion between the functor $S^{-1}()$ and the homological functors $Ext^n_A(M,-)$ and $Tor^A_n(M,-)$ which hav'nt been studied according to us.

In the case where the ring is not necessarily commutative, Maaouia has established in its works that the functor $S^{-1}()$ is adjoint to the functor $Hom_A(S^{-1}A-)$ (see [5], $chapI$ and [6].

We know that (see [10]) the functor $Ext^0_A(M,-)$ is equivalent to the functor $Hom_A(M,-)$, and $Tor^A_0(M,-)$ is equivalent to the functor $M \otimes_A -$ Which is isomorphic to the functor $S^{-1}()$.

So $Ext^0_{S^{-1}A}(S^{-1}A,-)$ is adjoint to $S^{-1}()$ and $Tor^{S^{-1}A}_0(S^{-1}A,-)$ is isomorphic to $S^{-1}()$. We generalize so this result for any non negative integer n as follows.

If B denotes a unitary ring, A a sub-ring of B, M a left finitely generated A-module, then $Ext^n_{(\overline{S})^{-1}B}(S^{-1}M,-)$ and $Tor^{S^{-1}A}_n(S^{-1}M,-)$ are adjoint functors for any non negative integer n.

Let us recall for this some useful cases for this work:
Let A, B and C be three rings.
-If M is an $(A-B)$-bimodule and N a left A-module, then $Hom_A(M,N)$ and $Ext^A_n(M,N)$ are left B-modules.
Thus we have established the following results:

(1) Isomorphisms of left $(\overline{S})^{-1}B$−modules: $(\overline{S})^{-1}Hom_A(M,N) \cong Hom_{S^{-1}A}$ $(S^{-1}M,S^{-1}N))$ where M is an $(A-B)$−-bimodule and N a left A-module;
(2) Isomorphisms of left $(\overline{S})^{-1}B$−modules: $(\overline{S})^{-1}Tor^A_n(M,N) \cong Tor^{S^{-1}A}_n$ $(S^{-1}M,S^{-1}N)$ and $Tor^A_n(M,N)_P \cong Tor^{A_P}_n(M_P,N_P)$;
(3) Isomorphisms of right $(\overline{S})^{-1}B$−modules: $(\overline{S})^{-1}Ext^n_A(M,N) \cong Ext^n_{S^{-1}A}$ $(S^{-1}M,S^{-1}N)$ and $Ext^n_A(M,N)_P \cong Ext^n_{A_P}(M_P,N_P)$;
(4) If B is noetherian, $_BM_A$ is a finitely generated free $(A\text{-}B)$-bimodule, then the covariant functors:
$$Ext^n_{(\overline{S})^{-1}B}(S^{-1}M,-) : (\overline{S})^{-1}B - Mod_{ff} \to S^{-1}A - Mod_{ff}$$
and $Tor^{S^{-1}A}_n(S^{-1}M,-) : S^{-1}A - Mod_{ff} \to (\overline{S})^{-1}B - Mod_{ff}$ are adjoint.

2   Recalls and Preliminary Results

2.1   Definition

Let Λ and Γ be two categories, $F : \Lambda \to \Gamma$ and $G : \Gamma \to \Lambda$ two covariant functors. It is said that the couple (F,G) is adjoint if for any $A \in Ob(\Lambda)$ and for any $B \in Ob(\Gamma)$, there is an isomorphism
$r_{A,B} : Hom_\Lambda(A,G(B)) \to Hom_\Gamma(F(A),B)$ so that:

(a) for any $f \in Hom_\Lambda(A', A)$, the following diagram is commutative :

$$Hom_\Lambda(A, G(B)) \xrightarrow{f_* = Hom(f, G(B))} Hom_\Lambda(A', G(B))$$
$$\downarrow r_{A,B} \qquad\qquad\qquad \downarrow r_{A',B}$$
$$Hom_\Gamma(F(A), B) \xrightarrow{F(f)_* = Hom(F(f), B)} Hom_\Gamma(F(A'), B)$$

where $f_* = Hom_\Lambda(f, G(B)) : Hom_\Lambda(A, G(B)) \longrightarrow Hom_\Lambda(A', G(B))$
such that for each $u \in Hom_\Lambda(A, G(B))$, $f_*(u) = u \circ f$ and $F(f)_* = Hom_\Gamma(F(f), B) : Hom_\Gamma(F(A), B) \longrightarrow Hom_\Gamma(F(A'), B)$ such that for each
$v \in Hom_\Gamma(F(A), B)$, $F(f)_*(v) = v \circ F(f)$.

(b) for any $g \in Hom_\Gamma(B, B')$, the following diagram is commutative:

$$Hom_\Lambda(A, G(B)) \xrightarrow{(G(g))^* = Hom(A, G(g))} Hom_\Lambda(A, G(B'))$$
$$\downarrow r_{A,B} \qquad\qquad\qquad \downarrow r_{A',B}$$
$$Hom_\Gamma(F(A), B) \xrightarrow{g^* = Hom(F(A), g)} Hom_\Gamma(F(A), B')$$

where $g^* = Hom_\Gamma(F(A), g) : Hom_\Gamma(F(A), B) \longrightarrow Hom_\Gamma(F(A), B')$
such that for each $u \in Hom_\Gamma(F(A), B)$, $g^*(u) = g \circ u$ and $G(g))^* = Hom_\Lambda(A, G(g)) : Hom_\Lambda(A, G(B)) \longrightarrow Hom_\Lambda(A, G(B'))$ such that for each
$v \in Hom_\Lambda(A, G(B))$, $G(g))^*(v) = G(g) \circ v$.

2.2 Definition

Let A be a ring and S a non empty subset of A non contenaing zero. Then, we say that S is a closed multiplicative subset of A if:

(i) $\forall s, t \in S$, $st \in S$ (we say that S is stable for multiplication)
(ii) $\forall s, t \in A$ such that $st \in S$, then $s \in S$ and $t \in S$ (we say that S is closed)

2.3 Definition

If S is a closed multiplicative subset of A, we say that S satisfies the left conditions of Ore (respectively right) if:

(i) $\forall a \in A, \forall s \in S$, there exist $t \in S$ and $b \in A$ such that $ta = bs$ (respectively $at = sb$). We say that S is left switchable (respectively right).
(ii) $\forall a \in A, \forall s \in S$ such that $as = 0$ (respectively $sa = 0$), then there exists $t \in S$ such that $ta = 0$ (respectively $at = 0$). We say that S is left reversible (respectively right).

2.4 Proposition

Let M be a left (respectively right) A - module and S a saturated multiplicative subset of A satisfying the left (respectively right) conditions of Ore. Then the binary relation R defined in $S \times M$ (respectively in $M \times S$) by:

$(s, m)R(s', m')$ if and only if there exist $x, y \in S$ such that $\begin{cases} xm = ym' \\ xs = ys' \end{cases}$

(respectively $(m, s)R(m', s')$ if and only if there exist $x', y' \in S$ such that
$\begin{cases} mx' = m'y' \\ sx' = s'y' \end{cases}$) is an equivalence relation (see [5] and [6]).

Proof. See ([5]).

2.5   Notation

The set $(S \times M)/R$ (respectively $(M \times S)/R$) of equivalence classes is denoted $S^{-1}M$, whose elements are denoted $\frac{m}{s}$ where $m \in M$ and $s \in S$.
If $M = A$, the set $(S \times A)/R$ is denoted $S^{-1}A$ which elements are denoted $\frac{a}{s}$ where $a \in A$ and $s \in S$.

2.6   Theorem

Let M be a left (respectively right) A - module and S a satured multiplicative subset of A satisfying the left (respectively right) conditions of Ore, then we have the following results:

(1) $S^{-1}A$ has a ring structure defined by the operations:
  $\cdot \ \frac{a}{t} + \frac{b}{s} = \frac{xa+yb}{ys}$ where $x, y \in S$ such that $xt = ys$ and
  $\cdot \ \frac{a}{t} \times \frac{b}{s} = \frac{zb}{wt}$ where $(w, z) \in S \times A$ such that $wa = zs$
  (respectively
  $\cdot \ \frac{a}{t} + \frac{b}{s} = \frac{ax+by}{sy}$ where $x, y \in S$ such that $tx = sy$ and
  $\cdot \ \frac{b}{s} \times \frac{a}{t} = \frac{bz}{tw}$ where $(w, z) \in S \times A$ such that $aw = sz$), (see [5] and [6])
(2) $S^{-1}M$ is a left $S^{-1}A-$module (respectively right $S^{-1}A-$module). Operations of $S^{-1}M$ are defined by:
  $\cdot \ \frac{m}{s} + \frac{m'}{s'} = \frac{xm+ym'}{ys'}$ where $x, y \in S$ such that $xs = ys'$ and
  $\cdot \ \frac{a}{t} \cdot \frac{m}{s} = \frac{zm}{wt}$ where $(w, z) \in S \times A$ such that $wa = zs$.
  (respectively
  $\cdot \ \frac{m}{s} + \frac{m'}{s'} = \frac{mx+m'y}{s'y}$ where $x, y \in S$ such that $sx = s'y$ and
  $\cdot \ \frac{m}{s} \cdot \frac{a}{t} = \frac{mz}{tw}$ where $(w, z) \in S \times A$ such that $aw = sz$) (see [5] and [6]).
(3) The relation
  $S^{-1}() : A - Mod \longrightarrow S^{-1}A - Mod$ (respectively $S^{-1}() : Mod - A \longrightarrow Mod - S^{-1}A$) which at all $M \in A - Mod$ (respectively $M \in Mod - A$) associates $S^{-1}M$ and for any morphism of left A-module (respectively right A-module) $f : M \longrightarrow N$ associates the morphism of left (respectively right) $S^{-1}A$-modules
  $$S^{-1}f : S^{-1}M \to S^{-1}N$$
  $$\frac{m}{s} \mapsto \frac{f(m)}{s}$$
  is covariant functor which is additive, exact and preserves projectivity (see [5] and [6]). In particular, this functor preserves injectivity if A is noetherian (see [2]).

2.7   Proposition and Definition

Let B be a ring, $A = Z(B)$, the center of B, S a central closed multiplicative subset of A satisfying the left (respectively right) conditions of Ore, note by $\overline{S}$, the saturated multiplicative subset of B generated by S. S is naturally the intersection of all saturated multiplicative subsets of B containing S. Then $\overline{S}$ satisfies the left (respectively right) conditions of Ore.

Proof. It's clear that a central closed multiplicative subset S satisfies the left and right conditions of Ore in A. S is also a central closed multiplicative subset of B, then it satisfies the left and right conditions of Ore in B. Therefore $\overline{S}$ the saturated multiplicative subset of B generated by S satisfies the left (respectively right) conditions of Ore (see [2], *chapI*).

2.8   Theorem

Let B be a ring, A a subring of B containing $Z(B)$, the center of B, S a central saturated multiplicative subset of A satisfying the left (respectively right) conditions of Ore, $\overline{S}$ the saturated multiplicative subset of B generated by S satisfying the left (respectively right) conditions of Ore, M an $(A-B)$--bimodule. Then for every left A-module N, we have the following isomorphism of left $(\overline{S})^{-1}B$--modules:
$$\psi : (\overline{S})^{-1}Hom_A(M,N) \to Hom_{S^{-1}A}(S^{-1}M, S^{-1}N))$$
$$\tfrac{f}{s} \mapsto \tfrac{1}{s}S^{-1}(f).$$

Proof. (a) Let us show that the correspondence ψ is a map.

Let $\tfrac{f}{s}$ and $\tfrac{g}{t}$ be two element of $\overline{S}^{-1}Hom_A(M,N)$, such that $\tfrac{f}{s} = \tfrac{g}{t}$. Let us show that $\psi(\tfrac{f}{s}) = \psi(\tfrac{g}{t})$. We have $\tfrac{f}{s} = \tfrac{g}{t}$ and $\overline{S}$ satisfies the right conditions of Ore in B, then there exist x and $y \in \overline{S}$ such that :
$$\begin{cases} fx = gy \\ sx = ty \end{cases}$$
$f.x = g.y \Rightarrow \forall m \in M, (f.x)(m) = (g.y)(m) \Rightarrow f(m.x) = g(m.y)$ Let $\tfrac{m}{t} \in S^{-1}(M)$ Let us show that $\psi(\tfrac{f}{s}) = \psi(\tfrac{g}{t})$, that means $[\tfrac{1}{s}S^{-1}(f)](\tfrac{m}{l}) = [\tfrac{1}{t}S^{-1}(g)](\tfrac{m}{l})$ for every $\tfrac{m}{l} \in S^{-1}(M)$.

We have $\psi(\tfrac{f}{s})(\tfrac{m}{l}) = \tfrac{1}{s}S^{-1}(f)(\tfrac{m}{l}) = S^{-1}(f)(\tfrac{m}{l}.\tfrac{1}{s}) = S^{-1}(f)(\tfrac{mz}{sw}) = \tfrac{f(mz)}{sw}$ where $(w,z) \in \overline{S} \times B$ such that $w.1 = z.l$;
$\psi(\tfrac{g}{t})(\tfrac{m}{l}) = \tfrac{1}{t}S^{-1}(g)(\tfrac{m}{l}) = S^{-1}(g)(\tfrac{m}{l}.\tfrac{1}{t}) = S^{-1}(g)(\tfrac{mz'}{tw'}) = \tfrac{g(mz')}{tw'}$ where $(w',z') \in \overline{S} \times B$ such that $w'.1 = z'.l$
Let us show that for every $\tfrac{m}{l} \in S^{-1}(M)$, there exist $(w,z) \in \overline{S} \times B$ and $(w',z') \in \overline{S} \times B$ such that $\tfrac{f(mz)}{sw} = \tfrac{g(mz')}{tw'}$
For every $m \in M$, there exist x and $y \in S$ such that:
$$\begin{cases} f(m.x) = g(m.y) \\ sx = ty \end{cases} \text{ Then we can take } \begin{cases} z = x \\ z' = y \end{cases},$$
then we have $f(mz) = g(mz')$.
And so $\begin{cases} sw = szl = sxl \\ tw' = tz'l = tyl \end{cases}$
So $\begin{cases} f(m.z) = g(m.z') \\ sxl = tyl \end{cases} \Rightarrow \begin{cases} f(m.z).1 = g(m.z').1 \\ sxl = tyl \end{cases} \Rightarrow \tfrac{f(mz)}{sx} = \tfrac{g(mz')}{ty}$

Then taking $\begin{cases} w = x \\ w' = y \end{cases}$, we have prooved that there exits $(w, z) \in \overline{S} \times B$ and

$(w', z') \in \overline{S} \times B$ such that $\frac{f(mz)}{sw} = \frac{g(mz')}{tw'}$.

Then $\psi(\frac{f}{s}) = \psi(\frac{g}{t})$, hence ψ is a map.

(b) Let us now show that ψ is a morphism of left $(\overline{S})^{-1}B-$module.

We have $\psi(\frac{f}{s} + \frac{g}{t}) = \psi(\frac{f.x+g.y}{ty})$ where $sx = ty$

$\Longrightarrow \psi(\frac{f}{s} + \frac{g}{t}) = \frac{1}{ty}S^{-1}(f.x + g.y)$.

For any morphism $h : M \to N$ of right $B-$module, the morphism $S^{-1}h : S^{-1}M \to S^{-1}N$ is $\overline{S}^{-1}B$ lineair and the functor $S^{-1}()$ is additive.

Then $\frac{1}{ty}S^{-1}(f.x + g.y) = \frac{S^{-1}(f).x+S^{-1}(g).y}{ty} = \frac{1}{ty}S^{-1}(f).x + \frac{1}{ty}S^{-1}(g).y = \psi(\frac{f}{s}) + \psi(\frac{g}{t})$

Therefore let us show that $\psi(\frac{f}{s}.\frac{a}{t}) = \frac{a}{t}.\psi(\frac{f}{s})$.

Let $\frac{m}{l} \in S^{-1}(M)$, we have $\psi(\frac{f}{s}.\frac{a}{t})(\frac{m}{l}) = \psi(\frac{f.z}{tw})(\frac{m}{l})$ where $(w, z) \in (\overline{S}) \times B$ such that $aw = sz$.

Then $\psi(\frac{f}{s}.\frac{a}{t})(\frac{m}{l}) = \frac{1}{tw}S^{-1}(fz)(\frac{m}{l}) = S^{-1}(fz)(\frac{m}{l}.\frac{1}{tw}) = S^{-1}(fz)(\frac{mz'}{tw.w'}) = \frac{f(m.z'.z)}{(tw).w'}$ where $(w', z') \in (\overline{S}) \times B$ such that $w'.1 = z'.l$.

Otherwise $\frac{a}{t}.\psi(\frac{f}{s})(\frac{m}{l}) = \frac{a}{t}.\frac{f(mz_1)}{sw_1}$ where $w_1.1 = z_1.l$.

Then $\frac{a}{t}.\psi(\frac{f}{s})(\frac{m}{l}) = \frac{z_2f(mz_1)}{w_2t}$ where $w_2.a = z_2(sw_1)$.

Then $\frac{a}{t}.\psi(\frac{f}{s})(\frac{m}{l}) = \frac{f(mz_1z_2)}{w_2t}$.

We have $t \in S$ then $w_2t = tw_2$.

Then $\frac{a}{t}.\psi(\frac{f}{s}(\frac{m}{l}) = \frac{f(mz_1z_2)}{tw_2}$.

To show that $\frac{f(m.z'.z)}{(tw).w'} = \frac{f(mz_1z_2)}{w_2t})$ we can take $\begin{cases} z_1 = z' \\ z_2 = z \\ w_2 = ww' \end{cases}$

So $\psi(\frac{f}{s}.\frac{a}{t}) = \frac{a}{t}.\phi(\frac{f}{s})$.

Then ψ is well a morphism of left $(\overline{S})^{-1}B-$module.

(c) Finally, let us show that ψ is bijective.

For this we consider canonical isomorphisms of groups:

$$\psi : \overline{S}^{-1}Hom_A(M, N) \to Hom_A(M, S^{-1}N)$$
$$\frac{f}{s} \mapsto f_\delta : M \to S^{-1}N$$
$$m \mapsto \frac{f(m)}{s} \text{ and}$$
$$\theta : Hom_A(M, S^{-1}N) \to Hom_{S^{-1}A}(\overline{S}^{-1}M, S^{-1}N)$$
$$g \mapsto \tilde{g} : S^{-1}M \to S^{-1}N$$
$$\frac{m}{t} \mapsto \frac{g(m)}{t}$$

By construction, we have $\psi = \theta \circ \varphi$

φ et θ are canonical isomorphisms of groups.

So $\psi = \theta \circ \varphi$ is bijective. Therefore, ψ is an isomorphism of left $(\overline{S})^{-1}B-$module.

2.9   Corollary

Let A be a ring (not necessarily commutative), S a non-empty closed multiplicative subset of A satisfying the left and right conditions of Ore, M an $(A-A)$--bimodule. Then for every left A-module N, we have the following isomorphism of left $S^{-1}A$--modules:
$$\psi : S^{-1}Hom_A(M, N) \to Hom_{S^{-1}A}(S^{-1}M, S^{-1}N))$$
$$\frac{f}{s} \mapsto \frac{1}{s}S^{-1}(f).$$

Proof. It suffices to take $B = A$ in Theorem 2.8. Indeed, we can localize A in the right because S satisfies the right conditions of Ore in A.

3   Functorial Isomorphisms with Functors Tor, Ext and $S^{-1}()$

In this section, B denotes an unit ring, $A = Z(B)$ the center of B, S a non-empty multiplicative subset of A satisfying the left and right conditions of Ore, $\overline{S}$ the multiplicative subset generated by S satisfying the left and right conditions of Ore. We show that the functor $S^{-1}()$ commutes with the functor Tor and Ext.

3.1   Theorem

let M be a left A-module (respectively a $(B-A)-$ bimodule), N a $(B-A)-$ bimodule (respectively a left A-module). Then for every non negative integer $n \geq 0$, we have the following isomorphism of left $(\overline{S})^{-1}B$-module :
$$(\overline{S})^{-1}Tor_n^A(M, N) \cong Tor_n^{S^{-1}A}(S^{-1}M, S^{-1}N).$$

Proof. It suffices to prove this theorem in the case where M is a left A-module and N a $(B-A)-$ bimodule. It's clear that $Tor_n^A(M, N)$ is a left B-module, so $(\overline{S})^{-1}Tor_n^A(M, N)$ is a left $(\overline{S})^{-1}B$-module. Also $S^{-1}M$ is a left $S^{-1}A$- module; $S^{-1}N$ is a $(S^{-1}B - S^{-1}A)-$ bimodule, so $Tor_n^{S^{-1}A}(S^{-1}M, S^{-1}N)$ is a left $(\overline{S})^{-1}B$-module.

(a) For $n = 0$ we have $Tor_0^A(M, N) \cong M \otimes_A N$ and
$Tor_0^{S^{-1}A}(S^{-1}M, S^{-1}N) \cong S^{-1}M \otimes_A S^{-1}N$ (see [9]).
Otherwise, $S^{-1}()$ is an exact and additive functor so according to [2], it commutes with tensor product functor $M \otimes_A -$ and then we have: $(\overline{S})^{-1}(M \otimes_A N) \cong S^{-1}M \otimes_A S^{-1}N$, hence the result.
(b) Let now P_N be a projective resolution of N
As the functor S^{-1} keep projectivity so $S^{-1}(P_N)$ is a projective resolution of $S^{-1}N$.
According to Theorem 2.8, Proving the existence of isomorphism ψ, we deduce the isomorphism of the complexes $(\overline{S})^{-1}(M \otimes_A P_N) \cong S^{-1}M \otimes_{S^{-1}A} S^{-1}(P_N)$.
Hence their homology groups are isomorphic and as the functor $S^{-1}()$ is exact and by definition of the functor Tor, we have $H_n(S^{-1}(M \otimes$

$_A P_N) \cong S^{-1} H_n(M \otimes {_A P_N}) \cong (\overline{S})^{-1} Tor_n^A(M, N)$. As well as $S^{-1}(P_N)$ is a projective resolution of $S^{-1}N$, so $H_n(\overline{S})^{-1}M \otimes {_{S^{-1}A}} S^{-1}(P_N)) \cong Tor_n^{S^{-1}A}(S^{-1}M, S^{-1}N)$.

3.2 Corollary

Let B be a ring, A the center of B, P a prime ideal of A, S the set of regular elements of $A - P$, M a left A-module and N an $(A - A)$−bimodule. Denotes respectively by A_P the ring of fractions $S^{-1}A$ of A with respect of S, M_P the modules of fractions $S^{-1}M$ and N_P the modules of fractions $S^{-1}N$
So we have the following isomorphism of left A_P-modules : $Tor_n^A(M, N)_P \cong Tor_n^{A_P}(M_P, N_P)$

Proof. The center A of B is a duo-ring, then according to [7], the ring of fractions A_P and the modules of fractions M_P and N exist. It is then sufficient to apply the Theorem 3.1 let M be a left A-module (respectively a $(A - B)-$ bimodule), N a $(A - B)-$ bimodule (respectively a left A-module). Then for every non negative integer $n \geq 0$, we have the following isomorphism of right(respectively a left) $(\overline{S})^{-1}B$-modules $(\overline{S})^{-1} Ext_A^n(M, N) \cong Ext_{S^{-1}A}^n(S^{-1}M, S^{-1}N)$.

3.3 Theorem

Let M be a left A-module (respectively a $(A - B)-$ bimodule), N a $(A - B)-$ bimodule (respectively a left A-module). Then for every non negative integer $n \geq 0$, we have the following isomorphism of right (respectively a left) $(\overline{S})^{-1}B$-modules $(\overline{S})^{-1} Ext_A^n(M, N) \cong Ext_{S^{-1}A}^n(S^{-1}M, S^{-1}N)$.

Proof. It suffices to proove this theorem in the case where M is a left A-module and N a $(A - B)-$ bimodule.
It's clear that $Ext_A^n(M, N)$ is a right B-module, so $(\overline{S})^{-1} Ext_A^n(M, N)$ is a right $(\overline{S})^{-1}B$-module. Also $S^{-1}M$ is a left $S^{-1}A$- module; $S^{-1}N$ is a $(S^{-1}A - S^{-1}B)$-bimodule, so $Ext_{S^{-1}A}^n(S^{-1}M, S^{-1}N)$ is a right $(\overline{S})^{-1}B$-module.
Now let P_M be a projective resolution of M.
According to **Theorem** 2.8, there is an isomorphism
$\psi : (\overline{S})^{-1} Hom_A(M, N) \to Hom_{S^{-1}A}(S^{-1}M, S^{-1}N)$.
We deduce the isomorphism of the complexes:
$(\overline{S})^{-1}(Hom_A(P_M, N)) \cong Hom_{S^{-1}A}(S^{-1}(P_M), S^{-1}N)$
By applying the functor homology H_n we have:
$H_n(\overline{S})^{-1}(Hom_A(P_M, N))) \cong (\overline{S})^{-1}H_n(Hom_A(P_M, N)) \cong (\overline{S})^{-1} Ext_A^n(M, N)$
since the functor $S^{-1}()$ is exact, therefore we have:
$H_n(Hom_{S^{-1}A}(S^{-1}P_M, S^{-1}N)) = Ext_{S^{-1}A}^n(S^{-1}M, S^{-1}N)$ because $S^{-1}(P_M)$ is a projective resolution.

3.4 Corollary

Let B be a ring, A the center of B, P a prime ideal of A, S the set of regular elements of $A - P$, M a left A-module and N an $(A - A)$−bimodule. Denotes

respectively by A_P the ring of fractions $S^{-1}A$ of A with respect of S, M_P the modules of fractions $S^{-1}M$ and N_P the modules of fractions $S^{-1}N$
So we have the following isomorphism of right A_P-modules : $Ext^n_A(M,N)_P \cong Ext^n_{A_P}(M_P, N_P)$

Proof. The center A of B is a duo-ring, this result come from [5] and from **Theorem 2.1**.

4 Adjoint Functors with Tor, Ext and $S^{-1}()$

In this section, Unless otherwise stated, B denotes a ring (not necessarily commutative), A the center of B, ${}_AM_B$ a finitely generated free $(A\text{-}B)$-bimodule, S a saturated multiplicative subset of A satisfying the left conditions of Ore, $S^{-1}A$ the ring of fractions of A with respect of S.
We know that (see [10]) the functor $Ext^0_A(M, -)$ Is equivalent to the functor $Hom_A(M-)$, $Ext^0_A(-, M)$ Is equivalent to the functor $Hom_A(-, M)$ and $Tor^A_0(M, -)$ Is equivalent to the functor $M \otimes_A -$ which is isomorphic to the functor $S^{-1}()$.
So $Ext^0_{S^{-1}A}(S^{-1}A, -)$ is adjoint to $S^{-1}()$ and $Tor^{S^{-1}A}_0(S^{-1}A, -)$ is isomorphic to $S^{-1}()$.
The aim of this section is to establish a generalization of these properties between the functor $S^{-1}()$ and functors $Ext^n_{S^{-1}B}(S^{-1}M, -)$ and $Tor^{S^{-1}A}_n(S^{-1}M, -)$ respectively for any non negative integer n.

4.1 Proposition

Let A and B be two rings, ${}_AM_B$ an (A, B) bimodule, ${}_BN_A$ an $(B, A)-$ bimodule, ${}_BP$ a left B-module, then we have the following isomorphism of left B-module :

$$Ext^n_A(M, Hom_B(N, P)) \cong Hom_B(Tor^A_n(M, N), P)$$

Proof. It's clear that with the conditions of this proposition, $Hom_B(N, P))$ is a left A-module, $Tor^A_n(M, N)$ is a $(B, B)-$ bimodule; then $Ext^n_A(M, Hom_B (N, P))$ and $Hom_B(Tor^A_n(M, N), P)$ are left B-modules.
For the rest of the proof, see [3] (chapter 6, Sect. 5: Duality homomorphisms).

4.2 Proposition

Let B be a noetherian duo-ring, A a subring of B, ${}_AM_B$ a finitely generated free $(A\text{-}B)$-bimodule, ${}_BN_A$ a finitely generated free $(B\text{-}A)$-bimodule, ${}_BP$ a left finitely generated free B-module, S a saturated multiplicative subset of A satisfying the left and right conditions of Ore, $\overline{S}$ the set of regular item of S satisfying the left and right conditions of Ore in B, then we have the following isomorphism of left $(\overline{S})^{-1}B$-modules: $Ext^n_{S^{-1}A}(S^{-1}M, Hom_{(\overline{S})^{-1}B}(S^{-1}N, S^{-1}P)) \cong Hom_{(\overline{S})^{-1}B}(Tor^A_n(S^{-1}M, S^{-1}N), S^{-1}P)$

Proof. It's clear that $_BP$ a left free finite type B-module is injective. As B is noetherian, the functor $S^{-1}()$ preserves injectivity (see [2]), then $S^{-1}P$ is a left injective $S^{-1}B$ module, This result is obtained by applying 2.8 and 4.1.

4.3   Proposition

Let A and B be two rings, $_BM_A$ a (B,A) bimodule, $_BN_A$ an $(B,A)-$ bimodule, $_BP$ a left B-module, then we have the following isomorphism of left B-module :

$$Tor_n^B(Hom_A(N,M),P) \cong Hom_B(Ext_A^n(M,N),P)$$

Proof. With these conditions, $Hom_A(N,M))$ and $Ext_A^n(M,N)$ are $(B,B)-$ bimodules; then $[Tor_n^B(Hom_A(N,M),P)$ and $Hom_B(Ext_A^n(M,N),P)$ are left B-modules. For the rest of the proof, see [3] (chapter 6, Sect. 5: Duality homomorphisms).

4.4   Theorem

Let B be a noetherian duo-ring, A a subring of B, $_BM_A$ and $_BN_A$ two free finite types (B,A) bimodules, $_BP$ a left free finite type B-module, S a saturated multiplicative subset of A satisfying the left conditions of Ore, $\overline{S}$ the set of regular item of S.
then we have the following isomorphism of left $(\overline{S})^{-1}B$-modules:

$$Tor_n^{(\overline{S})^{-1}B}(Hom_{S^{-1}A}(S^{-1}N,S^{-1}M),S^{-1}P) \cong Hom_{(\overline{S})^{-1}B}(Ext_{S^{-1}A}^n(S^{-1}M,S^{-1}N),S^{-1}P)$$

Proof. The proof is similar to that of Proposition 4.2.

4.5   Theorem

Let B be a noetherian ring, A a the center of B, $_BM_A$ a free finite type (B,A) bimodule, $_AN_B$ a free finite types (A,B) bimodule, $_BP$ a left free finite type B-module, S a saturated multiplicative subset of A satisfying the left conditions of Ore, $\overline{S}$ the set of regular item of S, $S^{-1}A$ the ring of fractions of A with respect of S. Then covariant functors:

$$Ext_{(\overline{S})^{-1}B}^n(S^{-1}M,-) : (\overline{S})^{-1}B - Mod_{ff} \to S^{-1}A - Mod_{ff}$$

and $Tor_n^{S^{-1}A}(S^{-1}M,-) : S^{-1}A - Mod_{ff} \to (\overline{S})^{-1}B - Mod_{ff}$ are adjoint.

Proof. It's clear that N is projective and P is injective, consequently the functors $Hom_A(-,P)$ and $Hom_B(N,-)$ are exact.

1. According to the adjoint isomorphism theorem, we have
   $Hom_A(N,Hom_B(M,P)) \cong Hom_B(M \otimes_A N,P)$, hence $(\overline{S})^{-1}Hom_A(N,$ $Hom_B(M,P)) \cong (\overline{S})^{-1}Hom_B(M \otimes_A N,P)$ and so $Hom_{S^{-1}A}(S^{-1}N,$ $Hom_{(\overline{S})^{-1}B}(S^{-1}M,S^{-1}P)) \cong Hom_{(\overline{S})^{-1}B}(S^{-1}M\otimes_{S^{-1}A}S^{-1}N,S^{-1}P)$ applying Theorem 2.8.

$S^{-1}N$ is projective module and $S^{-1}P$ is injective, therefore functors $Hom_{S^{-1}A}(S^{-1}N, -)$ and $Hom_{(\overline{S})^{-1}B}(-, S^{-1}P)$ are exacts.

Therefore, we have:

$H_n[Hom_{S^{-1}A}(S^{-1}N, Hom_{(\overline{S})^{-1}B}(S^{-1}M, S^{-1}P))] \cong H_n[Hom_{(\overline{S})^{-1}B}(S^{-1}M$

$\otimes_A S^{-1}N, S^{-1}P)]$ which equals to

$Hom_{S^{-1}A}[H_n(S^{-1}N, Hom_{(\overline{S})^{-1}B}(S^{-1}M, S^{-1}P))] \cong Hom_{(\overline{S})^{-1}B}[H_n(S^{-1}M$

$\otimes_{S^{-1}A} S^{-1}N, S^{-1}P)]$.

Let X_N be a projective resolution of $S^{-1}N$ and Y_P an injective resolution of $S^{-1}P$.

Then we have $Hom_A[H_n(X_N, Hom_B(S^{-1}M, S^{-1}P))] \cong Hom_B[H_n((\overline{S})^{-1}M$

$\otimes_A S^{-1}N, Y_P)]$ which equals to :

$Hom_{S^{-1}A}[S^{-1}N, Ext^n_{(\overline{S})^{-1}B}(S^{-1}M, S^{-1}P)] \cong Hom_{(\overline{S})^{-1}B}[Tor_n^{S^{-1}A}(S^{-1}M,$

$S^{-1}N), S^{-1}P]$.

2. Let us now prove by induction on $n \in IN$ than the following two diagrams, where $F = (Tor_n^{S^{-1}A}(S^{-1}M, -))$ and $G = Ext^n_{(\overline{S})^{-1}B}(S^{-1}M, -)$ are commutative:

Let $f : S^{-1}N' \longrightarrow S^{-1}N$ and $g : S^{-1}P \longrightarrow S^{-1}P'$

$$
\begin{array}{ccc}
Hom_{S^{-1}A}(S^{-1}N, G(S^{-1}P)) & \xrightarrow{f_* = Hom(f, G(S^{-1}P))} & Hom_{S^{-1}A}(S^{-1}N', G(S^{-1}P)) \\
\downarrow r_{S^{-1}N, S^{-1}P} & & \downarrow r_{S^{-1}N', S^{-1}P} \\
Hom_{(\overline{S})^{-1}B}(F(S^{-1}N), S^{-1}P) & \xrightarrow{(Ff)_* = Hom(F(f), S^{-1}P)} & Hom_{(\overline{S})^{-1}B}(F(S^{-1}N'), S^{-1}P)
\end{array}
\tag{1}
$$

$$
\begin{array}{ccc}
Hom_{S^{-1}A}(S^{-1}N, G(S^{-1}P)) & \xrightarrow{G(g)^* = Hom(S^{-1}N, G(g))} & Hom_{S^{-1}A}(S^{-1}N, G(S^{-1}P')) \\
\downarrow r_{S^{-1}N, (\overline{S})^{-1}B} & & \downarrow r_{S^{-1}N', S^{-1}P} \\
Hom_{(\overline{S})^{-1}B}(F(S^{-1}N), S^{-1}P) & \xrightarrow{g^* = Hom(F(S^{-1}N), g)} & Hom_{(\overline{S})^{-1}B}(F(S^{-1}N), S^{-1}P')
\end{array}
\tag{2}
$$

(a) For $n = 0$, we have $Ext^0_{(\overline{S})^{-1}B}(S^{-1}M, -) \cong Hom_{S^{-1}B}(S^{-1}M, -)$ and

$Tor_0^{(\overline{S})^{-1}B}(S^{-1}M, -) \cong S^{-1} \otimes_{S^{-1}A} -$ now (see [6] and [8], the functors $Hom_{(\overline{S})^{-1}B}(S^{-1}M, -) : (\overline{S})^{-1}B - Mod \to S^{-1}A - Mod$ and $M \otimes_{(\overline{S})^{-1}B} - : S^{-1}A - Mod \to (\overline{S})^{-1}B - Mod$ are adjoints. it results that the diagrams (1) and (2) are commutative in taking $F = M \otimes_{S^{-1}A} -$ and $G = Hom_{(\overline{S})^{-1}B}(S^{-1}M, -)$

(b) Suppose that for any non negative integer $k \leqslant n$, diagrams (1) and (2) are commutative where $F = Tor_k^{S^{-1}A}(S^{-1}M, -)$ and $G = Ext^k_{(\overline{S})^{-1}B}(S^{-1}M, -)$.

We show that (1) and (2) are commutative by taking

$F = Tor_{n+1}^{S^{-1}A}(S^{-1}M, -)$ and

$G = Ext^{n+1}_{S^{-1}B}(S^{-1}M, -)$.

Let $P_{S^{-1}M} : \cdots \to P_{n+1} \xrightarrow{d_{n+1}} P_n \xrightarrow{d_n} \cdots \longrightarrow P_2 \xrightarrow{d_2} P_1 \xrightarrow{d_1} P_0 \xrightarrow{\varepsilon} S^{-1}N \longrightarrow 0$ a projective resolution of $S^{-1}M$ of nth kernel $K_n = Ker(d_n)$.

Otherwise, we have (see [2]) :

$$Ext_{S^{-1}A}^{n+1}(S^{-1}M, S^{-1}P) \cong Ext_{S^{-1}A}^1(K_{n-1}, S^{-1}P)$$

and $Tor_{n+1}^{S^{-1}A}(S^{-1}M, S^{-1}N) \cong Tor_1^{S^{-1}A}(K_{n-1}, S^{-1}N)$. Hence, it is enough to apply the recurrence hypothesis to the functors $Ext_{S^{-1}A}^1(K_{n-1}, -)$ and $Tor_1^{S^{-1}A}(K_{n-1}, -)$.

(c) We conclude that for every non negative integer n, the functors $Ext_{S^{-1}B}^n(S^{-1}M, -)$ and $Tor_n^{S^{-1}A}(S^{-1}M, -)$ are adjoint.

4.6 Corollary

Let A be a noetherian ring S a saturated multiplicative subset of A satisfying the left conditions of Ore, $S^{-1}A$ The ring of fractions of A with respect of S. Then functors $Ext_{S^{-1}A}^n(S^{-1}A, -)$ and $Tor_n^{S^{-1}A}(S^{-1}A, -)$ are adjoint.

Proof. It resultes from Theorem 4.5 and in the fact that the ring A is an $(A-A)$-bimodule.

References

1. Anderson, F.W., Fuller, K.R.: Rings and Categories of Modules. GTM. Springer, New York (1973). https://doi.org/10.1007/978-1-4684-9913-1
2. Faye, D.: Thése Unique. Universitéé Cheikh Anta Diop, Dakar, Mai (2016)
3. MacLane, S.: Henri Cartan and Samuel Eilenberg, Homological Algebra. Princeton University Press, New Jersey (1956)
4. Ben Maaouia, M.: These 3me cycle. Université Cheikh Anta Diop, Dakar, Juillet (2003)
5. Ben Maaouia, M.: These d'Etat. Université Cheikh Anta Diop, Dakar (2011)
6. Ben Maaouia, M., Sanghare, M.: Anneaux et modules de fractions. Int. J. Math.-mai (2012)
7. Maaouia, M.B., Sanghare, M.: Localisation dans les duo - anneaux. Afrika Mathematika 3, 163–179 (2009)
8. Rotman, J.J.: Advanced Modern Algebra, 1st edn, pp. 898–921. Printice Hall, Upper Saddle River (2002)
9. Rotman, J.J.: Notes on homological algebra. University of Illinois, Urbana (1968)
10. Rotman, J.J.: An Introduction to Homological Algebra. Académic Press, New York (1972)

On the Computation of Minimal Free Resolutions with Integer Coefficients

Soda Diop$^{1(\boxtimes)}$, Guy Mobouale Wamba^2,
Andre Saint Eudes Mialebama Bouesso^2, and Djiby Sow^1

^{1} Faculté des Sciences et Techniques, Département de Mathématiques et
Informatique, Université Cheikh Anta Diop, BP: 5005, Dakar Fann, Senegal
sodettes@gmail.com, sowdjibab@yahoo.fr
^{2} Faculte des Sciences et Techniques, Universite Marien Ngouabi, BP: 69,
Brazzaville, Congo
wambastonn@gmail.com, andre.mialebama@umng.cg

Abstract. Let $I = \langle f_1, \ldots, f_s \rangle$ be an ideal of $R = \mathbb{Z}[x_1, \ldots, x_n]$. We introduce in this paper the concept of $\mathbb{Z}-$ideal $\mathbb{Z}(I)$ of I which is a proper ideal of R and we propose a technique for computing a weak Gröbner basis for $\mathbb{Z}(I)$. This result is central and leads to the computation of a minimal free resolution for $\mathbb{Z}(I)$ as an $R-$module.

Keywords: Special Gröbner bases · Weak Gröbner bases · Syzygies and free resolution

2010 Mathematics subject classification: 13P10 · 13C10 · 3P25

Introduction

The computation of free resolutions is a very important topic in commutative algebra. It is for example used to compute the cohomology group $\mathrm{Ext}_s^i(M, N)$ for further modules M, N. It has been widely studied for sub-modules with coefficients over fields (see [2,3,6,7]). In this paper we are interested in studying this problem in a polynomial ring over the integers. Denote by $R = \mathbb{Z}[x_1, \ldots, x_n]$ a polynomial ring over the integers, and given an ideal $I = \langle f_1, \ldots, f_s \rangle$ of R, we revisit the method introduced in [8–10] for computing special Gröbner bases for I in R, this notion is central and enables us to introduce the concept of $\mathbb{Z}-$ideal $\mathbb{Z}(I)$ of I which is our main object of work. Indeed, we prove in Lemma 1.3 that $\mathbb{Z}(I)$ is an ideal of R, this leads us to compute from a special Gröbner basis for I, a weak Gröbner basis for $\mathbb{Z}(I)$ in Theorem 1.4 (details on weak Gröbner bases can be found in [1]). The main goal of this paper is to propose a technique for computing a minimal free resolution for the $R-$module $\mathbb{Z}(I)$ from a given ideal I of R. Our method starts with an ideal $I = \langle f_1, \ldots, f_s \rangle$ of $R = \mathbb{Z}[x_1, \ldots, x_n]$, and computes a special Gröbner basis for I via the Algorithm 1, this is nothing but a strong Gröbner basis for the localized ideal $S^{-1}I$ of $S^{-1}R$ for some multiplicative subset S of $\mathbb{Z}$. The next step is the computation of a weak Gröbner basis for $\mathbb{Z}(I)$

C. T. Gueye et al. (Eds.): A2C 2019, CCIS 1133, pp. 51–72, 2019.
https://doi.org/10.1007/978-3-030-36237-9_3

using the Theorem 1.4 and the Algorithm 3. Since we are now able to compute special Gröbner bases which are strong Gröbner basis for the localized ideal $S^{-1}I$ of $S^{-1}R$ for some multiplicative subset S of $\mathbb{Z}$ (see details on strong Gröbner bases in [5,11]) and weak Gröbner bases for the ideal $\mathbb{Z}(I)$ of R, one moves to the syzyy theorems introduced in Sect. 1, these theorems are very useful for this method and work in the following way: if $G = \{g_1, \ldots, g_r\}$ is a special Gröbner basis for the ideal I i.e a strong Gröbner basis for the localized ideal $S^{-1}I$ of $S^{-1}R$, then the first syzygy theorem enables to compute a strong Gröbner basis $\{T_1, \ldots, T_l\}$ for the submodule $\mathrm{syz}(g_1, \ldots, g_r)$ in $(S^{-1}R)^r$, this result is used in the third syzygy theorem to obtain a weak Gröbner basis for a submodule in R^r. From here, it becomes easy to compute a minimal free resolution for the localized ideal $S^{-1}I$ of $S^{-1}R$ and at each stage, one uses some trick to obtain a minimal free resolution for the ideal $\mathbb{Z}(I)$ of R. This construction leads to the following diagram of minimal free resolutions.

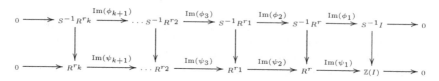

The basic notions for (special) Gröbner basis can be found in the Appendix A.

1   From Special to Weak Gröbner Bases

In this section we denote by $R = \mathbb{Z}[x_1, \ldots, x_n]$ the polynomial ring with integer coefficients and $I = \langle f_1, \ldots, f_t \rangle$ a finitely generated ideal of R.

1. A subset $G = \{g_1, \ldots, g_s\}$ of R is said to be a weak Gröbner basis for I if the leading ideal of I is $Lt(I) = \langle Lt(g_1), \ldots, Lt(g_s) \rangle$.
2. $G = \{g_1, \ldots, g_s\}$ is said to be a strong Gröbner basis for I if $\forall\ f \in I$ there exists $k \in \{1, \ldots, s\}$ such that $Lt(g_k) \mid Lt(f)$.
3. $G = \{g_1, \ldots, g_s\}$ is said to be a special Gröbner basis (see Appendix A) for I if it is a strong Gröbner basis for the ideal $S^{-1}I$ of $S^{-1}R$ for some multiplicative subset S of $\mathbb{Z}$.

With notations as above we have the following:

Definition 1.1 ($\mathbb{Z}$–ideal). Assume that $G = \{g_1, \ldots, g_s\}$ is a special Groebner basis for I and $q_1, \ldots, q_s$ are respectively the least common multiples of denominators occurring in $g_1, \ldots, g_s \in S^{-1}R$. By a $\mathbb{Z}$-ideal of I we mean the set
$$\mathbb{Z}(I) = \{f \in R \mid \exists\ v \in w\mathbb{N}^* : \frac{1}{v} \cdot f \in S^{-1}I\} \text{ where } w = \mathrm{lcm}(q_1, \ldots, q_s).$$

Remark 1.2. It is clear that $I \subsetneq \mathbb{Z}(I)$ for any ideal I of R. To see this, observe that if $f \in I$ then for any $v \in S$ we have $\frac{1}{v} \cdot f \in S^{-1}I$. For the converse, observe the following: if $I = \langle f_1, f_2 \rangle$ with $f_1 = 10xy + 1$, $f_2 = 6x^2 + 3$ is an ideal then $G = \{f_1, f_2, f_3, f_4\}$ (where $f_3 = \frac{3}{5}x - 3y$, $f_4 = 3y^2 + \frac{3}{50}$) is a strong Gröbner basis for $S^{-1}I$ with $S = 2^{\mathbb{N}} \cdot 5^{\mathbb{N}}$ (see Example A.6). Since $w = \text{lcm}(1, 1, 5, 50) = 50$, one can choose the polynomial $g = 30x - 150y \in R = \mathbb{Z}[x, y]$. It is easy to see that $\frac{1}{50}g = f_3 \in S^{-1}I$, that is $g \in \mathbb{Z}(I)$ while $g \notin I$.

Lemma 1.3. If I is an ideal of $R = \mathbb{Z}[x_1, \ldots, x_n]$ then so it is for $\mathbb{Z}(I)$.

Proof. The proof is straightforward.

Theorem 1.4 (FSTW). Let $I = \langle f_1, \ldots, f_t \rangle$ be a finitely generated ideal of $R = \mathbb{Z}[x_1, \ldots, x_n]$ and $G = \{g_1, \ldots, g_s\}$ be a special Gröbner basis for I. There exist $q_1, \ldots, q_s \in \mathbb{Z} \setminus \{0\}$ such that $G' = \{q_1 g_1, \ldots, q_s g_s\}$ is a weak Gröbner basis for $\mathbb{Z}(I) \subset R$.

Proof. Let $f \in \mathbb{Z}(I)$, we want to prove that $\mathbb{Z}(I) = \langle q_1 g_1, \ldots, q_s g_s \rangle$ and $Lt(f) \in \langle Lt(q_1 g_1), \ldots, Lt(q_s g_s) \rangle$. Since $\{g_1, \ldots, g_s\}$ is a special Gröbner basis for I, then there exists a multiplicative subset $S \subset \mathbb{Z}$ such that $\{g_1, \ldots, g_s\}$ is a strong Gröbner basis for $S^{-1}I$ in $S^{-1}R$. Let $q_1, \ldots, q_s$ be respectively least common multiples of denominators occurring in $g_1, \ldots, g_s$. Let $w = \text{lcm}(q_1, \ldots, q_s)$, since each $q_1, \ldots, q_s$ is invertible in $S^{-1}\mathbb{Z}$ then so it is for w, we can then find $v \in w\mathbb{N}^*$ such that $\frac{1}{v}f \in S^{-1}I$. By the division's algorithm of $\frac{1}{v}f$ by $\{g_1, \ldots, g_s\}$ there exist $h_1, \ldots, h_s \in S^{-1}R$ such that

$$\frac{1}{v}f = h_1 g_1 + \cdots + h_s g_s.$$

Let $r_1, \ldots, r_s$ be respectively least common multiples of denominators occurring in $h_1, \ldots, h_s$. Set $r = \text{lcm}(r_1, \ldots, r_s)$ and choose $v = wr$, it is then clear that

$$f = vh_1 g_1 + \cdots + vh_s g_s$$

that is

$$f = (rh_1)(wg_1) + \cdots + (rh_s)(wg_s).$$

Observe that $r = u_1 r_1 = \cdots = u_s r_s$ and $w = v_1 q_1 = \cdots = v_s q_s$, then

$$f = (v_1 u_1 r_1 h_1)(q_1 g_1) + \ldots + (v_s u_s r_s h_s)(q_s g_s) \in \langle q_1 g_1, \ldots, q_s g_s \rangle.$$

In addition, by the division's algorithm there exists $\{i_1, \ldots, i_k\} \subseteq \{1, \ldots, s\}$ such that $Lm(q_{i_j} g_{i_j}) \mid Lm(f)$ and $Lc(f) \in \langle Lc(q_{i_1} g_{i_1}), \ldots, Lt(q_{i_k} g_{i_k}) \rangle$, thus

$$Lt(f) \in \langle Lt(q_{i_1} g_{i_1}), \ldots, Lt(q_{i_k} g_{i_k}) \rangle.$$

Details on the computation of weak Gröbner bases for $\mathbb{Z}(I)$ can be found in the Algorithm 3.

Notation 1.5. Throughout this work by FSTW we mean from special to weak Gröbner bases.

2 Syzygy Theorems

For $r > 1$, denote by $(e_1, \ldots, e_r)$ the standard basis for $F = R^r$. Details on monomials, ordering, polynomial vectors of F can be found in the Definition A.1.

Definition 2.1 (Schreyer's ordering). Given a monomial ordering $>$ on R and non-zero polynomials $f_1, \ldots, f_r \in R$, we define the Schreyer's ordering $>_1$ on F induced by $>$ and $f_1, \ldots, f_r$ as follows: Let $x^\alpha e_i$ and $x^\beta e_j$ be monomials in F. One says that

$$x^\alpha e_i >_1 x^\beta e_j \text{ if and only if :}$$

1. $x^\alpha Lm(f_i) > x^\beta Lm(f_j)$ or;
2. $x^\alpha Lm(f_i) = x^\beta Lm(f_j)$ and $i > j \; \forall \; \alpha, \beta \in \mathbb{N}^n$.

Remark 2.2. Let $G = \{g_1, \ldots, g_s\}$ be a special Gröbner basis for the ideal I in R with multiplicative subset $S \subset \mathbb{Z}$. One knows by the Buchberger's criterion that for each pair (i, j) with $1 \leq j < i \leq s$, the remainder by the division's algorithm of the S-polynomial $S(g_i, g_j)$ (according to the Definition A.2) by G is equal to zero, in this case there exist $h_1, \cdots, h_s \in S^{-1}R$ such that

$$S(g_i, g_j) = \frac{Lc(g_j)}{Lc(g_i)} m_{ji} g_i - m_{ij} g_j = h_1 g_1 + \ldots + h_s g_s$$

(assuming that $Lc(g_i) \mid Lc(g_j)$), this means that

$$h_1 g_1 + \ldots + (h_j + m_{ij}) g_j + \ldots + \left(h_i - \frac{Lc(g_j)}{Lc(g_i)} m_{ji}\right) g_i + \ldots + h_s g_s = 0,$$

that is

$$G_{ij} = \left(h_1, \ldots, (h_j + m_{ij}), \ldots, \left(h_i - \frac{Lc(g_j)}{Lc(g_i)} m_{ji}\right), \ldots, h_s\right) \in (S^{-1}R)^s$$

is a syzygy (see Definition A.1) for $g_1, \ldots, g_s$ in $S^{-1}R$. Denoting $q_1, \ldots, q_s \in \mathbb{Z} \setminus \{0\}$ the least common multiple of denominators occurring in $g_1, \ldots, g_s$ respectively, we have seen in Theorem 1.4 that $G' = \{q_1 g_1, \ldots, q_s g_s\}$ is a weak Gröbner basis for $\mathbb{Z}(I)$ in R.

With notations as in the Remark 2.2, we have the following:

Remark 2.3 (Leading term of a syzygy). Observe by the Definition A.2 that $S(g_i, g_j) = \frac{Lc(g_j)}{Lc(g_i)} m_{ji} g_i - m_{ij} g_j$, this leads to $\frac{Lc(g_j)}{Lc(g_i)} m_{ji} Lt(g_i) - m_{ij} Lt(g_j) = 0$, that is $m_{ji} Lm(g_i) = m_{ij} Lm(g_j)$. Since $i > j$, then by the Schreyer's ordering $<_1$ induced by $<$ and G we have $m_{ji} e_i >_1 m_{ij} e_j$ therefore $Lt(G_{ij}) = -\frac{Lc(g_j)}{Lc(g_i)} m_{ji}$.

Denote by $syz(g_1, \ldots, g_s)$ the space of all syzygies of $g_1, \ldots, g_s$. With notations as in the Remark 2.2 we have the following:

Theorem 2.4 (First Syzygy Theorem). Let $G = \{g_1, \ldots, g_s\}$ be a special Gröbner basis for I (with multiplicative subset $S \subset \mathbb{Z}$) in R^r with respect to a monomial ordering $<$. The set $T = \{G_{ij} \in \text{Syz}(g_1, \ldots, g_s) / \ 1 \le j < i \le s\}$ form a strong Gröbner basis for $\text{syz}(g_1, \ldots, g_s)$ w.r.t the Schreyer's ordering $<_1$ induced by G and $<$.

Proof. Let $L = (L_1, \ldots, L_s) \in \text{syz}(g_1, \ldots, g_s)$, we need to prove that $\exists \, h \in T$ such that

$$\text{Lt}(h) \mid \text{Lt}(L).$$

By the division's algorithm in $(S^{-1}R)^s = (S^{-1}\mathbb{Z}[x_1, \ldots, x_n])^s$ for L by T (listed in some order) we have

$$L = \sum_{G_{ij} \in T, k} Q_k G_{ij} + U \quad (*)$$

where each $Q_k \in S^{-1}R = S^{-1}\mathbb{Z}[x_1, \ldots, x_n]$ and $U = (u_1, \ldots, u_s) \in (S^{-1}R)^s$.

Let $F = \begin{pmatrix} g_1 \\ \vdots \\ g_s \end{pmatrix} \in (S^{-1}R)^s$, multiplying $(*)$ by F we get $LF = \sum_{G_{ij} \in T, k} Q_k G_{ij} F + U F$ $(2*)$, since $L \in \text{syz}(g_1, \ldots, g_s)$ then $LF = 0$. Also, each $G_{ij} \in \text{syz}(g_1, \ldots, g_s)$ then each $G_{ij} F = 0$, it remains from $(2*)$ the expression: $u_1 g_1 + \ldots + u_s g_s = 0$ $(3*)$, this means that $U = (u_1, \ldots, u_s)$ is a syzygy for $g_1, \ldots, g_s$.

Assume that $U = (u_1, \ldots, u_s)$ is a non-trivial syzygy for $g_1, \ldots, g_s$, then there exists $K = \{i_1, \ldots, i_r\} \subseteq \{1, \ldots, s\}$ such that $g_{i_1} u_{i_1} + \ldots + g_{i_r} u_{i_r} = 0$ $(4*)$. We can transform $(4*)$ as follows

$$(5*): \quad 0 = \sum_{i=i_1}^{i_r} \text{Lt}(g_i)\text{Lt}(u_i) + \sum_{i=i_1}^{i_r} \text{tail}(g_i)u_i + \sum_{i=i_1}^{i_r} \text{Lt}(g_i)\text{tail}(u_i).$$

1. Assume that $\sum_{i=i_1}^{i_r} \text{Lt}(g_i)\text{Lt}(u_i) = 0$, then:

$$\text{Lt}(u_{i_r})\text{Lt}(g_{i_r}) = -\text{Lt}(g_{i_1})\text{Lt}(u_{i_1}) - \cdots - \text{Lt}(g_{i_{r-1}})\text{Lt}(u_{i_{r-1}}).$$

Since $\{g_1, \ldots, g_s\}$ is a strong Gröbner basis for $S^{-1}I$, then by the Buchberger's criterion one can compute the S-vector of each pair of polynomial vectors (g_i, g_j), this means that one always have $\text{Lc}(g_i) \mid \text{Lc}(g_j)$ or $\text{Lc}(g_j) \mid \text{Lc}(g_i)$ (see Definition A.3 or Algorithm 1). According to the Definition A.3 of S-vectors, this means that there exists $l \in \{i_1, \ldots, i_{r-1}\}$ such that $\text{Lc}(g_l) \mid \text{Lc}(g_j)$ for any $j \in \{i_1, \ldots, i_{r-1}\}$. Without loss of generalities we can assume that $l = i_1$, then each $\text{Lc}(g_j)$ is a multiple of $\text{Lc}(g_{i_1})$, hence

$$\text{Lc}(g_{i_r})\text{Lc}(u_{i_r}) \text{ is a multiple of } \text{Lc}(g_{i_1}) \quad (a).$$

In the other hand $Lm(g_{i_1}) \mid Lm(g_{i_r})Lm(u_{i_r})$, this means that $Lm(g_{i_r})Lm(u_{i_r})$ is a common multiple of $Lm(g_{i_1})$ and $Lm(g_{i_r})$, therefore it is a multiple of the $lcm(Lm(g_{i_1}), Lm(g_{i_r}))$ thus

$$\frac{lcm(Lm(g_l), Lm(g_{i_r}))}{Lm(g_{i_r})} \text{ divides } Lm(u_{i_r}) \text{ that is } Lm(G_{i_r i_1}) \mid Lm(u_{i_r}) \quad (b).$$

- If $Lc(g_{i_r}) \nmid Lc(g_{i_1})$ then $Lt(G_{i_r i_1}) = -\dfrac{lcm(Lm(g_{i_1}), Lm(g_{i_r}))}{Lm(g_{i_r})}$, in this case it is clear that $Lt(G_{i_r i_1}) \mid Lt(u_{i_r})$, which is a contradiction since by the division's algorithm no term occurring in the remainder is suppose to divide any of the $Lt(G_{ij})$.
- If $Lc(g_{i_r}) \mid Lc(g_{i_1})$ then from (a) we have $Lc(g_{i_1}) \mid Lc(g_{i_r})$ $Lc(u_{i_r})$, this means that $\dfrac{Lc(g_{i_1})}{Lc(g_{i_r})} \mid Lc(u_{i_r})$ therefore $Lt(G_{i_r i_1}) =$ $\dfrac{Lc(g_{i_1})}{Lc(g_{i_r})}\dfrac{lcm(Lm(g_{i_1}), Lm(g_{i_r}))}{Lm(g_{i_r})}$ divides $Lt(u_{i_r})$ which is again a contradiction.

2. Assume that $\displaystyle\sum_{i=i_1}^{i_r} Lt(g_i)Lt(u_i) \neq 0$, because of

$$(5*): \quad 0 = \sum_{i=i_1}^{i_r} Lt(g_i)Lt(u_i) + \sum_{i=i_1}^{i_r} tail(g_i)u_i + \sum_{i=i_1}^{i_r} Lt(g_i)tail(u_i).$$

this means that not every $Lt(g_i)Lt(u_i)$ occurs in $(5*)$ for $i \in K$, for this there exists $K_1 \subsetneq K = \{i_1, \ldots, i_r\}$ such that $\displaystyle\sum_{i \in K_1} Lt(g_i)Lt(u_i) = 0$, By (1) this leads again to a contradiction as seen in (1).

Thus the remainder U must be zero and $(*)$ becomes $L = \displaystyle\sum_{G_{ij} \in T, k} Q_k G_{ij}$ $(*)$, by the division's algorithm there exists $H \subseteq \{(i,j)/ 1 \leq j < i \leq s\}$ such that $\forall (k_1, k_2) \in H$, $Lt(G_{k_1, k_2}) \mid Lt(L)$ and we are done.

With notations as in the Example A.6 we have the following example.

Example 2.5. In the Example A.6 we have seen that $G = \{f_1, f_3, f_4\}$ is a strong Gröbner basis for $S^{-1}I$ in $S^{-1}R$ where $S = 2^{\mathbb{N}} \cdot 5^{\mathbb{N}}$. While computing the Gröbner basis G, we notice that:

$S(f_3, f_1) = yf_3 - \dfrac{3}{50}f_1 = f_4$ then $\dfrac{3}{50}f_1 - yf_3 + f_4 = 0$, this means that $G_{13} = (\dfrac{3}{50}, -y, -1)$ is a syzygy for f_1, f_3, f_4. In the same way, we have the following syzygies:

$G_{14} = (\dfrac{3}{10}y, \dfrac{1}{10}, -x)$ and $G_{34} = (0, y^2 + \dfrac{1}{50}, y - \dfrac{1}{5}x)$. By the Theorem 2.4 the set $T = \{G_{13}, G_{14}, G_{34}\}$ form a minimal strong Gröbner basis for $Syz(f_1, f_3, f_4)$ in $(\mathbb{Z}_{2.5}[x, y])^3$.

Theorem 2.6 (Second Syzygy Theorem). Let $G = \{g_1, \ldots, g_s\}$ be a special Gröbner basis for I with multiplicative subset S and $H = (h_1, \ldots, h_s) \in S^{-1}R$ be any syzygy for $g_1, \ldots, g_s$. Let $q_1, \ldots, q_s \in \mathbb{Z} \setminus \{0\}$ be respectively the least common multiples of denominators occurring in $g_1, \ldots, g_s$. Let $q = \mathrm{lcm}(q_1, \ldots, q_s)$, then There exist $r_1, \ldots, r_s \in \mathbb{Z} \setminus \{0\}$ such that

$$H' = (r_1 h_1, \ldots, r_s h_s) \in R^s$$

is a syzygy for $q_1 g_1, \ldots, q_s g_s$.

Proof. Since H is a syzygy for $g_1, \ldots, g_s$ then

$$(1): \quad h_1 g_1 + \cdots + h_s g_s = 0.$$

Observe that $q_1 g_1, \ldots, q_s g_s \in R = \mathbb{Z}[x_1, \ldots, x_n]$. Let $b_1, \ldots, b_s$ be least common multiples of denominators occurring in $h_1, \ldots, h_s$ respectively, we have $c_1 h_1, \ldots, c_s h_s \in R$. Denote by $q = \mathrm{lcm}(b_1 q_1, \ldots, b_s q_s)$, then multiplying (1) by q we have: (2) $q \cdot h_1 g_1 + \cdots + q \cdot h_s g_s = 0$. Observe that $q = v_1 \cdot b_1 q_1 = \cdots = v_s \cdot b_s q_s$ for some $v_1, \ldots, v_s \in \mathbb{Z} \setminus \{0\}$, then:

$$(3) \quad (v_1 b_1 h_1)(q_1 g_1) + \cdots + (v_s b_s h_s)(q_s g_s) = 0,$$

thus

$$(4) \quad (r_1 h_1)(q_1 g_1) + \cdots + (r_s h_s)(q_s g_s) = 0$$

where $r_1 = v_1 b_1, \ldots, r_s = v_s b_s$. $\qquad \square$

The Algorithm 4 deals with the computation of syzygies introduced in Theorem 2.6. With notations as in the Theorem 2.6 we have the following result.

Lemma 2.7. $S^{-1}\mathrm{syz}(q_1 g_1, \ldots, q_s g_s) = \mathrm{syz}(g_1, \ldots, g_s)$.

Proof. Let $y \in S^{-1}\mathrm{syz}(q_1 g_1, \ldots, q_s g_s) \subset (S^{-1}R)^s$ then there exists $z \in \mathrm{syz}(q_1 g_1, \ldots, q_s g_s) \subset R^s$ and $a = a_1 e_1 + \ldots + a_s e_s$ with $a_1, \ldots, a_s \in S$ such that $y = z \cdot \dfrac{1}{a}$. Since $z \in \mathrm{syz}(q_1 g_1, \ldots, q_s g_s)$ then $z = (z_1, \ldots, z_s) \in R^s$ such that $z_1(q_1 g_1) + \cdots + z_s(q_s g_s) = 0$ that is $(z_1 q_1)g_1 + \cdots + (z_s q_s)g_s = 0$ thus $z_1 q_1 e_1 + \cdots + z_s q_s e_s \in \mathrm{syz}(g_1, \cdots, g_s)$. Setting $a_1 = \dfrac{1}{q_1}, \cdots, a_s = \dfrac{1}{q_s}$ then we have $(z_1 e_1 + \cdots + z_s e_s)(\dfrac{1}{a_1} e_1 + \cdots + \dfrac{1}{a_s} e_s) \in \mathrm{syz}(g_1, \ldots, g_s)$ that is $z \cdot \dfrac{1}{a} = y \in \mathrm{syz}(g_1, \ldots, g_s)$.

Conversely let $y = y_1 e_1 + \ldots + y_s e_s \in \mathrm{syz}(g_1, \ldots, g_s)$ for some $y_1, \ldots y_s \in S^{-1}R$ then

$$y_1 g_1 + \ldots + y_s g_s = 0$$

this means that $(y_1 q_1^{-1})(q_1 g_1) + \ldots + (y_s q_s^{-1})(q_s g_s) = 0$ thus $y_1 q_1^{-1} e_1 + \ldots + y_s q_1^{-1} e_s \in \mathrm{syz}(q_1 g_1, \ldots, q_s g_s) \subset (S^{-1}R)^s$. Since

$$y = (y_1 q_1^{-1} e_1 + \ldots + y_s q_1^{-1} e_s)(q_1 e_1 + \ldots + q_s e_s) \text{ with } (q_1 e_1 + \ldots + q_s e_s) \in (S^{-1}\mathbb{Z})^s$$

then $y \in S^{-1}\mathrm{syz}(q_1 g_1, \ldots, q_s g_s)$.

Remark 2.8. By hypothesis $G = \{g_1, \ldots, g_s\}$ is a strong Gröbner basis for $S^{-1}I$ in $S^{-1}R$, by the Theorem 1.4 there exist $q_1, \ldots, q_s \in \mathbb{Z} \setminus \{0\}$ such that $\{q_1 g_1, \ldots, q_s g_s\}$ is a weak Gröbner basis for $\mathbb{Z}(I)$ in R. Also, $\{T_1, \ldots, T_l\}$ form a strong Gröbner basis for $\mathrm{syz}(g_1, \ldots, g_s) = S^{-1}\mathrm{syz}(q_1 g_1, \ldots, q_s g_s) \subset (S^{-1}R)^s$.

Set $T_1 = \begin{pmatrix} t_1^1 \\ \vdots \\ t_s^1 \end{pmatrix}, \ldots, T_l = \begin{pmatrix} t_1^l \\ \vdots \\ t_s^l \end{pmatrix}$ for some $t_i^j \in S^{-1}R$, $i = 1, \ldots, s$ and $j = 1, \ldots, l$. Denote by a_i^j the least common multiple of all denominators occurring in t_i^j and $a_1 = \mathrm{lcm}(a_1^1, \ldots, a_s^1), \ldots, a_l = \mathrm{lcm}(a_1^l, \ldots, a_s^l)$ and set $a = \mathrm{lcm}(a_1, \ldots, a_l)$. We have:

$$\mathbb{Z}(\mathrm{syz}(q_1 g_1, \ldots, q_s g_s)) = \{L = L_1 e_1 + \ldots + L_s e_s \in R^s \mid \exists\, v \in a\mathbb{N}^*$$
$$\text{such that } v^{-1} L \in S^{-1}\mathrm{syz}(q_1 g_1, \ldots, q_s g_s)\}.$$

Theorem 2.9 (Third Syzygy Theorem). Let $G = \{g_1, \ldots, g_s\}$ be a special Gröbner basis for an ideal $I \subset R$ with multiplicative subset $S \subset \mathbb{Z}$ w.r.t a monomial ordering $<$. Let $T_1, \ldots, T_l$ be elements of $(S^{-1}R)^s$ and assume that $\{T_1, \ldots, T_l\}$ form a strong Gröbner basis for $\mathrm{syz}(g_1, \ldots, g_s) \subset (S^{-1}R)^s$ w.r.t the Schreyer's ordering $<_1$ induced by $<$ and G. There exist $T_1', \ldots, T_l' \in R^s$ and $q_1, \ldots, q_s \in \mathbb{Z} \setminus \{0\}$ such that $T' = \{T_1', \ldots, T_l'\}$ form a weak Gröbner basis for $\mathbb{Z}(\mathrm{syz}(q_1 g_1, \ldots, q_s g_s))$ w.r.t to $<_1$.

Proof. Let $L = (L_1, \ldots, L_s) \in \mathbb{Z}(\mathrm{syz}(q_1 g_1, \ldots, q_s g_s))$, we need to prove that there exist $T_1', \ldots, T_s' \in R^s$ such that $Lt(L) \in \langle Lt(T_1'), \ldots, Lt(T_l') \rangle$.

$L \in \mathbb{Z}(\mathrm{syz}(q_1 g_1, \ldots, q_s g_s))$ implies there exists $v \in a\mathbb{N}^*$ such that

$$\frac{1}{v} \cdot L \in S^{-1}\mathrm{syz}(q_1 g_1, \ldots, q_s g_s) \subset (S^{-1}R)^s \quad (*).$$

Since $\{T_1, \ldots, T_l\}$ form a strong Gröbner basis for $\mathrm{syz}(g_1, \ldots, g_s) = S^{-1}\mathrm{syz}(q_1 g_1, \ldots, q_s g_s)$ then $S^{-1}\mathrm{syz}(q_1 g_1, \ldots, q_s g_s) = \langle T_1, \ldots, T_l \rangle$ and by the division's algorithm in $(S^{-1}R)^s$ of $\frac{1}{v}L$ by $\{T_1, \ldots, T_l\}$ there exist $Q_1, \ldots, Q_l \in S^{-1}R$ such that

$$\frac{1}{v}L = Q_1 T_1 + \cdots + Q_l T_l$$

that is,

$$v^{-1} \cdot L = Q_1 \begin{pmatrix} t_1^1 \\ \vdots \\ t_s^1 \end{pmatrix} + \cdots + Q_l \begin{pmatrix} t_1^l \\ \vdots \\ t_s^l \end{pmatrix} \quad (2*)$$

where $T_1 = \begin{pmatrix} t_1^1 \\ \vdots \\ t_s^1 \end{pmatrix}, \ldots, T_l = \begin{pmatrix} t_1^l \\ \vdots \\ t_s^l \end{pmatrix}$. Denote by $b_1, \ldots, b_l$ the least common multiples of denominators occurring in $Q_1, \ldots, Q_l$ respectively and denote by $b = \operatorname{lcm}(b_1, \ldots, b_l)$. By choosing $v = a \cdot b$ (since $a \cdot b \in a\mathbb{N}$) and multiplying $(2*)$ by v we get:

$$L = bQ_1 \begin{pmatrix} at_1^1 \\ \vdots \\ at_s^1 \end{pmatrix} + \cdots + bQ_l \begin{pmatrix} at_1^l \\ \vdots \\ at_s^l \end{pmatrix} \quad (3*).$$

$$L = bQ_1 T_1' + \cdots + bQ_l T_l' \quad (4*)$$

where

$T_1' = \begin{pmatrix} at_1^1 \\ \vdots \\ at_s^1 \end{pmatrix}, \ldots, T_l' = \begin{pmatrix} at_1^l \\ \vdots \\ at_s^l \end{pmatrix}$. Observe that $(4*)$ arises from the division's algorithm, there exists then $K = \{i_1, \ldots, i_r\} \subset \{1, \ldots, l\}$ such that $\operatorname{Lm}(T_j') \mid \operatorname{Lm}(L) \ \forall j \in K$ and $\operatorname{Lc}(L) \in \langle \operatorname{Lc}(T_j) \mid j \in K \rangle$ thus $\operatorname{Lt}(L) \in \langle \operatorname{Lt}(T_1'), \ldots, \operatorname{Lt}(T_l') \rangle$.

A weak Groebner basis for $\mathbb{Z}(Syz(q_1 g_1, \cdots, q_s g_s))$ is computed using the Algorithm 5.

3   Minimal Free Resolution in $\mathbb{Z}[x_1, \ldots, x_n]$

Given an ideal $I = \langle f_1, \ldots, f_s \rangle$ of $R = \mathbb{Z}[x_1, \ldots, x_n]$, we propose in this section a method for computing a minimal free resolution for $\mathbb{Z}(I)$ as an $R-$module. Our technique consists firstly to compute a minimal free resolution for $S^{-1}I$ as an $S^{-1}R-$module for some multiplicative subset $S \subset \mathbb{Z}$ arising from the computation of a special Gröbner basis for I, and then use syzygy theorems to obtain a minimal free resolution for $\mathbb{Z}(I)$.

3.1   Construction of a Free Resolution

Let $I = \langle f_1, \ldots, f_s \rangle$ be an ideal of R and $G = \{g_1, \ldots, g_r\}$ be a special minimal Gröbner basis for I w.r.t some monomial ordering $<$.

1. Theorem 1.4 states that there exist $q_1, \ldots, q_r \in \mathbb{Z} \setminus \{0\}$ such that $G' = \{q_1 g_1, \ldots, q_r g_r\}$ form a weak Gröbner basis for $\mathbb{Z}(I)$. We can then construct the following two maps:

$$\phi_1 : (S^{-1}R)^r \longrightarrow S^{-1}I = \langle g_1, \ldots, g_r \rangle, \ (h_1, \ldots, h_r) \longrightarrow \sum_{i=1}^r h_i g_i$$

and

$$\psi_1 : R^r \longrightarrow \mathbb{Z}(I) = \langle q_1 g_1, \ldots, q_r g_r \rangle, \ (h'_1, \ldots, h'_r) \longrightarrow \sum_{i=1}^{r} h'_i(q_i g_i).$$

Since these maps are surjective, we have the following exact sequences

$$(S^{-1}R)^r \longrightarrow S^{-1}I \longrightarrow 0 \quad (A_1)$$

and

$$R^r \longrightarrow \mathbb{Z}(I) \longrightarrow 0 \quad (A_2)$$

Observe that:

$$\ker(\phi_1) = \mathrm{syz}(g_1, \ldots, g_r) \ \text{and} \ \ker(\psi_1) = \mathrm{syz}(q_1 g_1, \ldots, q_r g_r) \quad (*).$$

(A_1) and (A_2) can be represented as follows:

$$(S^{-1}R)^r \xrightarrow{\mathrm{Im}(\phi_1)} S^{-1}I \longrightarrow 0$$

$$R^r \xrightarrow{\mathrm{Im}(\psi_1)} \mathbb{Z}(I) \longrightarrow 0$$

2. Since $\{g_1, \ldots, g_r\}$ form a strong minimal Gröbner basis for $S^{-1}I$, by the First Syzygy Theorem (see Theorem 2.4) one can find a strong minimal Gröbner basis $\{T_1, \ldots, T_{r_1}\}$ for $\mathrm{syz}(g_1, \ldots, g_r) \subset (S^{-1}R)^{r_1}$, and then by the Third Syzygy Theorem (see Theorem 2.9) one have the corresponding weak Gröbner basis $\{T'_1, \ldots, T'_{r_1}\}$ for $\mathbb{Z}(\mathrm{syz}(q_1 g_1, \ldots, q_r g_r)) \subset R^{r_1}$. Therefore

$$\mathrm{syz}(g_1, \ldots, g_r) = \langle T_1, \ldots, T_{r_1} \rangle \ \text{and} \ \mathbb{Z}(\mathrm{syz}(q_1 g_1, \ldots, q_r g_r)) = \langle T'_1, \ldots, T'_{r_1} \rangle.$$

Now consider the maps

$$\alpha_1 : (S^{-1}R)^{r_1} \longrightarrow \ker(\phi_1) = \langle g_1, \ldots, g_r \rangle$$

and

$$\beta_1 : R^{r_1} \longrightarrow \ker(\psi_1) = \langle q_1 g_1, \ldots, q_r g_r \rangle,$$

Using the inclusion maps i_1 and i_2 we have the following commutative diagrams:

$$
\begin{array}{ccc}
\ker(\phi_1) \xrightarrow{i_1} (S^{-1}R)^r & \text{and} & \ker(\psi_1) \xrightarrow{i_2} R^r \\
\alpha_1 \uparrow \ \ \nearrow_{\phi_2 = i_1 \circ \alpha_1} & & \beta_1 \uparrow \ \ \nearrow_{\psi_2 = i_2 \circ \beta_1} \\
(S^{-1}R)^{r_1} & & R^{r_1}
\end{array}
$$

adding these to (A_1) and (A_2) we obtain the following exact sequences

$$(S^{-1}R)^{r_1} \xrightarrow{\phi_2} (S^{-1}R)^r \xrightarrow{\phi_1} S^{-1}I \longrightarrow 0, \quad (A3)$$

and

$$R^{r_1} \xrightarrow{\psi_2} R^r \xrightarrow{\psi_1} \mathbb{Z}(I) \longrightarrow 0 \quad (A4)$$

3. Now let us compute $\ker(\psi_2)$ and $\ker(\phi_2)$. Since $\{T_1, \ldots, T_{r_1}\}$ forms a strong Gröbner basis for $\mathrm{syz}(g_1, \ldots, g_r)$ w.r.t the Schreyer's ordering $<_1$ induced by $<$ and $\{g_1, \ldots, g_r\}$, then the remainder of the division's algorithm of $S(T_i, T_j)_{1 \le j < i \le r_1}$ by $\{T_1, \ldots, T_{r_1}\}$ is zero. For each such S-vector, one can collect the corresponding syzygy T_{ij} (playing the role of G_{ij} seen in Remark 2.2) and using the Theorem 2.4 one sees that the set $\{T_{ij} \in \mathrm{syz}(T_1, \ldots, T_{r_1})/1 \le j < i \le r_1\}$ forms a strong Gröbner basis for $\mathrm{syz}(T_1, \ldots, T_{r_1}) = \mathrm{syz}(\mathrm{syz}(g_1, \ldots, g_s))$ w.r.t the Schreyer's ordering $<_2$ induced by $<_1$ and $\{T_1, \ldots, T_{r_1}\}$. Assume without loss of generalities that $\langle T_{ij} \in \mathrm{syz}(T_1, \ldots, T_{r_1})/i > j \rangle = \langle G_1, \ldots, G_{r_2} \rangle$ and the set $\{G_1, \ldots, G_{r_2}\}$ form a minimal Gröbner basis for $\mathrm{syz}(\mathrm{syz}(g_1, \ldots, g_s))$, then:

$$\ker(\phi_2) = \mathrm{syz}(\mathrm{syz}(g_1, \ldots, g_r)) = \langle G_1, \ldots, G_{r_2} \rangle$$

and using the Third Syzygy Theorem (see Theorem 2.9) there exists a corresponding weak Gröbner basis $\{G_1', \ldots, G_{r_2}'\}$ for $\mathrm{syz}(T_1', \ldots, T_{r_1}') = \mathrm{syz}(\mathrm{syz}(q_1 g_1, \ldots, q_r g_r))$ and

$$\ker(\psi_2) = \mathbb{Z}(\mathrm{syz}(\mathrm{syz}(q_1 g_1, \ldots, q_r g_r))) = \langle G_1', \ldots, G_{r_2}' \rangle.$$

Consider the maps

$$\alpha_2 : (S^{-1}R)^{r_2} \longrightarrow \ker(\phi_2) \quad \text{and} \quad \beta : R^{r_2} \longrightarrow \ker(\psi_2),$$

using the inclusion maps i_3 and i_4 we get the following commutative diagrams:

$$
\ker(\phi_2) \xrightarrow{i_3} (S^{-1}R)^{r_1} \quad \text{and} \quad \ker(\psi_2) \xrightarrow{i_4} R^{r_1}
$$

$$\alpha_2 \uparrow \quad \nearrow \phi_3 = i_3 \circ \alpha_2 \qquad\qquad \beta_2 \uparrow \quad \nearrow \psi_3 = i_4 \circ \beta_2$$

$$(S^{-1}R)^{r_2} \qquad\qquad\qquad\qquad R^{r_2}$$

Adding these to $(A3)$ and $(A4)$ we get the following exact sequences

$$S^{-1}R^{r_2} \xrightarrow{\mathrm{Im}(\phi_3)} S^{-1}R^{r_1} \xrightarrow{\mathrm{Im}(\phi_2)} S^{-1}R^s \xrightarrow{\mathrm{Im}(\phi_1)} S^{-1}I \longrightarrow 0. \quad (A5)$$

and

$$R^{r_2} \xrightarrow{\mathrm{Im}(\psi_3)} R^{r_1} \xrightarrow{\mathrm{Im}(\psi_2)} R^s \xrightarrow{\mathrm{Im}(\psi_1)} \mathbb{Z}(I) \longrightarrow 0. \quad (A6)$$

Continuing this way and so forth, these leads to two free resolutions not necessarily finite. To obtain finite and minimal free resolution, we use the technique introduced in [3] which is based on the Hilbert's syzygies theorem. The degree reverse lexicographic ordering is the best choice for this method.

Theorem 3.1 (Hilbert's syzygies theorem). Let $R = \mathbb{Z}[x_1, \ldots, x_n]$, then any finitely generated R−module has a minimal free resolution of length at most $\dim(R) = n + 1$.

Since R and $S^{-1}R$ have Krull dimension equal to $n + 1$, we know in advance the length of our minimal free resolution.

Method for Computing a Minimal Free Resolution

1. Choose an integer $k \in \{1, \ldots, n\}$ such that none of the leading terms $Lt(f_i)$ (w.r.t the degree reverse lexicographic order) involves the variables $x_{k+1}, \ldots, x_n$;
2. choose $k = n$ if one of the $Lt(f_i)$ involves x_n;
3. In the Buchberger's test, let the f_i be arranged such that, for $j < i$, the exponent of x_k in $Lt(f_j)$ is smaller or equal than that of x_k in $Lt(f_i)$.
4. To compute a minimal free resolution, we apply the previous process by respecting this rule at each step, this will lead to the following exact sequences written in the diagram:

$$S^{-1}R^{r_k} \xrightarrow{\mathrm{Im}(\phi_{k+1})} \cdots S^{-1}R^{r_2} \xrightarrow{\mathrm{Im}(\phi_3)} S^{-1}R^{r_1} \longrightarrow S^{-1}R^r \xrightarrow{\mathrm{Im}(\phi_1)} S^{-1}I \longrightarrow 0$$

$$R^{r_k} \xrightarrow{\mathrm{Im}(\psi_{k+1})} \cdots R^{r_2} \xrightarrow{\mathrm{Im}(\psi_3)} R^{r_1} \xrightarrow{\mathrm{Im}(\psi_2)} R^r \xrightarrow{\mathrm{Im}(\psi_1)} \mathbb{Z}(I) \longrightarrow 0$$

with $\ker(S^{-1}R^{r_k} \longrightarrow S^{-1}R^{r_{k-1}}) = 0$ and using the Theorem 2.4 we have the following exact sequences:

$$0 \longrightarrow S^{-1}R^{r_k} \xrightarrow{\mathrm{Im}(\phi_{k+1})} \cdots S^{-1}R^{r_2} \xrightarrow{\mathrm{Im}(\phi_3)} S^{-1}R^{r_1} \xrightarrow{\mathrm{Im}(\phi_2)} S^{-1}R^r \xrightarrow{\mathrm{Im}(\phi_1)} S^{-1}I \longrightarrow 0$$

$$0 \longrightarrow R^{r_k} \xrightarrow{\mathrm{Im}(\psi_{k+1})} \cdots R^{r_2} \xrightarrow{\mathrm{Im}(\psi_3)} R^{r_1} \xrightarrow{\mathrm{Im}(\psi_2)} R^r \xrightarrow{\mathrm{Im}(\psi_1)} \mathbb{Z}(I) \longrightarrow 0$$

Example 3.2. Let us compute a minimal free resolution for $\mathbb{Z}(I)$ in $R = \mathbb{Z}[x, y]$ where $I = \langle f_1 = 10xy + 1, f_2 = 6x^2 + 3 \rangle$ is an ideal of R. Since $\dim(R) = 3$, we know in advance that $\mathbb{Z}(I)$ has a minimal free resolution of length at most 3.

– We have seen in the Example A.6 that $G = \{f_1, f_3, f_4\}$ is a special Gröbner basis for I and by the Example A.9 we have seen that $G' = \{g_1, g_3, g_4\}$ is a weak Gröbner basis for $\mathbb{Z}(I)$ in R. This leads to the exact sequences

$$S^{-1}R^3 \longrightarrow S^{-1}I \longrightarrow 0$$

$$R^3 \longrightarrow \mathbb{Z}(I) \longrightarrow 0$$

where $S = 2^N \cdot 5^N$, $f_1 = 10xy + 1$, $f_3 = \dfrac{3}{5}x - 3y$, $f_4 = 3y^2 + \dfrac{3}{50}$ and $g_1 = 6x^2 + 3$, $g_3 = 3x - 15y$, $g_4 = 150y^2 + 3$.

- In the Example 2.5 we have seen that $T = \{G_{13}, G_{14}, G_{34}\}$ is a strong minimal Gröbner basis for $\mathrm{syz}(f_1, f_3, f_4)$ w.r.t the Schreyer's ordering induced by $<$ and f_1, f_3, f_4. In the other hand, in the Example A.12 we have seen that $T' = \{G'_{13}, G'_{14}, G'_{34}\}$ is a weak minimal Gröbner basis for $\mathbb{Z}(\mathrm{syz}(q_1 f_1, q_3 f_3, q_4 f_4))$ w.r.t the Schreyer's ordering induced by $<$ and f_1, f_3, f_4. These lead to the exact sequences

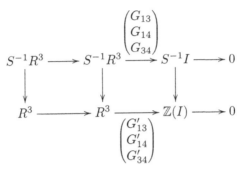

- Observe that $Lm(G_{13}) = ye_2$, $Lm(G_{14}) = xe_3$, $Lm(G_{34}) = xe_3$. Let us arrange these polynomials as follows: $t_1 = G_{14}, t_2 = G_{34}, t_3 = G_{13}$. Since $T = \{t_1, t_2, t_3\}$ is a strong minimal Gröbner basis for $\mathrm{Syz}(f_1, \ldots, f_4)$ w.r.t $<_1$ induced by d_p and f_1, f_3, f_4, then the remainder of all the S-vectors by division's algorithm is zero. Observe that:
$S(t_3, t_1) = S(t_3, t_2) = 0$; and $S(t_2, t_1) = 5t_2 - t_2 = -5yt_3 \implies H_{12} = (-1, 5, 5y)$ is a syzygy for t_1, t_2, t_3 and by the First Syzygy Theorem (see Theorem 2.4), the set $H = \{H_{12}\}$ form a strong Gröbner basis for $\mathrm{syz}(t_1, t_2, t_3)$ w.r.t the Schreyer's ordering $<_2$ induced by $<_1$ and t_1, t_2, t_3. This means that

$$\mathrm{syz}(t_1, t_2, t_3) = \mathrm{syz}(\mathrm{syz}(f_1, f_3, f_4)) = \langle H_{12} \rangle.$$

By the Third Syzygy Theorem (see Theorem 2.9) the set $H' = \{H'_{12} = (-1, 5, 5y)\}$ form a weak minimal Gröbner basis for $\mathbb{Z}(\mathrm{syz}(t'_1 t'_2, t'_3))$ (where $t'_1, t'_2 t'_3$ are like in the Algorithm 5) w.r.t the Schreyer's ordering $<_2$ induced by $<_1$ and G_{13}, G_{14}, G_{34}. These lead to the exact sequences:

$$
\begin{array}{ccccccc}
S^{-1}R^3 & \xrightarrow{H_{12}} & S^{-1}R^3 & \xrightarrow{\left(\begin{smallmatrix} G_{13} \\ G_{14} \\ G_{34} \end{smallmatrix}\right)} & S^{-1}I & \longrightarrow & 0 \\
\downarrow{\scriptstyle FSTW} & & \downarrow{\scriptstyle FSTW} & & \downarrow & & \\
R^3 & \xrightarrow{H'_{12}} & R^3 & \xrightarrow{\left(\begin{smallmatrix} G'_{13} \\ G'_{14} \\ G'_{34} \end{smallmatrix}\right)} & \mathbb{Z}(I) & \longrightarrow & 0
\end{array}
$$

Since $\text{syz}(H_{12}) = 0$ (because the ring has no zero divisors), then $\text{syz}(H'_{12}) = 0$. Thus we have the following minimal free resolution

$$
\begin{array}{ccccccccc}
0 & \longrightarrow & S^{-1}R & \xrightarrow{(0)} & S^{-1}R^3 & \xrightarrow{H_{12}} & S^{-1}R^3 & \xrightarrow{\begin{pmatrix} G_{13} \\ G_{14} \\ G_{34} \end{pmatrix}} & S^{-1}I & \longrightarrow & 0 \\
& & \downarrow{\scriptstyle \text{FSTW}} & & \downarrow{\scriptstyle \text{FSTW}} & & \downarrow{\scriptstyle \text{FSTW}} & & \downarrow & & \\
0 & \longrightarrow & R & \xrightarrow{(0)} & R^3 & \xrightarrow{H'_{12}} & R^3 & \xrightarrow{\begin{pmatrix} G'_{13} \\ G'_{14} \\ G'_{34} \end{pmatrix}} & \mathbb{Z}(I) & \longrightarrow & 0
\end{array}
$$

4   Conclusion

The problem treated in this paper was inspired by the idea introduced in [3] for computing minimal free resolutions with coefficients over a field. Our goal was to adapt this technique for any ideal I of $\mathbb{Z}[x_1, \ldots, x_n]$ but the First Syzygy Theorem which is central for this technique failed to be adapted, to get round this situation, we introduce the ideal $\mathbb{Z}(I)$ of I which is a suitable ideal of $\mathbb{Z}[x_1, \ldots, x_n]$ for this technique. We encourage interested people to propose a technique for computing minimal free resolutions for any ideal of $\mathbb{Z}[x_1, \ldots, x_n]$.

A   Appendix

Basic notions

Definition A.1 [4]. Let $R = \mathbb{Z}[x_1, \ldots, x_n]$ and $F = R^r$ be an R–free module with canonical basis $e_1, \ldots, e_r$ for some $r > 1$.

1. A monomial in R is a multivariate polynomial of the form $x_1^{\alpha_1} \cdots x_n^{\alpha_n}$ where $\alpha_i \in \mathbb{N}$. We denote $x^\alpha = x_1^{\alpha_1} \cdots x_n^{\alpha_n}$ where $\alpha = (\alpha_1, \ldots, \alpha_n)$. We denote by $\text{Mon}(x_1, \ldots, x_n)$ the set of all monomials in R.
2. A monomial in F is a product of the form $x^\alpha e_i$ where x^α is a monomial in R and e_i an element of the basis. We denote by $\text{Mon}(F)$ the set of all monomials in F.
3. Accordingly, a term in F is the product of a monomial in F with an element of $\mathbb{Z}$.
4. A term $c_1 x^\alpha e_i$ divides a term $c_2 x^\beta e_j$ if the following conditions are satisfied:
   - $i = j$;
   - c_1 divides c_2 in $\mathbb{Z}$;
   - x^α divides x^β in R.
   We also say that a monomial $x^\alpha \in \text{Mon}(x_1, \ldots, x_n)$ divides a monomial $x^\beta e_j \in \text{Mon}(F)$ if x^α divides x^β and in this case we write $(\frac{x^\beta}{x^\alpha}) e_j \in \text{Mon}(F)$.

5. Let $x^\beta e_i, x^\beta e_j \in \mathrm{Mon}(F)$, we define the product as follows:

$$x^\alpha e_i \cdot x^\beta e_j = \begin{cases} x^{\alpha+\beta} e_i & \text{if } i = j \\ 0 & \text{otherwise.} \end{cases}$$

In the same way, we define the least common multiple as follows: if $c_1, c_2 \in \mathbb{Z}$, then

$$\mathrm{lcm}(c_1 x^\alpha e_i, c_1 x^\beta e_j) = \begin{cases} \mathrm{lcm}(c_1 x^\alpha, c_1 x^\beta) e_i = (\mathrm{lcm}(c_1, c_2) \cdot \mathrm{lcm}(x^\alpha, x^\beta)) e_i & \text{if } i = j \\ 0 & \text{otherwise.} \end{cases}$$

6. A monomial ordering on F is a total ordering $<$ on $\mathrm{Mon}(F)$ such that if $x^\alpha e_i$ and $x^\beta e_j$ are monomials in $\mathrm{Mon}(F)$, and x^γ any monomial in R, then

$$x^\alpha e_i < x^\beta e_j \implies (x^{\alpha+\gamma}) e_i < (x^{\beta+\gamma}) e_j.$$

Moreover, $x^\alpha e_i < x^\beta e_i$ if and only if $x^\alpha e_j < x^\beta e_j \; \forall \, i, j$.

7. Let $<$ be a monomial ordering on F, let $f \in F \setminus \{0\}$, and let $f = c x^\alpha e_i + f^*$ be the unique decomposition of f with $c \in \mathbb{Z} \setminus \{0\}, x^\alpha e_i \in \mathrm{Mon}(F)$, and $x^\alpha e_i > x^\beta e_j$ for any non-zero term $c^* x^\beta e_j$ occurring in f^*. We define the leading monomial, the leading coefficient, the leading term, and the tail of f as $Lm(f) := x^\alpha e_i, Lc(f) := c, Lt(f) := c x^\alpha e_i, \mathrm{tail}(f) := f - Lt(f)$ respectively.

8. For any sub-module $I \subset F$ (or ideal in the case $r = 1$), we define $Lt(I) := \langle Lt(f)/f \in F \rangle \subset F$ the leading module (or leading ideal) of F.

9. An element $(h_1, \ldots, h_s) \in R^s$ is called syzygy for $f_1, \ldots, f_r \in R$ if $h_1 f_1 + \ldots, h_s f_s = 0$.

Definition A.2 (S-vector). Let $f, g \in F = R^r = (\mathbb{Z}[x_1, \ldots, x_n])^r$ be non-zero polynomial vectors and $<$ be a monomial ordering. We define the S-vector of f and g in F as follows:

$$S(f, g) = \begin{cases} \dfrac{Lc(g)}{Lc(f)} m_{gf} \cdot f - m_{fg} \cdot g & \text{if } Lc(f) \mid Lc(g) \\[2mm] m_{gf} \cdot f - \dfrac{Lc(f)}{Lc(g)} m_{fg} \cdot f & \text{if } Lc(g) \mid Lc(f) \text{ and } Lc(f) \nmid Lc(g). \end{cases}$$

Where $m_{gf} = \dfrac{\mathrm{lcm}(Lm(f), Lm(g))}{Lm(f)}$ and $m_{fg} = \dfrac{\mathrm{lcm}(Lm(f), Lm(g))}{Lm(g)}$.

If $F = R$ we call an S-vector by S-polynomial.

Special Gröbner bases

In this section we denote by $I = \langle f_1, \ldots, f_s \rangle$ an R–submodule of $F = R^r = (\mathbb{Z}[x_1, \ldots, x_n])^r$. We revisit the technique for computing special Gröbner bases for I introduced in [8–10]. Let $f, g \in F$ and $<$ be a monomial ordering. If $Lc(f) = \gcd(Lc(f), Lc(g)) \cdot a$ and $Lc(g) = \gcd(Lc(f), Lc(g)) \cdot b$ with $\gcd(a, b) = 1$, then by the fundamental theorem of arithmetic there exist prime numbers $p_1, \ldots, p_r, q_1, \ldots, q_s$ with $p_i \neq q_j \; \forall \, i, j$ and integers $l_1, \ldots, l_r, m_1, \ldots, m_s$ such that $a = p_1^{l_1} \cdots p_r^{l_r}$ and $b = q_1^{m_1} \cdots q_r^{m_r}$.

With these notations, we have the following:

Definition A.3 (Special S-vector). By a special S-vector of f and g we mean an S-vector $S(f, g)$ defined over the localized ring $\mathbb{Z}_{[p_1 \cdots p_r]}$ or $\mathbb{Z}_{[q_1 \cdots q_s]}$ where
$$\mathbb{Z}_{[p_1 \cdots p_r]} = \{\frac{c}{p_1^{n_1} \cdots p_r^{n_r}} \mid c \in \mathbb{Z} \text{ and } n_1, \ldots, n_r \in \mathbb{N}\}$$

Remark A.4. The idea of the previous definition is described by the following tree

$$\mathbb{Z}$$
$$\mathbb{Z}_{[p_1 \cdots p_r]} \qquad\qquad \mathbb{Z}_{[q_1 \cdots q_s]}$$

which means that if one of the leading coefficient divides the other, use directly the Definition A.2 with coefficients in $\mathbb{Z}$. If none of the leading coefficient divides the other, apply the Definition A.2 in any of $\mathbb{Z}_{[p_1 \cdots p_r]}$ or $\mathbb{Z}_{[q_1 \cdots q_s]}$ since in any of these localized ring, one of the leading coefficient divides the other.

Algorithm 1. SPECIAL S-VECTORS

Input: $f_1, f_2 \in F = R^r = (\mathbb{Z}[x_1, \ldots, x_n])^r$ and a monomial ordering $>$ on $\mathrm{Mon}(F)$.

Output: Special S-vectors of f and g.

1 if $Lc(f_1) \mid Lc(f_2)$ then

2 $\quad$ $p := S(f_2, f_1) = m_{12} \cdot f_2 - \dfrac{Lc(f_2)}{Lc(f_1)} m_{21} \cdot f_1$

3 $\quad$ return p.

4 end

5 if $Lc(f_2) \mid Lc(f_1)$ then

6 $\quad$ $p := S(f_2, f_1) = \dfrac{Lc(f_1)}{Lc(f_2)} m_{12} \cdot f_2 - m_{21} \cdot f_1$

7 $\quad$ return p.

8 end

9 write $Lc(f_2) = \gcd(Lc(f_1), Lc(f_2)) \cdot a_2$ and $Lc(f_1) = \gcd(Lc(f_1), Lc(f_2)) \cdot a_1$;

10 write $a_2 = p_1^{l_1} \cdots p_r^{l_r}$;

11 compute $p := S(f_2, f_1) = \dfrac{Lc(f_1)}{Lc(f_2)} m_{12} \cdot f_2 - m_{21} \cdot f_1$ in $F' = (\mathbb{Z}_{[p_1 \cdots p_r]}[x_1, \ldots, x_n])^r$;

12 return p.

Example A.5. Let $f = \begin{pmatrix} 12x^2y + y^2 \\ 5xy^2 + 9 \end{pmatrix}$, $g = \begin{pmatrix} 15x^3 + 7y^3 \\ 4xy \end{pmatrix} \in R^2 = (\mathbb{Z}[x, y])^2$.
Let $<$ be the degree reverse lexicographic ordering, then we have: $f = f_1 e_1 + f_2 e_2$, $g = g_1 e_1 + g_2 e_2$ with $f_1 = 12x^2y + y^2$, $f_2 = 5xy^2 + 9$, $g_1 = 15x^3 + 7y^3$ and $g_2 = 4xy$. Observe that $Lt(f_1) = 12x^2y > Lt(f_2) = 5xy^2$ then $Lt(f) = 12x^2ye_1$ and $Lt(g) = 15x^3e_1$. Since $Lc(f) = 12$, $Lc(g) = 15$ and $\gcd(Lc(f), Lc(g)) = 3$ then $Lc(f) = 3 \cdot 4$, $Lc(g) = 3 \cdot 5$. In this case

$$\mathbb{Z}$$
$$\diagup \diagdown$$
$$\mathbb{Z}_{[5]} \qquad\qquad \mathbb{Z}_{[2]}$$

where $\mathbb{Z}_{[5]} = \{\dfrac{a}{5^{n_2}} \mid a \in \mathbb{Z} \text{ and } n_2 \in \mathbb{N}\}$ and $\mathbb{Z}_{[2]} = \{\dfrac{a}{2^n} \mid a \in \mathbb{Z} \text{ and } n \in \mathbb{N}\}$.

1. Observe that 5 is a unit in $\mathbb{Z}_{[5]}$, in this ring 15 divides 12 and we have:

$$S(f,g) = x \cdot f - \frac{4}{5}y \cdot g;$$

2. 2 is a unit in $\mathbb{Z}_{[2]}$ and in this ring 12 divides 15 and we have:

$$S(f,g) = \frac{5}{4}x \cdot f - y \cdot g.$$

Algorithm 2. SPECIAL GRÖBNER BASES: BUCHBERGER'S ALGORITHM

Input: An ideal $I = \langle f_1, \ldots, f_s \rangle \subset F = R^r = (\mathbb{Z}[x_1, \ldots, x_n])^r$ and a monomial ordering $>$ on $\mathrm{Mon}(F)$.
Output: Gröbner basis $G = \{g_1, \ldots, g_t\}$ for $S^{-1}I$ in $F' = (S^{-1}\mathbb{Z}[x_1, \ldots, x_n])^r$ where S a suitable multiplicative subset of $\mathbb{Z}$.

1   $k := s$;
2   Compute the set P of all special S-vectors $P := \{S(f_i, f_j)_{1 \leq j < i \leq k}\}$ together with the list L of integers representing the multiplicative subset on which one work;
3   **while** $P \neq \emptyset$ **do**
4   choose $h \in P$ and denote by r the remainder of the division's algorithm of h by $\{f_1, \ldots, f_k\}$ in the localization of $\mathbb{Z}$ at the multiplicative subset given by L;
5   **if** $r = 0$ **then**
6        $\mid$ $P := P \setminus \{h\}$
7   **end**
8   **else**
9        $\mid$ $k := k + 1$;
10        $\mid$ $f_k := r$;
11        $\mid$ Compute the set of all special S-vectors $P := (P \setminus \{h\}) \cup \{S(f_k, f_i), S(f_i)/\ 1 \leq i \leq k\}$ together with the new set of integers M
12        $\mid$ $L := L \cup M$
13   **end**
14 **end**
15 **return** $\{f_1, \ldots, f_k\}$ together with L

Example A.6. Let $I = \langle f_1 = 10xy + 1, f_2 = 6x^2 + 3 \rangle$ be an ideal of $\mathbb{Z}[x, y]$. We wish to compute a special Gröbner basis for I w.r.t the degree reverse lexicographic ordering d_p. Observe that $\mathrm{Lc}(f_1) \nmid \mathrm{Lc}(f_2)$ and $\mathrm{Lc}(f_2) \nmid \mathrm{Lc}(f_1)$, in this case we write: $\mathrm{Lc}(f_1) = 2 \cdot 5$ and $\mathrm{Lc}(f_2) = 2 \cdot 3$, we have:

$$\mathbb{Z}$$
$$\diagup \diagdown$$
$$\mathbb{Z}_{[5]} = (5^{\mathbb{N}})^{-1}\mathbb{Z} \qquad\qquad \mathbb{Z}_{[3]} = (3^{\mathbb{N}})^{-1}\mathbb{Z}$$

Observe that in $\mathbb{Z}_{[5]}$, 5 is invertible and $Lc(f_1) \mid Lc(f_2)$, we have in $R_1 = \mathbb{Z}_{[5]}[x,y]$:

$$S(f_1, f_2) = \frac{3}{5}xf_1 - yf_2 = \frac{3}{5}x - 3y.$$

Since $\frac{3}{5}x - 3y$ is the remainder of the division of $S(f_1, f_2)$ by $\{f_1, f_2\}$, we set $f_3 = \frac{3}{5}x - 3y$. Observe that $Lc(f_1)$ and $Lc(f_3)$ are not compatible under the division in R_1. Since $Lc(f_1) = 2 \cdot 5$ and $Lc(f_3) = 3 \cdot \frac{1}{5}$, we can localize either by $2^{\mathbb{N}}$ or $3^{\mathbb{N}}$. Let us localize R_1 by $2^{\mathbb{N}}$, this gives $R_2 = \mathbb{Z}_{[5.2]}[x,y]$. Note that in R_2 we have $Lc(f_1) \mid Lc(f_3)$ and:

$$S(f_1, f_3) = \frac{3}{50}f_1 - yf_3 = 3y^2 + \frac{3}{50}.$$

Since $3y^2 + \frac{3}{50}$ is the remainder of $S(f_1, f_3)$ by $\{f_1, f_2, f_3\}$ then we set $f_4 = 3y^2 + \frac{3}{50}$.

Observe that:

$$S(f_1, f_4) = \frac{3}{10}yf_1 - xf_4 = -\frac{1}{10}f_3;$$

$$S(f_2, f_3) = 10xf_3 - f_2 = -3f_1;$$

$$S(f_2, f_4) = y^2 f_2 - 2x^2 f_4 = -\frac{1}{50}f_2 + f_4$$

$$S(f_3, f_4) = y^2 f_3 - \frac{1}{5}xf_4 = -\frac{1}{50}f_3 - yf_4$$

Thus $G = \{f_1, f_2, f_3, f_4\}$ is a strong Gröbner basis for $S^{-1}I$ in R_2 where $S = 2^{\mathbb{N}} \cdot 5^{\mathbb{N}}$, that is, G is a special Gröbner basis for I in R.

Example A.7. Let $N = \langle f_1, f_2 \rangle$ be an R–submodule of $R^2 = (\mathbb{Z}[x,y])^2$ with $f_1 = \begin{pmatrix} 12x^2y + y^2 \\ 5xy^2 + 9 \end{pmatrix}$, $f_2 = \begin{pmatrix} 15x^3 + 7y^3 \\ 4xy \end{pmatrix} \in R^2$. Let $<$ be the degree reverse lexicographic ordering and $<_M$ be a module ordering with priority given to monomials. We have:
$Lt(f) = 12x^2y e_1$ and $Lt(f_2) = 15x^3 e_1$. Since $Lc(f_1) = 12$, $Lc(f_2) = 15$ and none of 12 and 15 divides the other in $\mathbb{Z}$ then consider: $\gcd(Lc(f_1), Lc(f_2)) = 3$ with $Lc(f_1) = 3 \cdot 4$ and $Lc(f_2) = 3 \cdot 5$. In this case we have

$$\mathbb{Z}$$
$$\diagup \diagdown$$
$$\mathbb{Z}_{[5]} \qquad\qquad \mathbb{Z}_{[2]}$$

where $\mathbb{Z}_{[5]} = \{\frac{a}{5^{n_2}} \mid a \in \mathbb{Z} \text{ and } n_2 \in \mathbb{N}\}$ and $\mathbb{Z}_{[2]} = \{\frac{a}{2^n} \mid a \in \mathbb{Z} \text{ and } n \in \mathbb{N}\}$.

1. Since 2 is a unit in $\mathbb{Z}_{[2]}$, then in this ring 12 divides 15 and we have:

$$S(f_1, f_2) = \frac{5}{4}x \cdot f_1 - y \cdot f_2 = \begin{pmatrix} -7y^4 + \dfrac{5}{4}xy^2 \\ \dfrac{25}{4}x^2y^2 - 4xy^2 + \dfrac{5}{4}xy \end{pmatrix}.$$

Since the remainder of the division's algorithm is $S(f_1, f_2)$ then set

$$f_3 = \begin{pmatrix} -7y^4 + \dfrac{5}{4}xy^2 \\ \dfrac{25}{4}x^2y^2 - 4xy^2 + \dfrac{5}{4}xy \end{pmatrix}.$$

2. Observe that $Lt(f_3) = \dfrac{25}{4}x^2y^2 e_3$ and $S(f_1, f_3) = S(f_2, f_3) = 0$. This means that $G = \{f_1, f_2, f_3\}$ is a strong Gröbner basis for $S^{-1}N$ as an $S^{-1}R$−submodule of $(S^{-1}\mathbb{Z}[x, y])^2$ with $S = \{2^0, 2, 2^2, 2^3, \dots, \}$. In the other words, G is a special Gröbner basis for N in R^2.

Remark A.8. The Theorem 1.4 can be generalized for sub-modules of R^s. The following algorithm computes a weak Gröbner basis for $\mathbb{Z}(I)$.

Algorithm 3. FSTW

Input: A submodule $I = \langle f_1, \dots, f_s \rangle$ of R^r and a monomial ordering $>$.
Output: A weak Gröbner basis for $\mathbb{Z}(I)$ in R^r.
1 Use the algorithm 2 to compute a special Gröbner basis $G = \{g_1, \dots, g_r\}$ for I w.r.t some multiplicative subset $S \subset \mathbb{Z}$;
2 **for** $i = 1, \dots, r$ **do**
3 $\quad$ compute q_i the least common multiple of all denominators occurring in g_i;
4 **end**
5 **return** $\{q_1 g_1, \dots, q_r g_r\}$

Example A.9. Let $I = \langle f_1 = 10xy + 1, f_2 = 6x^2 + 3 \rangle$ be an ideal of $R = \mathbb{Z}[x, y]$. Let us compute a weak Gröbner basis for $\mathbb{Z}(I)$ w.r.t d_p. We have seen in the Example A.6 that $G = \{10xy + 1, f_2 f_3 = \dfrac{3}{5}x - 3y, f_4 = 3y^2 + \dfrac{3}{50}\}$ form a special Gröbner basis for I, using the Algorithm 3, the set $G' = \{f_1, f_2, 5f_3, 50f_4\}$ for a weak Gröbner basis for $\mathbb{Z}(I)$ in R w.r.t d_p.

Remark A.10. From Theorem 2.6 we give the following algorithm in the general case of modules. Observe that if $G = \{g_1, \dots, g_s\}$ is a special Gröbner basis for a sub-module $I \subset R^r$ with multiplicative subset S then each $g_i \in (S^{-1}R)^r$ that is, $g_i = (g_i^1, \dots, g_i^r)$ where $g_i^j \in S^{-1}R$ for $j = 1, \dots, r$ and $i = 1, \dots, s$.

Algorithm 4. Computing a syzygy of a weak Gröbner basis

Input: A special Gröbner basis $G = \{g_1, \ldots, g_s\}$ for a sub-module $I \subset R^r$
with multiplicative subset S w.r.t a monomial ordering $>$; and a
syzygy $H = (h_1, \ldots, h_s) \in (S^{-1}R)^s$ for $g_1, \ldots, g_s$.
Output: A syzygy H' for the corresponding weak Gröbner basis
$\{q_1 g_1, \ldots, q_s g_s\}$ for $\mathbb{Z}(I)$.

1 **for** $k = 1, \ldots, s$ **do**
2      $g_k = (g_k^1, \ldots, g_k^r)$ and denote by $q_k^1, \ldots, q_k^r$ the least common multiple
     of denominators occurring in $g_k^1, \ldots, g_k^r$;
3   compute b_k the least common multiple of denominators of terms
     occurring in h_k;
4   Set $q_k = \text{lcm}(q_k^1, \ldots, q_k^r)$;
5   Denote by b_k the least common multiple of denominators occurring in
     h_k;
6 **end**
7 Compute $q = \text{lcm}(b_1 q_1, \ldots, b_s q_s)$.
8 **for** $k = 1, \ldots, s$ **do**
9   Compute $r_k = \dfrac{q}{q_k}$;
10 **end**
11 **return** $H' = (r_1 h_1, \ldots, r_s h_s)$

Example A.11. In the Example 2.5 we have computed a strong Gröbner basis
T for $\text{Syz}(f_1, f_3, f_4)$ in $S^{-1}\mathbb{Z}[x, y]$ where $S = 2^{\mathbb{N}} \cdot 5^{\mathbb{N}}$. In the Example A.9 we
have seen that $G' = \{g_1, g_3, g_4\}$ form a a minimal weak Gröbner basis for $\mathbb{Z}(I)$
in $R = \mathbb{Z}[x, y]$ with $g_1 = f_1, g_3 = 5f_3$ and $g_4 = 50f_4$. Our gaol is to compute from
each G_{ji} obtained in the Example 2.5, the corresponding syzygies for g_1, g_3, g_4
using the Algorithm 4. We have:

- for G_{13}, we have $\dfrac{3}{50}f_1 - yf_3 - f_4 = 0$. Since $q_1 = 1, q_3 = 5, q_4 = 50$ are least
  common multiples of denominators of f_1, f_3, f_4 respectively and $b_1 = 50, b_3 = b_4 = 1$ are least common multiples of denominators of components of G_{13},
  then by the Algorithm 4 we have $q = \text{lcm}(q_1 b_1, q_3 b_3, q_4 b_4) = 50$ and $r_1 = \dfrac{q}{q_1} = 50, r_3 = \dfrac{q}{q_3} = 10$ and $r_4 = \dfrac{q}{q_4} = 1$. Thus $G'_{13} = (r_1 h_1, r_3 h_3, r_4 h_4) = (3, -10y, -1)$ form a syzygy for g_1, g_3, g_4 in R since:

$$3g_1 - 10yg_3 - g_4 = 3(10xy + 1) - 10(3x - 15y) - (3y^2 + \dfrac{3}{50}) = 0.$$

- For each other G_{ji} one can apply the same technique to obtain the corresponding syzygies in R.

Algorithm 5. COMPUTING A WEAK GRÖBNER BASIS FOR $\mathbb{Z}(\text{SYZ}(q_1 g_1, \ldots, q_s g_s))$

Input: A special Gröbner basis $G = \{g_1, \ldots, g_s\}$ for $I \subset R$; a multiplicative subset $S \subset \mathbb{Z}$; a strong Gröbner basis $\{T_1, \ldots, T_l\}$ for $\text{syz}(g_1, \ldots, g_s) \subset (S^{-1}R)^s$ w.r.t the Schreyer's ordering $<_1$ induced by $<$ and G.

Output: $T_1', \ldots, T_l' \in R^s$ and $q_1, \ldots, q_s \in \mathbb{Z} \setminus \{0\}$ such that $\{T_1', \ldots, T_l'\}$ form a weak Gröbner basis for $\mathbb{Z}(\text{syz}(q_1 g_1, \ldots, q_s g_s))$ w.r.t to $<_1$.

1   **for** $k = 1, \ldots, s$ **do**
2 | Compute a_k, the least common multiple of denominators of all polynomials occurring in T_k;
3 | Compute q_k, the least common multiple of denominators occurring in g_k;
4   **end**
5   Compute $a = \text{lcm}(a_1, \ldots, a_s)$;
6   **for** $k = 1, \ldots, s$ **do**
7 | Compute $T_k' = a \cdot T_k$;
8   **end**
9   **return** $\{T_1', \ldots, T_s'\}$

With notations as in the Example A.11 we give the following example

Example A.12. We have seen in the Example 2.5 that $T = \{G_{13}, G_{14}, G_{34}\}$ form a a minimal strong Gröbner basis for $\text{syz}(f_1, f_3, f_4)$ in $(\mathbb{Z}_{2.5}[x, y])^3$. To obtain a weak Gröbner basis for $\text{syz}(g_1, g_3, g_4)$ observe that: $G_{13} = (\frac{3}{50}, -y, -1)$ and denote by $a_1 = 50$ the least common multiple of all denominator occurring in G_{13}, $G_{14} = (\frac{3}{10}y, \frac{1}{10}, -x)$ and denote $a_2 = 10$, and $G_{34} = (0, y^2 + \frac{1}{50}, y - \frac{1}{5}x)$ and denote $a_3 = 50$. Let $a = \text{lc}(a_1, a_2, a_3) = 50$ and according to the Algorithm 5, the set $\{aG_{13}, aG_{14}, aG_{34}\} = \{(3, -50y, -50), (15y, 5, -50x), (0, 50y^2 + 1, 50y - 10)\}$ form a weak Gröbner basis for $\mathbb{Z}(\text{Syz}(g_1, g_3, g_4))$ w.r.t the Schreyer's ordering induced by d_p and f_1, f_3, f_4.

References

1. Adams, W.W., Laustaunau, P.: An introduction to Gröbner bases. Graduate studies in Mathematics, vol. 3, American Mathematical Society, Providence, RI (1994)
2. Berkesch, C., Schreyer, F.-O.: Syzygies, finite length modules, and random curves. arXiv. 1403.0581 (2014)
3. Erocal, B., Motsak, O., Schreyer, F.-O., Steenpass, A.: Refined Algorithms to Compute Syzygies. arxiv:1502.01654v2 [math.AC] (2016)
4. Greuel, G.-M., Pfister, G.: A Singular Introduction to Commutative Algebra, 2nd edn. Springer, Heidelberg (2008). https://doi.org/10.1007/978-3-540-73542-7
5. Popescu, A.: Signature standard bases over principal ideal rings. Ph.D. thesis, Kaiserslautern University (2016)

6. Schreyer, F.-O.: A standard basis approach to syzygies of canonical curves. J. Reine Angew. Math. **421**, 83–123 (1991)
7. Schreyer, F.-O.: Die Berechnung von Syzygien mit dem verallgemeinerten weier-strasschen divisionssatz. Diplomarbeit, Hamburg (1980)
8. Yengui, I.: Dynamical Gröbner bases. J. Algebra **301**, 447–458 (2006)
9. Yengui, I.: Corrigendum to "Dynamical Gröbner bases" [J. Algebra 301 (2) (2006) pp. 447–458] & to "Dynamical Gröbner bases over Dedekind rings" [J. Algebra 324 (1) (2010) pp. 12–24]. J. Algebra **339**, pp. 370–375 (2011)
10. Yengui, I.: Constructive commutative algebra: projective module over polynomial rings and dynamical Gröbner bases. https://doi.org/10.1007/978-3-319-19494-3
11. Wienand, O.: Algorithm for symbolic computation and their applications: standard bases over rings and rank tests in statistics. Ph.D. thesis, Kaiserslautern University (2011)

On the Splitting Field of Some Polynomials with Class Number One

Abdelmalek Azizi$^{(\boxtimes)}$ 🆔

Faculty of Sciences, Mohammed Premier University, 60000 Oujda, Morocco
ab.azizi@ump.ac.ma

Abstract. Let $P(X)$ be an irreducible monic polynomial of $\mathbb{Z}[X]$, d be the discriminant of $P(X)$ and L be the splitting field of $P(X)$. In this paper, we study the class number one problem for the splitting field L or the condition for which the class number of L is equal to 1 using an algebraic approach based on the Hilbert class field towers of some fields.

Keywords: The class number one problem · The splitting field of a polynomials · The Hilbert class field towers

1 Introduction

The class number one problem was studied, since the time of Gauss, for number fields as the quadratic fields, the cubic fields, the quartic and the bicyclic fields, the CM-fields and other fields. Almost of this studies were based on analytic methods using some approximations of different L-functions as the Dirichlet L-functions, the L-functions of elliptic curves, ... (see for example [4,5,14,17] or [1]). In this paper, we study the class number one problem for the splitting field of some polynomials but using an algebraic approach based on the Hilbert class field towers of some fields.

Let n be an integer greater than or equal to 3, $P(X)$ be an irreducible monic polynomial with coefficients in $\mathbb{Z}$ and of degree n, K a field generated by a root of $P(X)$, L the normal closure of K, and d be the discriminant of $P(X)$. The discriminant d is equal to

$$d = \prod_{i<j}(\alpha_i - \alpha_j)^2$$

where the α_i are the roots of $P(X)$. Then d is a square in L. So, if d is not square in $\mathbb{Z}$, then the quadratic field $F = \mathbb{Q}(\sqrt{d})$ is included in L. Several authors have studied the ramification in the extension L/F. In particular, Movahhedi [10] gave necessary and sufficient conditions for the non ramification of finite ideals in L/F. On the other hand, it was shown (see [3,12,16]) that if d is a squarefree (or less, the discriminant of K is squarefree (see [11])), then the Galois group of

Supported by ACSA Laboratory FSO Mohammed Premier University.

C. T. Gueye et al. (Eds.): A2C 2019, CCIS 1133, pp. 73–80, 2019.
https://doi.org/10.1007/978-3-030-36237-9_4

$L/\mathbb{Q}$ is equal to the symmetric group S_n while the Galois group of L/F is equal to A_n. Moreover, the extension L/F is unramified for the finite primes.

The last results was generalized by Takeshi KONDO in [7] as followed: let $d(K)$ be the discriminant of K and $F = \mathbb{Q}(\sqrt{d}) = \mathbb{Q}(\sqrt{d(K)})$; if the **main condition: $d(K)$ is not square in $\mathbb{Q}$ and is equal to the discriminant of** $\mathbb{Q}(\sqrt{d(K)})$ is satisfied, then the Galois group of $L/\mathbb{Q}$ is equal to the symmetric group S_n while the Galois group of L/F is equal to A_n. Moreover, the extension L/F is unramified for the finite primes. It implies that under the main condition, L/F is unramified at all primes in the case where $d < 0$ or in the case where L/F is a real extension.

Assuming the main condition and the condition L/F is unramified at the infinite primes satisfied, one asks under what other conditions the class number of L is equal to 1 or the class number one problem for L is solved?

To resolve this problem (the class number one problem for the splitting field L), one uses the notion of class field tower of number fields. Let k be an algebraic number field, C_k be its class group and $k^{(1)}$ be the Hilbert class field of k. For a nonnegative integer i, we define inductively $k^{(i)}$ as $k^{(0)} = k$ and $k^{(i+1)} = (k^{(i)})^{(1)}$. Then $k^{(0)} \subseteq k^{(1)} \subseteq k^{(2)} \subseteq ... \subseteq k^{(i)} \subseteq ...$ is called the Hilbert class field tower of k. If m is the minimal integer such that $k^{(m)} = k^{(m+1)}$, then m is called the length of the tower. If no such m exists, then the tower is said to be infinite.

The infinite class field towers has a very important applications in coding theory and Cryptography (see [6, 8, 15]) and the finite class field towers are very useful as in the study of the main question of this paper.

If we assume that the class number of L is equal to 1; then F has a finite class field tower. Indeed, if $F^{(i+1)}$ (resp. $L^{(i+1)}$) is the Hilbert class field of $F^{(i)}$ (resp. $L^{(i)}$) then since $F \subset L$ we obtain that $F^{(i)} \subset L^{(i)}$.

In other hand, since the class number of L is equal to 1, then $L^{(i)} = L$. It follows then, that in this case, the class field tower of $F^{(i)}$ is finite. Moreover we prove that if the class number of L is equal to 1 then $L = F^{(m)}$ where m is the length of the tower $F^{(i)}$. In particular, we prove the following main result:

The Main Theorem 1. *Let $P(X) \in \mathbb{Z}[X]$ be a monic irreducible polynomial of degree n greater than or equal to 3 and with discriminant d, K a field generated by a root of $P(X)$, $d(K)$ the discriminant of K and L the splitting field of $P(X)$. We suppose that $d(K)$ is not square in $\mathbb{Q}$ and is equal to the discriminant of $\mathbb{Q}(\sqrt{d(K)})$ and L/F is unramified at the infinite primes. Let $F = \mathbb{Q}(\sqrt{d}) = \mathbb{Q}(\sqrt{d(K)})$ be the quadratic subfield of L and $h(d)$ it's class number. Then we have:*

1. *if $n = 3$, then the class number of L is equal to 1 if and only if $h(d) = 3$ and the class number of $F^{(1)}$ is equal to 1 (the Hilbert class field tower of F stops at $F^{(1)}$). In this case $L = F^{(1)}$.*
2. *if $n = 4$, then the class number of L is equal to 1 if and only if $h(d) = 3$, the class number of $F^{(1)}$ is equal to 4 and the Hilbert class field of F stops at $F^{(2)}$. In this case $L = F^{(2)}$.*
3. *if $n \geq 5$ and $h(d) \neq 1$, then the class number of L is not equal to 1.*

Under the conditions of the main theorem, if the class number of L is equal to 1 then the squarefree part of d is divisible by one or two primes.

In the case where $d < 0$, there exists a finite square-free parts of d such that $h(d) = 3$ (the square-free part of d must be $\in \{-23, -31, -59, -83, -107, -139, -211, -307, -379, -499, -547, -883, -907\}$); so using PARI/GP we prove that for $n = 4$ the class number of the splitting field L is not equal to 1. In addition, from the last case, we found some examples where the class number of L is not equal to 1; and which are examples of imaginary quadratic fields for which the class field towers stop only after the third round.

2   Proof of the Main Theorem

2.1   Case Where $n = 3$

In this paragraph we suppose that the degree n of $P(X)$ is equal to 3.

Theorem 1. *We keep the previous notations. We suppose that $d(K)$ is not square in $\mathbb{Q}$ and is equal to the discriminant of $\mathbb{Q}(\sqrt{d(K)})$ and L/F is unramified. The class number of L is equal to 1 if and only if $h(d) = 3$ and the Hilbert class field tower of F stops at $F^{(1)}$ (the class number of $F^{(1)}$ is equal to 1). In this case $L = F^{(1)}$.*

Proof. Let $P(X)$ be a polynomial of degree 3 such that $d(K)$ is not square in $\mathbb{Z}$ and is equal to the discriminant of $\mathbb{Q}(\sqrt{d(K)})$ and L/F is unramified. Then the Galois group of L/F is equal to A_3. So L/F is an unramified extension of degree 3. Thus, the class number of F is divisible by 3.

If the class number of L is equal to 1, then we have $F \subsetneq F^{(1)} \subset L$. It follows that the class number of $F^{(1)} = L$ is equal to 1 and the class number of F is equal to 3. Conversely, since L/F is an unramified abelian extension then $L \subset F^{(1)}$; moreover if the class number of F is equal to 3 then $F^{(1)}$ and L have the same degree on F, which leads to $F^{(1)} = L$. Since the tower stops at $F^{(1)}$, then the class number of L is equal to 1.

Theorem 2. *Let $P(X) = X^3 + lX - 1$ where $l \in 2\mathbb{Z}$, $l \geq 2$, α be a root of $P(X)$, $K = \mathbb{Q}(\alpha)$ and L be the splitting field of $P(X)$. It is assumed that its discriminant $d = -(4l^3 + 27)$ is squarefree. Let $F = \mathbb{Q}(\sqrt{d})$ then the class number of L is equal to 1 if and only if K is principal and the class number of F, $h(F)$ is equal to 3.*

Proof. Let $K = \mathbb{Q}(\alpha)$ be the cubic field generated by a root α of the equation $P(X) = X^3 + lX - 1 = 0$ where $l \in \mathbb{Z}$, $l \geq 2$. Let L be the splitting field of $P(X)$. If $4l^3 + 27 = -d$ is squarefree, then α is a fundamental unit of K. Moreover if l is even, then two roots of this equation form a fundamental system of L the splitting field of $P(X)$ (see [9]).

In addition, the class number of L is given by the formula

$$h(L) = a(h(K))^2 h(F)/3$$

with $a = [E_L : E_K E_{K'}]$, K' is the conjugate of K, $F = \mathbb{Q}(\sqrt{d})$, $h(L)$ is the class number of the field L and E_L, E_K and $E_{K'}$ are the unit groups modulo the unity roots group of the fields L, K and K' respectively (see [9]). Since L/F is unramified of degree 3, then 3 divides $h(F)$.

If the class number of L is equal to 1 ($h(L) = 1$), then $h(F) = 3$. Since two roots of $P(X)$ form a fundamental system of units of L, then $a = 1$; and $h(L) = (h(K))^2$; and since $h(L) = 1$ then $h(K) = 1$.

Conversely, if $h(K) = 1$ then $h(L) = ah(F)/3$. Moreover if $h(F) = 3$, then $h(L) = a$. From [9], we have that $a = 1$ or $a = 3$. If $a = 3$, then $h(L) = 3$ and $[L^{(1)} : F] = 9$. It follows that $L^{(1)}/F$ is an abelian unramified extension, so the class number $h(F)$ of F is divisible by 9 which is impossible since $h(F) = 3$. Then the class number of L is equal to 1.

Example 1. Let be $P(X) = X^3 + aX^2 + bX + c \in \mathbb{Z}[X]$ an irreducible polynomial, d its discriminant, K the field generated by a root of $P(X)$ and $h(K)$ the class number of K. Using the program PARI we have: Through the previous theorem;

Table 1. Examples with negative discriminant.

a	b	c	$d = \text{disc}(P(X))$	disc(K)	h(d)	h(K)
0	2	−1	−59	−59	3	1
0	4	−1	−283	−283	3	2
0	8	−1	$-2075 = -5^2 \times 83$	−83	3	1

we easily check that the class number of L is equal to 1 in the first example and in the third one wiles it is equal to 1 in the second example (we found the first and the second examples also in [9]). The Hilbert class field $F^{(1)}$ coincides with L and the class field tower of F stops at $F^{(1)}$ in the first example and in the third example. In the second example since $h(K) = 2$, then it is obvious that $h(L) = 4$ and so the tower does not stop at $F^{(1)}$ (Table 1).

Example 2. Real examples. Let be $P(X) = X^3 + aX^2 + bX + c \in \mathbb{Z}[X]$ an irreducible polynomial, d its discriminant, K the field generated by a root of $P(X)$ and $h(K)$ the class number of K. Using the program PARI we have: For all these real examples, the class number of the Hilbert class field F^1 is equal to 1 and F^1 is of degree 3 over F; so it is equal to L (Table 2).

Remarque 1. *The condition $d(K)$ is equal to the discriminant of $\mathbb{Q}(\sqrt{d(K)})$ implies that $d(K) \equiv 1 \bmod 4$. In addition to the conditions on the discriminant, the condition $h(d) = 3$ leads, in the imaginary case, to the fact that the squarefree*

Table 2. Examples with positive discriminant.

a	b	c	$d = disc(P(X))$	$h(d)$	$h(K)$	$h(F^1)$
0	−4	−1	229	3	1	1
−1	−4	3	257	3	1	1
−1	−4	1	$321 = 3 \times 107$	3	1	1
−1	−5	4	$469 = 7 \times 67$	3	1	1
0	−5	−1	$473 = 11 \times 43$	3	1	1

part of the discriminant is divisible by only one prime number $p \equiv -1 \bmod 4$; while in the real case, the squarefree part of the discriminant is divisible by only one prime number $p \equiv 1 \bmod 4$ or two prime numbers p and q such that $p \equiv -1 \bmod 4$ and $q \equiv -1 \bmod 4$. Otherwise, by genus theory the class number of F must be divisible by 2.

2.2   Case Where $n = 4$

In this paragraph we suppose that the degree n of $P(X)$ is equal to 4.

Theorem 3. *We keep the previous notations. We suppose that $d(K)$ is not square in $\mathbb{Q}$ and is equal to the discriminant of $\mathbb{Q}(\sqrt{d(K)})$ and L/F is unramified. The splitting field L has a subfield F' of degree 3 over F such that $Gal(L/F')$ is isomorphic to $\mathbb{Z}/2\mathbb{Z} \times \mathbb{Z}/2\mathbb{Z}$ (of type (2.2)). It implies that L/F' is an abelian unramified extension.*

Proof. As the Galois group L/F is equal to A_4 and contains a normal subgroup $V_4 = \{\sigma \in A_4 \mid \sigma^2 = id\}$ which is of type (2.2); then the subfield F' of fixed elements of L by V_4 is an abelian extension of F of degree 3 and $Gal(L/F')$ is isomorphic to V_4 which is isomorphic to $\mathbb{Z}/2\mathbb{Z} \times \mathbb{Z}/2\mathbb{Z}$.

Theorem 4. *We keep the previous notations. We suppose that $d(K)$ is not square in $\mathbb{Q}$ and is equal to the discriminant of $\mathbb{Q}(\sqrt{d(K)})$ and L/F is unramified. The class number of L is equal to 1 if and only if the class number of F is exactly equal to 3, the class number of $F^{(1)}$ is equal to 4 and the Hilbert class field tower of F stops at $F^{(2)}$. In addition, the class group of $F^{(1)}$ is of type $(2,2)$. In this case $L = F^{(2)}$.*

Proof. If the class number of L is equal to 1, then $F^{(1)}$ and $F^{(2)}$ are included in L. From the last theorem, there exists a subfield F' of L which is unramified and of degree 3 over F so $F' \subset F^{(1)}$. Since V_4 is the unique non trivial normal subgroup of A_4 the Galois group of L/F; then the extension F'/F is the maximal abelian sub-extension of L/F. It follows that $F' = F^{(1)}$. Moreover, as the Galois group of L/F' is abelian and $F^{(2)}$ is the maximal unramified abelian extension of $F^{(1)}$, then $L = F^{(2)}$.

Conversely, if the class number of F is equal to 3 and F'/F is the sub-extension of L/F of degree 3, then $F^{(1)} = F'$ so $F^{(1)}$ is a subfield of L. On the other hand, $L/F^{(1)}$ is abelian unramified extension of degree 4 and it is the same for $F^{(2)}/F^{(1)}$. As a result, $L = F^{(2)}$ and therefore the class number of L is equal to 1.

Theorem 5. *Let p be a prime congruent to -1 modulo 4 and $P(X)$ be an irreducible polynomial of degree 4 whose discriminant is equal to $-p$. Then the class number $h(-p)$ of $F = \mathbb{Q}(\sqrt{-p})$ is divisible by 3. Moreover if the class number of F is equal to 3, then the splitting field N of $P(X)$ is cyclic of order 2. It coincides with $F^{(2)}$ and contains $F^{(1)}$ which is of type $(2,2)$.*

Proof. In this case $d = -p$ is squarefree and then is equal to $d(K)$; moreover $d(K) = -p$ is equal to the discriminant of $\mathbb{Q}(\sqrt{d(K)})$ and L/F is unramified. Using the PARI Programm [13], we check that for $p \in \{23, 31, 59, 83, 107,$ $139, 211, 283, 307, 331, 379, 499, 547, 643, 883, 907\}$ the set of all prime numbers p congruent to -1 modulo 4 such that $h(-p) = 3$ (see [2]), the class number of $F^{(1)}$ is equal to 1 except for $p \in \{283, 331, 643\}$ where the class group of $F^{(1)}$ is of type $(2, 2)$ and for the last values $-p$ is the discriminant of an irreducible polynomial of degree 4.

Also, if $P(X)$ is an irreducible polynomial of degree 4 over $\mathbb{Q}$ with discriminant $-p$ and α is a root of $P(X)$, then $\mathbb{Q}(\alpha)F^{(1)}$ is of degree 24. In addition, $F^{(1)} \subset L$, $\mathbb{Q}(\alpha) \subset L$ and $[L : Q] = 24$ then $L = \mathbb{Q}(\alpha)F^{(1)} = F^{(2)}$. Following a computation with PARI, the class number of $N = F^{(2)}$ is equal to 2 for $p \in \{283, 331, 643\}$. Consequently, the class number of L is not equal to 1 and the Hilbert class field tower of F does not stop at $F^{(2)}$.

Remarque 2. *The fields $F = \mathbb{Q}(\sqrt{-p})$ for $p \in \{283, 331, 643\}$ are examples of imaginary quadratic fields for which the length of their Hilbert class field tower is at least 3 (the tower will stop after the third floor). It is a tower for which the first three levels are respectively cyclic of order 3, of type (2.2) and cyclic of order 2. It is not known if the fourth level exists or no; but the 2-part does not exists.*

Example 3. Real examples. Let $P(X) = X^4 + aX^3 + bX^2 + cX + e$ be an irreducible polynomial; the following examples are for the splitting field $L = F^2$ where the class number of L is equal to 1 or is equal to 2 (Table 3).

2.3   Case Where $n \geq 5$

Theorem 6. *We keep the previous notations. We suppose that $d(K)$ is not square in $\mathbb{Q}$ and is equal to the discriminant of $\mathbb{Q}(\sqrt{d(K)})$ and L/F is unramified. Assume that $n \geq 5$.*

1. *The extension L/F does not contain any Galois sub-extension F'/F.*
2. *Assume that L/F is unramified. If the class number of L is equal to 1, then the class number of F is also equal to 1.*

Table 3. Examples for the splitting field L where the class number is equal to 1 or to 2.

a	b	c	e	d = disc(P(X))	h(d)	$h(F^1)$	$h(F^2)$
0	−4	1	1	1957	3	4	1
−1	−4	2	1	3981 = 3 × 1327	3	4	1
−2	−4	3	3	7053 = 3 × 2351	3	4	1
0	−5	−1	4	2777	3	4	2
0	−7	−5	3	7537	3	4	1

Proof. Assume that $n \geq 5$.

1. The group A_n has no nontrivial group quotient and therefore L/F has no Galois sub-extension.
2. We assume that L/F is unramified. If the class number of F is not equal to 1, then there exists an abelian unramified extension k/F which is nontrivial. Since, according to 1., L/F does not have a Galois sub-extension, then $L \cap k = F$. As a result $L.k/L$ is a non-trivial extension which is abelian and unramified, thus the class number of L is not equal to 1. This completes the proof of this theorem.

Corollary 1. *We keep the previous notations. We suppose that $n \geq 5$, $d(K)$ is not square in $\mathbb{Q}$ and is equal to the discriminant of $\mathbb{Q}(\sqrt{d(K)})$ and L/F is unramified. If the class number $h(d)$ of F is not equal to 1, then the class number of L can never be equal to 1.*

2.4 Conclusions

The main theorem is then proved through the three last sections (Theorems 1, 4 and 6).

In the case where $d < 0$; let $I_1 = \{-3, -4, -7, -8, -11, -19, -43, -67, -163\}$, $I_2 = \{-23, -31, -59, -83, -107, -139, -212, -283, -307, -331, -379, -499, -547, -643, -883, -907\}$ and $F = \mathbb{Q}(\sqrt{d})$ where $d \in \mathbb{Z}$ is squarefree. It is very known that the class number of F is equal to 1 if and only if $d \in I_1$ and it is equal to 3 if and only if $d \in I_2$ (see [2]).

Corollary 2. *We keep the previous notations. Let be $P(X) \in \mathbb{Z}[X]$ an irreducible monic polynomial of degree greater than or equal to 3 and with discriminant d supposed negative. We suppose that $d(K)$ is not square in $\mathbb{Q}$ and is equal to the discriminant of $\mathbb{Q}(\sqrt{d(K)})$. Let d_0 be the squarefree part of $d(K)$. Then we have:*

1. *if $n \geq 5$ and $d_0 \notin I_1$, then the class number of L is not equal to 1.*
2. *if $n \leq 4$ and $d_0 \notin I_2$, then the class number of L is not equal to 1.*

Proof. Let n be the degree of the polynomial $P(X)$. If $n \geq 5$ and $d_0 \notin I_1$, then the class number of F is not equal to 1. Using the last theorem, we have that the class number of L is not equal to 1.

If $n \leq 4$ and $d_0 \notin I_2$, then the class number of F is not equal to 3, using the last theorem then the class number of L is not equal to 1.

References

1. Ahn, J.-H., Boutteaux, G., Kwon, S.-H., Louboutin, S.: The class number one problem for some non-normal CM-fields of degree 2p. J. Number Theory **132**, 1793–1806 (2012)
2. Arno, S., Robinson, M.L., Weeler, F.S.: Imaginary quadratic field with small odd class number. Acta Arithmetica **I.XXXIII**(4), 295–330 (1998)
3. Elstodt, J., Grunewal, F., Mennicke, J.: On unramified A_n-extension of quadratic number fields. Glasgow Math. J. **27**, 31–37 (1985). J. Math. **4**, 367–369 (1974)
4. Goldfeld, D.: The Gauss Class Number Problem for Imaginary Quadratic Fields. Heegner Points and Rankin L-Series MSRI Publications, vol. 49 (2004)
5. Goldfeld, D.: Gauss class number problem for imaginary quadratic fields. Bull. (New Ser.) Am. Math. Soc. **13**(1), 23–37 (1985)
6. Guruswami, V.: Constructions of codes from number fields. In: AAECC-14, Melbourne, Australia, 26–30 November 2001
7. Kondo, T.: Algebraic number fields with the discriminant equal to that of a quadratic number field. J. Math. Soc. Japan **47**(1), 31–36 (1995)
8. Lenstra, W.: Codes from algebraic number fields. In: Hazewinkel, M., Lenstra, J.K., Meertyens, L.G.L.T. (eds.) Mathematics and Computer Science II, Fundamental Contributions in the H. Netherlands since 1945, CWI Monograph 4, pp. 95–104. North-Holland, Amsterdam (1986)
9. Moser, N.: Unités et nombre de classes d'une extension galoisienne diédrale de Q. Séminaire de théorie des nombres de Grenoble, tome 3, exp. no. 4, pp. 1–22 (1973–1974)
10. Movahhedi, A.: Sur une classe d'extension non ramifiées. Acta Arithmetica, LIX.1 (1991)
11. Nakagawa, J.: on the Galois group of number field with square free discriminant, comment. Math. Univ. St. Paul **37**(1), 95–98 (1988)
12. Osada, H.: The Galois group of the polynomials $X^n + aX + b$. J. Number Theory **25**, 230–238 (1987)
13. PARI/GP (version 2.7.5). http://pari.math.u-bordeaux.fr/
14. Stark, H.M.: The gauss class-number problems. In: Clay Mathematics Proceedings, vol. 7 (2007)
15. Temkine, A.: Tours de corps de classes de Hilbert pour les corps globaux et applications. Thèse de Doctorat, Faculté des sciences de Luminy, Université de la Méditerranée, Aix-Marseille II, France (2000)
16. Yamamura, K.: On unramified Galois extensions of real quadratic number field. Osaka J. Math. **23**, 471–478 (1986)
17. Watkins, M.: Class numbers of imaginary quadratic fields. Math. Comput. **73**(246), 907–938 (2003)

Code, Cryptology and Information
Security

Group Codes

Santos González[1], Victor Markov[2], Olga Markova[2], and Consuelo Martínez[1($\boxtimes$)]

[1] Department of Mathematics, Oviedo University, Oviedo, Spain
`santos@uniovi.es`, `cmartinez@uniovi.es`
[2] Department of Mechanics and Mathematics, Moscow State University,
Moscow, Russia
`ov_markova@mail.ru`

Abstract. Mathematical objects in this paper are group codes. In the first part of the paper we present a survey with some of the main results about group codes, mainly the existence of group codes that are not abelian group codes, the minimal length and the minimal dimension of such codes and the existence of a non-abelian group code that has better parameters than any abelian group code. In particular, in a previous paper [1], we have shown that the minimal dimension of a group code that is not abelian group code is 4. However, all known examples of group codes of dimension 4 that are non-abelian group codes are constructed using groups that are not p-groups.

We do not know if such codes exist for the case of p-groups, but in the second part of this paper we prove that, under some restrictions on the base field, all four-dimensional G-codes for an arbitrary finite p-group G are abelian.

Keywords: Group codes · Length · Dimension · Groups · Non-abelian groups · Abelian groups

Introduction

Error correcting codes play a key role to guarantee the reliability of (digital) information that is sent through a channel with noise. During the transmission process some errors may appear and it is essential for the recipient to be able to detect that some errors have indeed been produced and, eventually, to correct them. This process of detection and correction of errors can take place thanks to error correcting codes *with good properties*.

Usually linear codes are the most widely used, since Linear Algebra provide some powerful tools to deal with them. For general information about Coding Theory we refer readers to [2] and [3]. So all codes considered here are, in particular, vector subspaces of dimension k of an F-vector space of dimension n. We will refer to n as the length of the code and k is its dimension.

Granted by MTM2017-83506-C2-2-P, FD-GRUPIN-IDI/2018/000193 and RFBR grant 17-01-00895 A.

It is worth to mention that there are some important and useful non-linear codes, for instance, binary Hamming codes. However, Hamming codes can be seen as linear codes over the ring $\mathbb{Z}_4$ as proved by A. Nechaev in 1989 (see [4]) and Hammons et al. in 1994 (see [5]).

At present, many algebraic structures (groups, rings, modules, ...) are used in Coding Theory.

All groups and fields considered in what follows are supposed to be finite and p denotes a prime.

Cyclic codes have nice properties and efficient decodifying algorithms have been developed for them. A code $\mathcal{C}$ is cyclic if it satisfies that a word $\mathbf{c} = (c_1, \ldots, c_n) \in \mathcal{C}$ if and only if $(c_2, \ldots, c_n, c_1) \in \mathcal{C}$.

A cyclic code can be seen as an ideal in the quotient ring $F[x]/(x^n - 1)$ and it is generated by a (unique) monic polynomial $g(x)$ satisfying $g(x)|x^n - 1$.

The notion of group code extends in a natural way the one of cyclic code. A cyclic code is a group code when the associated group is cyclic.

In this paper we will survey on group codes constructed using non-abelian groups. It is clear that work with non-abelian groups is sensibly more complicated than work with abelian groups. So the aim of that survey is to prove that the use of non-abelian groups has interest and opens new possibilities. We will include also some new results about group codes of dimension 4.

1   Group Codes

From now on F will denote a finite field and $G = \{g_0 = e, g_1, \ldots, g_{n-1}\}$ will denote a finite group of order n. The set of all formal linear combinations of elements of G,

$$FG = \left\{ \sum_{g \in G} \alpha_g g \mid \alpha_g \in F \right\}$$

has a well known structure of algebra over F and is called the group algebra (or group ring) of G over F.

Following [6] we say that a linear code $\mathcal{C}$ over F is a (left) G-code if its length is equal to $n = |G|$ and there exists a one-to-one mapping $\nu : \{1, \ldots, n\} \to G$ such that

$$\left\{ \sum_{i=1}^{n} a_i \nu(i) : (a_1, \ldots, a_n) \in \mathcal{C} \right\}$$

is a (left) ideal in FG. We will also say that this (left) ideal is permutation equivalent to the code $\mathcal{C}$.

A code $\mathcal{C}$ is called an (abelian) group code if there exists an (abelian) group A such that $\mathcal{C}$ is an A-code.

So a cyclic code is a G-code, where G is a cyclic group.

The question of how to distinguish group codes among linear codes was addressed in [6] using the *automorphism permutation group* of the code. Given a code $\mathcal{C}$ and a permutation $\sigma \in S_n$, for $\mathbf{c} = (\alpha_0, \alpha_1, \ldots, \alpha_{n-1}) \in \mathcal{C}$, we denote $\sigma(\mathbf{c}) = (\alpha_{\sigma(0)}, \alpha_{\sigma(1)}, \ldots \alpha_{\sigma(n-1)})$. The automorphism permutation group of the code $\mathcal{C}$, PAut($\mathcal{C}$), is the subgroup of S_n that consists of all permutations $\sigma \in S_n$ satisfying that $\sigma(\mathcal{C}) = \mathcal{C}$.

In [6] authors prove that a code $\mathcal{C}$ of length n is a left group code (resp. a group code) if and only if PAut($\mathcal{C}$) contains H (resp. $H \cup C_{S_n}(H)$), where H is a transitive subgroup of order n. The code $\mathcal{C}$ is an abelian group code if and only if PAut($\mathcal{C}$) contains a regular abelian subgroup $A \leq S_n$.

These characterizations, important from a theoretical point of view, are not very helpful in applications. For instance, they could not help to address the question about the existence of group codes that cannot be realized as abelian group codes (we will call those codes *non-abelian group codes*, for short). Let us emphasise that a given code $\mathcal{C}$ can be seen as group code over two different groups. It is possible also that one of them is abelian and the other is not. So, in order to justify the study of non-abelian group codes, the first question that was needed to answer was: Do they exist?

We started studying this question in [7], where we proved that if G is a non-abelian group and there is a G-code that is a non-abelian group code, then $|G| \geq 24$. Then we found the first examples of non-abelian group codes using the group S_4. We started with the semisimple case, constructing for $G = S_4$ and $F = \mathbb{Z}_5$ a G-code whose weight distribution does not coincide with the weight distribution of any abelian group code of length 24. To check the weight distributions of all those codes we needed the help of a computer. The result turned out to be the same in the non-semisimple case and we could prove the existence of G-codes over $\mathbb{Z}_2$ and $\mathbb{Z}_3$ that are non-abelian group codes. It is worth to mention that case $F = \mathbb{Z}_3$ follows the same lines of case $F = \mathbb{Z}_5$. However, when $F = \mathbb{Z}_2$ some interesting differences appear, since the considered code has the same parameters and the same weight distribution of some abelian code of length 24.

However those codes constructed using the group $G = S_4$ have worse parameters than abelian group codes of the same length. So, in that point it was still unclear that the study of non-abelian group codes is worth. But finally we were able to construct some non-abelian group codes that achieve better properties than abelian group codes of the same length.

Using the group $G = \mathrm{SL}(2, \mathbb{Z}_3)$ we could construct a binary code having dimension 6 and minimal weight 10. So such code is, in some sense, optimal.

Indeed, in [8] we prove that a binary code of length 24 and dimension 6 has minimal distance ≤ 10. Furthermore, the distance 10 can not be reached using an abelian group code. And, what is also important, in this construction we did not have to use computers. All results were proved in a pure algebraic way and they allow to say that the study of non-abelian group codes is worth.

Still we can ask if the existence of non-abelian group codes is some exceptional fact and appears only in some cases or, by the contrary, it happens in all characteristics.

V. Markov checked the existence of non-abelian group codes for every prime field $\mathbb{Z}_p$, $p < 100$ and conjectured that there are non-abelian group codes of length 24 over every finite field.

In [9] we proved that for every $p \geq 3$ there are G-codes over $\mathbb{Z}_p$ of dimension 9, where $G = S_4$, that are not abelian. In a similar way, we could prove that there are G-codes over $\mathbb{Z}_p$ that are not abelian and have dimension 4, with $G = \mathrm{GL}(2, \mathbb{Z}_3)$.

Now we can give a positive answer to Markov's conjecture using a previous result in [7]:

Theorem 1. *If E/F is an extension of fields, G is a finite group and every G-code over E is abelian, then every G-code over F is abelian.*

Nothing is known about the converse of the assertion in the above mentioned theorem.

2   Dimension of Non-abelian Group Codes

In the results mentioned until now we have paid attention specially to the length of the group code. So we know that 24 is the minimal length of a non-abelian group code and this length is achieved, that is, there are G-codes over any finite field with length 24 and that are non-abelian group codes.

In this section we will pay attention to the dimension of those codes. What is the minimal dimension of a non-abelian group code?

It was shown in [6] that any one-dimensional group code over a field F is an abelian group code (moreover it is a C-code for a cyclic group C). And as we have mentioned in the previous section there are non-abelian group codes of dimension 4. What about codes of dimensions 2 and 3?

This question was addressed in [1] were we proved the following main result:

Theorem 2. *Let C be a G-code over a finite field F, where G is a finite group. If $\dim_F C \leq 3$, then C is an abelian group code.*

So now we can say that the minimal length of a non-abelian group code is 24 and the minimal dimension is 4 and there are non-abelian group codes of dimension 4 and length 24. However, in none of the known examples there is a non-abelian group code linked to a p-group. We can not claim that if G is a p-group then every G-code is an abelian group code, but we can prove that, with some additional restrictions on the base field, this is the case.

Theorem 3. *Let p be a prime, and let G be a finite p-group. If F is a field with $|F| < p^3$ then any G-code C over F with $\dim_F C = 4$ is an abelian group code.*

In what follows we will prove this result, considering before the semisimple case and then the modular case in a separate way.

3 Semisimple Case

Given a field F, we denote its multiplicative group by F^*. Let $M_k(F)$ be the algebra of all $k \times k$-matrices over F. For any integer $n \geq 1$ we use the notation $\mathrm{GL}_n(F)$, $\mathrm{T}_n(F)$ and $\mathrm{UT}_n(F)$ respectively for the group of all invertible $n \times n$-matrices, the group of all invertible upper triangular $n \times n$-matrices and the group of all upper unitriangular $n \times n$-matrices, i.e. upper triangular matrices with diagonal elements equal to 1, over the field F.

As usual write $A \leq B$ to express that A is a subgroup of the group B, while $A \triangleleft B$ means that A is a normal subgroup in B. $Z(G)$ and $Z(R)$ will denote the center of the group G and of the ring R, respectively.

Let's remember a useful sufficient condition for all G-codes to be abelian.

Theorem 4 ([6, theorem 3.1]). *Let G be a finite group. Assume that G has two abelian subgroups A and B such that every element of G can be written as ab with $a \in A$ and $b \in B$. Then every G-code is an abelian group code.*

We say that a group G *has an abelian decomposition* if it satisfies the condition of this theorem.

For any finite group G and any subgroup $N \leq G$ we consider an element $N_\Sigma = \sum_{u \in N} u \in FG$. We will use the following properties of N_Σ: $N_\Sigma = uN_\Sigma = N_\Sigma u$ for every $u \in N_\Sigma$, and $N_\Sigma \in Z(FG)$ iff $N \triangleleft G$.

Given two finite groups G, H of the same order n and a one-to-one mapping $\varphi : G \to H$ we define its natural extension $\tilde{\varphi} : FG \to FH$ by the rule

$$\tilde{\varphi}\left(\sum_{g \in G} a_g g\right) = \sum_{g \in G} a_g \varphi(g).$$

If I, J are left (right, two-sided) ideals in the group rings FG and FH, respectively, and there exists a one-to-one mapping $\varphi : G \to H$ such that $\tilde{\varphi}(I) = J$, we say that I and J are *permutation equivalent*.

We say that a subgroup $U \leq G$ *acts trivially* (from the left) on some set $X \subseteq FG$ if $ux = x$ for every $u \in U$ and $x \in X$. Our proofs are based on Theorem 4 and on the following simple observation.

Lemma 1 ([1, Lemma 1.3]). *Let F be a field and let G, H be two groups of the same order $n < \infty$. Suppose that there exist two normal subgroups $N \triangleleft G$ and $K \triangleleft H$ such that $G/N \cong H/K$, and that N acts trivially on some (left, right, two-sided) ideal $I \in FG$. Then I is permutation equivalent to some (left, right, two-sided) ideal of the ring FH.*

Evidently Lemma 1 remains valid is we consider the right action instead of the left action.

We add here the following remark which we will use in what follows.

Lemma 2. *Let e be a central idempotent of a group ring $R = FG$. Then the mapping $f : g \mapsto ge$ for any $g \in G$ is a group homomorphism from G to the group of invertible elements of the ring Re and its kernel acts trivially on $I = Re$ from the left and from the right.*

Proof. Straightforward.

In this section we assume that F is a finite field such that $|F| = q < p^3$ with $(p, q) = 1$ and that G is a finite p-group.

By Maschke's theorem the ring $R = FG$ decomposes into the direct sum of matrix rings over some extensions of the base field F. It follows that any ideal I of the ring R has the form $I = Re$ for some central idempotent $e \in R$. Suppose that $I = Re$ is an ideal in R such that $\dim_F I = 4$. Then either I is commutative or I is isomorphic to the ring $M_2(F)$. If I is commutative then G' acts trivially on I and one can take $H = G/G' \times A$ in Lemma 1 where A is an arbitrary abelian group with $|A| = |G'|$ to show that I defines an abelian group code.

Now consider the case $I \cong M_2(F)$. Let $f : G \rightarrow \mathrm{GL}_2(F)$ be the homomorphism defined in Lemma 2 and let K be its kernel. Then $|G/K| = p^r$ for some integer $r > 0$ (otherwise G acts trivially on I hence any subspace of I is an ideal which is impossible since I is a simple ring) and p^r divides $|\mathrm{GL}_2(F)| = (q^2 - 1)(q^2 - q) = q(q - 1)^2(q + 1)$.

Next we prove that $r \leq 4$ for $p > 2$ and $r \leq 5$ for $p = 2$.

If $p = 2$ then only the values $q = 3, 5, 7$ are possible since $q < p^3 = 8$ and $(q, p) = 1$. In these cases $|\mathrm{GL}_2(\mathbb{Z}_3)| = 48 = 16 \cdot 3$, $|\mathrm{GL}_2(\mathbb{Z}_5)| = 480 = 32 \cdot 15$, $|\mathrm{GL}_2(\mathbb{Z}_7)| = 2016 = 32 \cdot 63$, correspondingly, thus $r \leq 5$.

Suppose now that $p > 2$. Then either $p^r | (q + 1)$ or $p^r | (q - 1)^2$. If $p^r | (q - 1)^2$ then $r = 2k$ where $p^k | (q - 1)$. Since $q < p^3$ then $k \leq 2$ thus $r \leq 4$. At last consider the case $p^r | (q + 1)$. By condition we have $q + 1 \leq p^3$, hence $r \leq 3$.

Propositions [7, 3.1] and [7, 4.2] imply that if $|G/K| = p^r$ where $r \leq 4$ or $|G/K| = 2^r$ where $r \leq 5$ then the group G/K has an abelian decomposition. Application of Lemmas 1, 2 and Theorem 4 finishes the proof of Theorem 3 in the semisimple case.

4   Modular Case

In this section we assume that F is a finite field with $\mathrm{char} F = p$ and $|F| = q = p^s$ with $s \leq 2$. Let G be a finite p-group.

Let I be an ideal in FG with $\dim I = 4$. Arguing as in [1], we consider the left and right action of G on I. We fix two group homomorphisms $\varphi, \psi : G \rightarrow \mathrm{GL}(I) = \mathrm{GL}_4(F)$ defined as follows:

$$\forall g \in G, \ v \in I, \varphi(g)(v) = gv, \ \psi(g)(v) = vg^{-1}.$$

Let $K = \varphi(G)$ and $L = \psi(G)$. Then $K, L \leq \mathrm{GL}_4(F)$ $AB = BA$ for any $A \in K$ and $B \in L$ by associativity. Our aim is to prove that at least one of these groups has an abelian decomposition and then to apply Lemma 1 and Theorem 4 which would imply that I defines an abelian group code.

In what follows we will denote the elements of a matrix by the same letter as the matrix itself but in the lower case, for example, $A = (a_{i,j})$.

As in [1] we can assume that the subgroup K has no abelian decomposition and is contained in $UT_4(F)$ since $UT_4(F)$ is a Sylow p-subgroup in $GL_4(F)$. First note that the group $UT_4(F)$ has the following abelian decomposition:

$$UT_4(F) = \left\{ \begin{pmatrix} 1 & \alpha & 0 & 0 \\ 0 & 1 & 0 & 0 \\ 0 & 0 & 1 & \beta \\ 0 & 0 & 0 & 1 \end{pmatrix} : \alpha, \beta \in F \right\} \left\{ \begin{pmatrix} 1 & 0 & a & b \\ 0 & 1 & c & d \\ 0 & 0 & 1 & 0 \\ 0 & 0 & 0 & 1 \end{pmatrix} : a, b, c, d \in F \right\}.$$

So $K \neq UT_4(F)$.

4.1 Case $|F| = p = \mathrm{char} F$

If $p = 2$ and K is a proper subgroup of $UT_4(F)$ then either $|K| = 32$ and K has an abelian decomposition by [7, Proposition 4.2] or $|K| \in \{1, 2, 4, 8, 16\}$ and K has an abelian decomposition by [7, Proposition 3.1]. A contradiction shows that the statement is valid for $p = 2$. Suppose now that p is odd. Then the only possible value of $|K|$ is p^5. Consider the natural homomorphism $\xi : UT_4(F) \to F^3$ defined as follows:

$$\xi : \begin{pmatrix} 1 & a & b & c \\ 0 & 1 & d & f \\ 0 & 0 & 1 & g \\ 0 & 0 & 0 & 1 \end{pmatrix} \mapsto (a, d, g)$$

for all $a, b, c, d, f, g \in F$, and let $N = \ker \xi \cap K$, $V = \xi(K)$. Since $|\ker \xi| = p^3$, $|V| \geq p^2$. But if $|V| = p^3$ then K contains a generating system of the group $UT_4(F)$ modulo its Frattini subgroup $\ker \xi = UT_4(F)'$ (cf. [10, Corollary 10.4.3]) so $K = UT_4(F)$, a contradiction. If $|V| = p^2$ then $|N| = p^3 = |\ker \xi|$. An easy calculation shows that the centralizer of $\ker \xi$ is contained in the commutative group

$$\left\{ \begin{pmatrix} 1 & 0 & b & c \\ 0 & 1 & d & f \\ 0 & 0 & 1 & 0 \\ 0 & 0 & 0 & 1 \end{pmatrix} : b, c, d, f \in F \right\}.$$

This implies commutativity of the group L and the theorem in this case is valid.

4.2 Case $|F| = p^2$

Now we assume that $|F| = p^2$. Let F_0 be the subfield in F containing p elements. Then F is a two-dimensional space over F_0 with a basis $1, \theta$.

Lemma 3. *By some choice of basis in I we can assume that the groups K and L are contained in the group H of block upper-triangular matrices of the form*

$$\begin{pmatrix} 1 & \bar{v} \\ \hline 0 & \\ 0 & A \\ 0 & \end{pmatrix}, \tag{1}$$

where $A \in GL_3(F)$ and $\bar{v} \in F^3$.

Proof. Let $\Delta = \omega G = \left\{ \sum_{g \in G} a_g g \mid \sum_{g \in G} a_g = 0 \right\}$ be the augmentation ideal of the group ring FG. It is well known (see e.g. [11, Lemma 3.1.6]) that this ideal is nilpotent and that its left annihilator coincides with its right annihilator and is the one-dimensional ideal $V = F(\sum_{g \in G} g)$. It follows that any non-zero left (right, two-sided) ideal in FG contains V. Taking a basis of I with the first vector in V we see that any operator of left (right) action of an element $g \in G$ has a matrix of the form (1) with respect to this basis.

As in the previous proof we can assume that the subgroup K is contained in $\mathrm{UT}_4(F)$ since $\mathrm{UT}_4(F)$ is a Sylow p-subgroup in the group of matrices having the form (1).

We need some auxiliary results.

Lemma 4. *If M, N are two groups, M is commutative and every subgroup of N has an abelian decomposition then every subgroup of the group $M \times N$ has an abelian decomposition.*

Proof. Consider the natural projection $\pi : M \times N \to N$. For any $S \leq M \times N$ let $\pi(S) = AB$, where A, B are abelian subgroups in $\pi(S)$. If $\tilde{A} = \pi^{-1}(A) \cap S$ then $\tilde{A}$ is commutative since $\tilde{A} \subseteq M \times A$ and the same is true for $\tilde{B} = \pi^{-1}(B) \cap S$. Note now that $S = \tilde{A}\tilde{B}$: if $x \in S$ then $\pi(x) = ab$ with $a \in A$ and $b \in B$, so $x = \tilde{a}\tilde{b}c$ for some $\tilde{a} \in \tilde{A}, \tilde{b} \in \tilde{B}$ and $c \in \ker \pi = M$. Then $\tilde{b}c \in \tilde{B}$ since $c \in M \cap S$.

Lemma 5. *Let $W = F^2$ be a two-dimensional space over F and let V be any F_0-subspace of W with $\dim_{F_0} V = 3$. Then there exist two linearly independent (over F_0) elements $v_1, v_2 \in V$ such that $v_2 = \theta v_1$.*

Proof. Since $|V| = p^3 > |F|$, it follows that $\dim_F FV \geq 2$ and there exist two vectors $e_1, e_2 \in V$ linearly independent over F and a vector $e_3 \in V$, such that e_1, e_2, e_3 is a basis for V over F_0. As a vector of W

$$e_3 = (\alpha_1 1 + \beta_1 \theta)e_1 + (\alpha_2 1 + \beta_2 \theta)e_2$$

for some $\alpha_i, \beta_i \in F_0$, $i = 1, 2$. Then define

$$v_1 = \beta_1 e_1 + \beta_2 e_2 \neq 0, \quad v_2 = e_3 - \alpha_1 e_1 - \alpha_2 e_2 = \theta(\beta_1 e_1 + \beta_2 e_2) \neq 0,$$

since e_1, e_2, e_3 are linearly independent over F_0. By definition $v_2 = \theta v_1$.

Lemma 6. *Let $W = F^2$ be a two-dimensional space over F. Then there exists a basis of W over F_0 of the form $v_1, \theta v_1, v_2, \theta v_2$.*

Proof. Take a basis v_1, v_2 of W over F and note that $v_1, \theta v_1, v_2, \theta v_2$ are linearly independent over F_0. Hence they generate a 4-dimensional space over F_0, which necessary coincides with W.

Lemma 7. *If $|F| = p^2$ with a prime p then every subgroup in $\mathrm{UT}_3(F)$ has an abelian decomposition.*

Proof. Let S be a subgroup in $G = \mathrm{UT}_3(F)$. First note that $\mathrm{UT}_3(F) = AB$ where

$$A = \left\{ \begin{pmatrix} 1 & \alpha & \beta \\ 0 & 1 & 0 \\ 0 & 0 & 1 \end{pmatrix} : \alpha, \beta \in F \right\}, \qquad B = \left\{ \begin{pmatrix} 1 & 0 & \beta \\ 0 & 1 & \alpha \\ 0 & 0 & 1 \end{pmatrix} : \alpha, \beta \in F \right\}.$$

So we have to prove that S has an abelian decomposition only for $S \neq G$. Then $|S| \in \{p^i : 0 \leq i \leq 5\}$. If $p = 2$ then the result follows from [7, Proposition 4.2]. Assume that $p > 2$ and that $|S| = p^5$.

Now consider a group homomorphism $\xi : \mathrm{UT}_3(F) \rightarrow (F^2, +)$ defined as follows:

$$\xi : \begin{pmatrix} 1 & \alpha & \gamma \\ 0 & 1 & \beta \\ 0 & 0 & 1 \end{pmatrix} \mapsto (\alpha, \beta) \quad \forall \alpha, \beta, \gamma \in F.$$

Clearly $\ker \xi = G' = Z(G)$. If $\dim_{F_0} \xi(S) < 3$ then $|S| = |\xi(S)| \cdot |S \cap \ker \xi| \leq p^2 \cdot p^2 = p^4$ which contradicts the assumption $|S| = p^5$. If $\dim_{F_0} \xi(S) = 4$ then $SG' = G$ and $S = G$ by virtue of [10, Corollary 10.3.3], which also leads to a contradiction. So we must have $\dim_{F_0} \xi(S) = 3$, so $|S/(S \cap G')| = p^3$, hence $|S \cap G'| = p^2$ and $S \supset G'$. By Lemma 5 there exist two linearly independent (over F_0) vectors $v_1 = (\alpha_1, \beta_1)$ and $v_2 = (\alpha_2, \beta_2)$ in $\xi(S)$ such that $(\alpha_1, \beta_1) = \lambda(\alpha_2, \beta_2)$ for some $\lambda \in F$. There exist matrices $x, y \in S$ such that

$$x = \begin{pmatrix} 1 & \alpha_1 & \gamma_1 \\ 0 & 1 & \beta_1 \\ 0 & 0 & 1 \end{pmatrix}, \qquad y = \begin{pmatrix} 1 & \alpha_2 & \gamma_2 \\ 0 & 1 & \beta_2 \\ 0 & 0 & 1 \end{pmatrix} \qquad \text{for some } \gamma_1, \gamma_2 \in F.$$

Direct calculation shows that $[x, y] = 1$, so x, y and $G' = Z(G)$ generate an abelian subgroup A of the group S. Since $|A| = p^4$, we obtain $S = A\langle z \rangle$ for any element $z \notin A$.

Now we proceed with the proof of Theorem 3. First note that if $a_{1,2} = a_{3,4} = 0$ for any matrix $A = (a_{i,j}) \in K$, then K is abelian, which contradicts our assumption.

Case 1. Suppose that $a_{2,3} = a_{3,4} = 0$ for any matrix $A = (a_{i,j}) \in K$. Then there exists an inclusion $\xi : K \rightarrow \mathrm{UT}_2(F) \times \mathrm{UT}_3(F)$ given by the rule

$$\xi(A) = \left(\begin{pmatrix} 1 & a_{1,3} \\ 0 & 1 \end{pmatrix}, \begin{pmatrix} 1 & a_{1,2} & a_{1,4} \\ 0 & 1 & a_{2,4} \\ 0 & 0 & 1 \end{pmatrix} \right).$$

Lemmas 4 and 7 imply that K has an abelian decomposition, a contradiction.

Case 2. If $a_{2,3} = a_{1,2} = 0$ for any matrix $A \in K$, then K is anti-isomorphic to a group from the previous case with the anti-isomorphism given by the rule $A \rightarrow P^{-1}A^T P$, where P is the backward identity matrix and A^T is the transposed matrix A.

Case 3. Suppose that $a_{3,4} = 0$ for any matrix $A \in K$, but there exist matrices $A^{(1)}$, $A^{(2)} \in K$ with $a_{1,2}^{(1)} \neq 0$ and $a_{2,3}^{(2)} \neq 0$.

Consider a matrix $X \in H$ (see Lemma 3). Let $C = [A^{(2)}, X] = A^{(2)}X - XA^{(2)}$, then $c_{3,3} = -a_{2,3}^{(2)}x_{3,2}$ and $c_{4,3} = -a_{2,3}^{(2)}x_{4,2}$. Suppose now that $X \in L$. Then $C = 0$, thus $x_{3,2} = x_{4,2} = 0$ and $x_{2,2} = 1$ since L is a p-group.
Therefore matrices $\left\{ \begin{pmatrix} x_{3,3} & x_{3,4} \\ x_{4,3} & x_{4,4} \end{pmatrix} \mid X \in L \right\}$ form a p-subgroup L_2 of $GL_2(F)$ and there exists a matrix $T \in GL_2(F)$ such that $TL_2T^{-1} \leq UT_2(F)$. If we take $U = \begin{pmatrix} E & 0 \\ 0 & T \end{pmatrix} \in GL_4(F)$ with E standing for the identity matrix in $M_2(F)$, then $ULU^{-1} \leq UT_4(F)$. Note that $UKU^{-1} \leq UT_4(F)$ and $b_{3,4} = 0$ for any matrix $B \in UKU^{-1}$. If $b_{2,3} = 0$ for every matrix $B \in UKU^{-1}$ then we can apply the argument of case 1. Therefore without loss of generality we assume further that $K, L \leq UT_4(F)$ and that there exists a matrix $B \in K$ such that $b_{2,3} \neq 0$. In this case the relation $[B, X] = 0$ implies that $x_{3,4} = 0$ for any $X \in L$. Since K is not abelian, there exist $A', A'' \in K$ with $[A', A''] \neq 0$, which is equivalent to $a'_{1,2}a''_{2,3} - a''_{1,2}a'_{2,3} \neq 0 \vee a'_{1,2}a''_{2,4} - a''_{1,2}a'_{2,4} \neq 0$. Then

$$\begin{cases} [A', X] = 0 \\ [A'', X] = 0 \end{cases}$$

is equivalent to

$$\begin{cases} a'_{1,2}x_{2,3} - a'_{2,3}x_{1,2} = 0 \\ a'_{1,2}x_{2,4} - a'_{2,4}x_{1,2} = 0 \\ a''_{1,2}x_{2,3} - a''_{2,3}x_{1,2} = 0 \\ a''_{1,2}x_{2,4} - a''_{2,4}x_{1,2} = 0 \end{cases}. \tag{2}$$

Consider the case when $a'_{1,2}a''_{2,3} - a''_{1,2}a'_{2,3} \neq 0$. Then the first and the third equations of (2) imply that $x_{1,2} = x_{2,3} = 0$. Since $a'_{1,2} \neq 0$ or $a''_{1,2} \neq 0$, the second or the fourth equation show that $x_{2,4} = 0$, thus L is commutative. The case when $a'_{1,2}a''_{2,4} - a''_{1,2}a'_{2,4} \neq 0$ is analogous.

The anti-isomorphism from case 2 can be applied to reduce the case when $a_{1,2} = 0$ for any matrix $A \in K$ to the considered case.

Case 4. Suppose that $a_{2,3} = 0$ for any matrix $A \in K$, but there exist matrices $A^{(1)}, A^{(2)} \in K$ with $a_{1,2}^{(1)} \neq 0$ and $a_{3,4}^{(2)} \neq 0$. Consider a matrix $X \in H$. Let again $C = [A^{(2)}, X]$, then $c_{3,2} = a_{3,4}^{(2)}x_{4,2}$ and $c_{3,3} = a_{3,4}^{(2)}x_{4,3}$. Suppose now that $X \in L$. Then $C = 0$, thus $x_{4,2} = x_{4,3} = 0$ and $x_{4,4} = 1$ since L is a p-group. Therefore matrices $\left\{ \begin{pmatrix} x_{2,2} & x_{2,3} \\ x_{3,2} & x_{3,3} \end{pmatrix} \mid X \in L \right\}$ form a p-subgroup in $GL_2(F)$. Arguing as in case 3, we can assume that $K, L \leq UT_4(F)$. In this case the relation $[A^{(2)}, X] = 0$ implies that $x_{2,3} = 0$. For a matrix $X \in L$ having the aforementioned structure the system of equations $[A, X] = 0 \ \forall A \in K$ is equivalent to the system of linear equations

$$a_{1,2}x_{2,4} + a_{1,3}x_{3,4} - a_{2,4}x_{1,2} - a_{3,4}x_{1,3} = 0 \ \forall A \in K. \tag{3}$$

Since K is not abelian, the rank over F of the coefficient matrix in (3) is at least 2. If this rank is at least 3, then the space of its solutions over F is at most 1-dimensional.

It is easy to check that L is abelian.

Consider the case when the system (3) has rank 2. Then the set of vectors $V = \{(a_{1,2}, a_{1,3}, a_{2,4}, a_{3,4})| \ A \in K\}$ form a linear space over F_0 of dimension 2, 3 or 4. If $\dim_{F_0} V = 2$, then $|K| \leq p^4$ (since we have no restrictions on matrix elements in the position $(1,4)$), therefore K has an abelian decomposition by [7, Proposition 3.1], a contradiction. If $\dim_{F_0} V = 3$, by Lemma 5 we have a basis over F_0 for V of the form $v_1, v_2 = \theta v_1, v_3$. Consider matrices $A_1, A_2, A_3 \in K$ corresponding to the vectors v_1, v_2, v_3 and subgroups $K_1 = \langle A_1, A_2 \rangle$ and $K_2 = \langle A_3 \rangle$ in K. Since v_1 and v_2 generate a 1-dimensional linear space over F, thus K_1 is abelian and $K = (K_1 Z(K))(K_2 Z(K))$, a contradiction.

If $\dim_{F_0} V = 4$, similarly by Lemma 6 we have a basis over F_0 for V of the form $v_1, v_2 = \theta v_1, v_3, v_4 = \theta v_3$. Again we have matrices $A_1, A_2, A_3, A_4 \in K$ corresponding to these vectors and subgroups $K_1 = \langle A_1, A_2 \rangle$ and $K_2 = \langle A_3, A_4 \rangle$ in K. Both K_1 and K_2 are abelian and $K = (K_1 Z(K))(K_2 Z(K))$, a contradiction.

Case 5. Suppose that there exist matrices $A^{(1)}, A^{(2)}, A^{(3)} \in K$ with $a_{1,2}^{(1)} \neq 0$, $a_{2,3}^{(2)} \neq 0$ and $a_{3,4}^{(3)} \neq 0$.

A subgroup of K generated by $A^{(1)}, A^{(2)}, A^{(3)}$ contains a matrix A with nonzero entries $a_{1,2}, a_{2,3}$ (A may be taken equal to $A^{(1)}$, or $A^{(2)}$ or $A^{(1)} A^{(2)}$) and a matrix A' with nonzero entries $a'_{2,3}, a'_{2,4}$. If it is possible to choose $A = A'$, i.e. $a_{1,2} \neq 0, a_{2,3} \neq 0$ and $a_{3,4} \neq 0$, then A is a non-derogatory matrix, that is its characteristic polynomial coincides with its minimal polynomial. It follows from [12, Theorem 1.3.5] that all matrices commuting with A form a commutative F-algebra. In particular, $C(A) = C_{GL_4(F)}(A)$ is commutative, hence is $L \subset C(A)$. Assume next $a_{3,4} = a'_{1,2} = 0$. Consider a matrix $X \in L$ and the relations $[A, X] = [A', X] = 0$. As in case 3, the relation $[A, X] = 0$ implies $x_{3,2} = x_{4,2} = 0$, $[A', X] = 0$ implies $x_{4,3} = 0$. Thus X is an upper-triangular matrix, consequently, $X \in UT_4(F)$ since L is a p-group. Then $[A, X] = [A', X] = 0$ imply $x_{1,2} = x_{3,4} = 0$, that is, L is abelian.

This finishes the proof of Theorem 3.

There exists an alternative proof of Theorem 3 for the case $|F| = p$, namely, it can be deduced from the case $|F| = p^2$ using a slightly modified Theorem 5.1 [7].

Proposition 1. *Let F be a subfield of a field E and G be a group. If all G-codes over E of given dimension k are abelian then all G-codes over F of given dimension k are abelian.*

Proof. For any ideal $I \lhd FG$, $\dim_E(EI) = \dim_F(I)$ and $\mathrm{PAut}(I) = \mathrm{PAut}(EI)$ by [7, Lemma 5.1], so the proof of [7, Theorem 5.1] remains valid.

5   An Example

Here we show that the method used in the proof of Theorem 3 does not work if $|F| \geq p^3$. First we prove

Lemma 8. *For any prime p and any field F such that $\mathrm{char} F = p$ and $|F| > p^2$ the group $\mathrm{UT}_3(F)$ contains a subgroup S that does not have an abelian decomposition.*

Proof. It is easy to see that there exists an element $t \in$ such that $1, t, t^2$ are linearly independent over the subfield F_0 of the field F with $|F_0| = p$ (for instance it is true if t generates F^*). Indeed, if F_0 is a finite field then one can take as t any generating element of F over F_0, otherwise the set of all roots in F of all non-zero polynomials over the field F_0 of degree ≤ 2 is finite, and it is sufficient to take t outside of this set. Consider the following three matrices:

$$x = \begin{pmatrix} 1 & 1 & 0 \\ 0 & 1 & 1 \\ 0 & 0 & 1 \end{pmatrix}, \quad y = \begin{pmatrix} 1 & t & 0 \\ 0 & 1 & 1 \\ 0 & 0 & 1 \end{pmatrix}, \quad z = \begin{pmatrix} 1 & 1 & 0 \\ 0 & 1 & t^2 \\ 0 & 0 & 1 \end{pmatrix}.$$

Let S be the subgroup generated by z, y, z. It is clear that $(\mathrm{UT}_3(F))' = Z(\mathrm{UT}_3(F))$, so $S' \subseteq Z(S)$ and the group S' is generated by $[x, y], [y, x]$ and $[x, z]$. By direct calculation we obtain

$$[x, y] = \begin{pmatrix} 1 & 0 & \alpha \\ 0 & 1 & 0 \\ 0 & 0 & 1 \end{pmatrix}, \quad [x, z] = \begin{pmatrix} 1 & 0 & \beta \\ 0 & 1 & 0 \\ 0 & 0 & 1 \end{pmatrix}, \quad [y, z] = \begin{pmatrix} 1 & 0 & \gamma \\ 0 & 1 & 0 \\ 0 & 0 & 1 \end{pmatrix},$$

where $\alpha = 1 - t$, $\beta = t^2 - 1$, $\gamma = t^3 - 1$. α, β, γ are linearly independent over F_0 (otherwise the elements $1, t+1$ and $t^2 + t + 1$ would be linearly dependent over F_0) so α, β, γ generate a subgroup of order p^3 of the additive group $(F, +)$, and $|S'| = p^3$. Since the vectors $(1,1)$, $(t,1)$ and $(1,t^2)$ are linearly independent over F_0, we get $|S/S'| = p^3$ so $|S| = p^6$.

Let us prove that the centralizer of each non-central element $g \in S$ in the group S is a commutative group of order p^4. To do this, fix an element $a = x^k y^l z^m c$ with $(k, l, m) \in \mathbb{Z}_0 \backslash \{(0,0,0)\}$ and $c \in Z(S)$ and consider an arbitrary element $x^u y^v z^w d \in S$ where $(u, v, w) \in \mathbb{Z}_0^3$ and $d \in Z(S)$. Computing the commutator $[a, b]$ we see that $ab = ba$ if and only if $(k + lt + m)(u + v + wt^2) = (k + l + mt^2)(u + vt + w)$ (here we identify the elements of $\mathbb{Z}_p$ with the correspondent elements of F_0). In matrix form we obtain

$$(k\ l\ m) \begin{pmatrix} 0 & 1-t & t^2-1 \\ t-1 & 0 & t^3-1 \\ 1-t^2 & 1-t^3 & 0 \end{pmatrix} \begin{pmatrix} u \\ v \\ w \end{pmatrix} = 0.$$

Dividing by $t - 1 \neq 0$ we get

$$(k\ l\ m) \begin{pmatrix} 0 & -1 & t+1 \\ 1 & 0 & t^2+t+1 \\ -t-1 & -t^2-t-1 & 0 \end{pmatrix} \begin{pmatrix} u \\ v \\ w \end{pmatrix} = 0.$$

Computing coefficients with respect to the basis $\{1, t, t^2\}$ of F over F_0 we obtain the following system of equations:

$$\begin{pmatrix} l-m & -k-m & k+l \\ -m & -m & k+l \\ 0 & -m & l \end{pmatrix} \begin{pmatrix} u \\ v \\ w \end{pmatrix} = \begin{pmatrix} 0 \\ 0 \\ 0 \end{pmatrix}.$$

Subtracting the second equation from the first one and the third equation from the second one gives an equivalent system

$$\begin{pmatrix} l & -k & 0 \\ -m & 0 & k \\ 0 & -m & l \end{pmatrix} \begin{pmatrix} u \\ v \\ w \end{pmatrix} = \begin{pmatrix} 0 \\ 0 \\ 0 \end{pmatrix}.$$

Note that the matrix of the last system contains 2×2 submatrices with determinants $-k^2$, m^2 and l^2, hence for any $(k, l, m) \neq (0, 0, 0)$ the rank of this matrix equals 2. It follows that the last system of equations has $|F_0| = p$ solutions, so the centralizer of a in S is $\langle a \rangle S'$ thus it is a commutative group of p^4 elements. Then it is a maximal abelian subgroup in S since any abelian subgroup containing a must be contained in $C_S(a)$. So if $S = AB$ for some abelian subgroups A and B then without loss of generality we can assume that A and B contain $Z(S)$ and that $A = C_S(a)$ and $B = C_S(b)$ for any elements $a \in A\backslash Z(S)$ and $b \in B\backslash Z(S)$. But then $|S| = |A||B|/|A \cap B| \leq p^4 p^4/p^3 = p^5 < p^6 = |S|$. The contradiction proves our claim.

Proposition 2. *If F is a field with* $\mathrm{char} F = p$ *and* $|F| > p^2$ *then there exist two subgroups* $K, L \subseteq UT_4(F)$ *such that* $AB = BA$ *for any* $A \in K$, $B \in L$, *but neither* K *nor* L *has an abelian decomposition.*

Proof. Consider the groups

$$K_1 = \left\{ \begin{pmatrix} 1 & a & 0 & c \\ 0 & 1 & 0 & g \\ 0 & 0 & 1 & 0 \\ 0 & 0 & 0 & 1 \end{pmatrix} : a, c, g \in F \right\}$$

and

$$L_1 = \left\{ \begin{pmatrix} 1 & 0 & b & c' \\ 0 & 1 & 0 & 0 \\ 0 & 0 & 1 & h \\ 0 & 0 & 0 & 1 \end{pmatrix} : b, c', h \in F \right\}.$$

It is easy to check that elements of K_1 commute with those of L_1, and that there exist isomorphisms

$$\begin{pmatrix} 1 & a & 0 & c \\ 0 & 1 & 0 & g \\ 0 & 1 & 0 & 0 \\ 0 & 0 & 0 & 1 \end{pmatrix} \longmapsto \begin{pmatrix} 1 & a & c \\ 0 & 1 & g \\ 0 & 0 & 1 \end{pmatrix}, \quad \begin{pmatrix} 1 & 0 & b & c' \\ 0 & 1 & 0 & 0 \\ 0 & 1 & h & 0 \\ 0 & 0 & 0 & 1 \end{pmatrix} \longmapsto \begin{pmatrix} 1 & b & c' \\ 0 & 1 & h \\ 0 & 0 & 1 \end{pmatrix}$$

between K_1, L_1 and $UT_3(F)$. The inverse images $K \subseteq K_1$ and $L \subseteq L_1$ of the group S from Lemma 8 under these isomorphisms have the desired properties.

It follows only that the technique used in Sect. 3 cannot be applied to the case when $|F| > p^2$. Of course this does not answer in negative the question of existence for the non-abelian codes of dimension 4 in the case of p-groups.

Acknowledgements. Authors want specially remember Victor Markov and acknowledge his valuable contribution to this paper before he passed away in July 2019.

References

1. García Pillado, C., González, S., Markov, V., Markova, O., Martínez, C.: Group codes of dimension 2 and 3 are abelian. Finite Fields Appl. **55**, 167–176 (2019)
2. Macwilliams, F.J., Sloane, N.J.A.: The Theory of Error-Correcting Codes. North-Holland, Amsterdam (1977)
3. Pless, V.S., Huffman, W.C. (eds.): Handbook of Coding Theory I, vol. II. Elsevier, New York (1998)
4. Nechaev, A.A.: Kerdock code in a cyclic form. Diskret. Mat. **1**(4), 123–139 (1989). (in Russian); English transl. in: Discrete Math. Appl. 2(6), 659–683 (1992)
5. Hammons Jr., A.R., Kumar, P.V., Calderbank, A.R., Sloane, N.J.A., Solé, P.: The Z_4-linearity of Kerdock, Preparata, Goetals and related codes. IEEE Trans. Inform. Theory **40**, 301–319 (1994)
6. Bernal, J.J., del Río, Á., Simón, J.J.: An intrinsical description of group codes. Des. Codes Crypt. **51**(3), 289–300 (2009)
7. García Pillado, C., González, S., Markov, V., Martínez, C., Nechaev, A.A.: Group codes over non-abelian groups. J. Algebra Appl. **12**, 1350037-1–1350037-20 (2013)
8. García Pillado, C., González, S., Markov, V., Martínez, C., Nechaev, A.A.: New examples of non-abelian group codes. Adv. Math. Commun. **10**(1), 1–10 (2016)
9. García Pillado, C., González, S., Markov, V., Martínez, C.: Non-abelian group codes over an arbitrary finite field. Fundementalnaya i prikladnaya matematika **20**(1), 17–22 (2015). (in Russian)
10. Hall, M.: The Theory of Groups. Harper and Row, New york (1968)
11. Passman, D.S.: The Algebraic Structure of Group Rings. Wiley, New York (1977)
12. Suprunenko, D.A., Tyshkevich, R.I.: Commutative Matrices. Academic Pr, New York (1968)

Designing a Public Key Cryptosystem Based on Quasi-cyclic Subspace Subcodes of Reed-Solomon Codes

Thierry P. Berger[1]([⊠]), Cheikh Thiécoumba Gueye[2]([⊠]), Jean Belo Klamti[2]([⊠]), and Olivier Ruatta[1]([⊠])

[1] XLIM UMR 7252 Université de Limoges - CNRS, Limoges, France
thierry.berger@unilim.fr, olivier.ruatta@unilim.fr
[2] Université Cheikh Anta Diop de Dakar - Faculté des Sciences et Techniques, Department de Mathématiques et Informatique LACGAA, Dakar, Senegal
cheikht.gueye@ucad.edu.sn, jklamty@gmail.com

Abstract. In this paper we introduce a code-based cryptosystem using quasi-cyclic generalized subspace subcodes of Generalized Reed-Solomon codes in order to reduce the public key size. In our scheme the underlying Generalized Reed-Solomon code is not secret, so the classical attacks such as square code or folding attacks have no more purpose against it. In addition one part of the security of this scheme is based on hard problems in coding theory like *Equivalence Subcodes (ES) Problem*. We propose some parameters to reach at least a security level of 128 and 192 bits. We make a public key size comparison with some well established code-based public key encryption schemes. We also see that for the 128 bits security level the key size of our proposals are often better than the code-based schemes in competition for NIST's second round.

Keywords: Mceliece public key cryptosystem · Subspace subcodes · Reed-Solomon codes · Quasi-cyclic codes

1 Introduction

The quantum computer is a major threat against public key cryptography since *Shor* introduced a quantum algorithm [30] for factoring numbers and computing the discrete logarithm problem. That is why recently in 2000s the *Post-Quantum* cryptography was introduced [9]. To protect against quantum attacks NIST launched a call for *Post-Quantum* cryptography standardization in 2016 [26].

Introduced in 1978 by *McEliece* [21], code-based cryptography is one of promising candidates for post-quantum cryptography. Its security relies on the hardness of decoding a random-like linear code without any visible structure. This random decoding problem is known to be a NP-complete problem [8] for which if an algorithm needed 2^w binary operations with classical computers, then using quantum computers and Grover's algorithm they will need $2^{(w/2)}$

© Springer Nature Switzerland AG 2019
C. T. Gueye et al. (Eds.): A2C 2019, CCIS 1133, pp. 97–113, 2019.
https://doi.org/10.1007/978-3-030-36237-9_6

binary operations. The main drawback of code-based cryptography is the size of the public key which is very large. In order to reduce the key size, a long track of research works has been devoted to variants of these systems characterized by smaller public keys [4,24,25]. However most of these variants based on codes with non trivial automorphism group like in [4,24] and they have been seriously dented. This is why the design of a secure McEliece-like cryptosystem with a small public key remains a major issue.

Our Contribution

In this paper we design code-based Public Key Encryptions (PKEs) using structured generalized subspace subcodes of Generalized Reed-Solomon codes (GSS-GRS) [7,17]. The main originality of our approach is the fact that the use of subspace subcodes that are not subfield subcodes. This destroys the finite field extension structure, which is the core of all known attacks against codes derived from Generalized Reed-Solomon codes. The underlying Generalized Reed-Solomon (GRS) code is public so the classical attacks such as square code attacks and folding attacks have no more purpose against it. We give some practical instantiations of our cryptosystems and we give some security analysis of our scheme. We conclude our paper with a comparison of our proposals with related previous works.

Organization of Paper

This paper is organized as follows: in Sect. 2 we give some definitions and properties of generalized subspace subcodes. In Sect. 3 we specify the notion of quasi-cyclic generalized subspace subcodes of Reed-Solomon codes (RS codes). In Sect. 4 we introduce a code-based PKEs using quasi-cyclic/cyclic GSS-RS codes. In Sect. 5 we give the rationale of designing our scheme. In Sect. 6 we give a security analysis of the cryptosystem. In Sect. 7 we introduce some instantiations of our scheme. The Sect. 8 concludes the paper with a comparison of our proposals with previous works.

2   Background on Block Codes

We will use the following notations:

- $GF(2)$ is the finite field with 2 elements.
- $GF(2^m)$ is an extension of $GF(2)$ of degree m.
- $GL_2(m)$ is the General Linear Group of the $m \times m$ invertible matrices with entries in $GF(2)$.

2.1   Binary m-block Codes

For more details on block codes, the reader can refer to [3,7].

Definition 1. *A binary m-block code of length n over* $\mathbb{E} = GF(2)^m$ *is an additive code over the additive group* $(\mathbb{E}, +)$, *i.e. a subgroup of* $(\mathbb{E}^n, +)$. *The integer m is the size of the blocks.*

A binary m-block code of length n is clearly a binary linear code of length nm, however, we are interested in its block properties.

For the comparison of m-block codes with linear codes over $\mathbb{F}_{2^m}$, we use the following notation for parameters as a code C which is either an m-block code or an $\mathbb{F}_{2^m}$-linear code: $[n, k, d]_{2^m}$, where n is the block-length of the code, $k = \log_{2^m}(|C|)$ is the pseudo-dimension of C relative to the size of the alphabet 2^m and d is its block-minimum distance. If a code is $\mathbb{F}_{2^m}$-linear, its pseudo-dimension is its dimension.

In the sequel, if $x \in E^n$ is an m-block codeword, we set $x = (\overline{x}_1, ..., \overline{x}_n)$ and $\overline{x}_i = (x_{i,1}, ..., x_{i,m}) \in E$.

For the construction of m-block codes from $GF(2^m)$-linear codes we choose a basis $\mathcal{B} = (b_1, ..., b_m)$ of $GF(2^m)$ considered as a vector space over $GF(2)$. We then replace the coefficients of codewords in $GF(2^m)$ by their m-block coordinates over $GF(2)$. Note that the parameters of this 2-ary image seen as an m-block code are the same than the original $GF(2^m)$-linear code.

2.2   Isometries of m-block Codes

An isometry for an m-block code is a $GF(2)$-linear map on E^n which preserves the m-block distance.

If we permute the coordinates (at $\mathbb{E}$-level) of an m-block code C of length n, we obtain another m-block code C' with same length, pseudo-dimension and m-block distance.

Let $GL_2(m)$ be the General Linear Group of the $m \times m$ invertible binary matrices. Let $L = (\Lambda_1, ..., \Lambda_n) \in GL_2(m)^n$ be an ordered set of elements of $GL_2(m)$. It is easy to check that the map $x = (\overline{x}_1, ..., \overline{x}_n) \mapsto L(x) = (\overline{x}_1 \Lambda_1, ..., \overline{x}_n \Lambda_n)$ is a linear isometry for the m-block distance. From a matrix point of view, it consists in multiplying on the right the elements of $\mathbb{E}^n = \mathbb{F}_2^{nm}$ by the n-block diagonal matrix $\text{Diag}(L) = \text{Diag}(\Lambda_1, ..., \Lambda_n)$.

Such a map L is called a *multiplier*. The following theorem gives a full characterization of isometries for m-block distance.

Theorem 1. (Theorem 1 of [7]). *The $\mathbb{F}_2$-isometries of $\mathbb{E}^n$ (i.e. linear isomorphisms preserving the Hamming block-weight) is the group generated by the m-block permutations and the multipliers.*

We are now able to define some notions of equivalence of block codes.

Definition 2. *Let C and C' be two m-block codes of length n over $\mathbb{E}$.*

- *C and C' are equivalent if there is an isometry $f = L \circ \pi$, (where L is a multiplier and π a permutation) such that $C' = f(C)$.*
- *C and C' are equivalent by permutation if there is a permutation at block level $\pi \in \text{Sym}(n)$ such that $C' = \pi(C)$.*
- *C and C' are equivalent by multiplier if there is a multiplier $L \in GL_2(m)^n$ such that $C' = L(C)$.*

2.3 Subspace Subcodes

The notion of subspace subcode was introduced in [17] and developed in [7].

Let C be a binary m-block code of length n and V be an $GF(2)$-subspace of $GF(2^m)$ of dimension $\mu \leq m$. The subspace subcode over V of C is the $GF(2)$-linear code $C_{|V^n} = C \cap V^n$.

To obtain an effective representation of the code $C_{|V^n}$, it is necessary to choose a basis of V as a subspace of $GF(2)^m$ of dimension μ. It can be done for instance by choosing a generator matrix M_V of V viewed as a linear code of length m. The elements of V are replaced by their μ coordinates on this basis. By this identification, we obtain a μ-block code of length n, denoted $SS_V(C)$. In the sequel, the notion of μ-subspace subcode means its μ-block representation $SS_V(C)$.

Let ψ be a $GF(2)$-projection of rank μ of $GF(2^m)$ onto a $GF(2)$-subspace $V \subset GF(2^m)$ of dimension μ. Let Ψ be the action of ψ on each coordinate of a word in $GF(2^m)^n$. Let C be an $GF(2^m)$-linear. The (μ-)projected code relative to ψ of C is the code $P_\psi(C) = \Psi(C)$.

The following proposition describes the link between subspace subcodes and projected codes as μ-block codes of length n.

Proposition 1 (Proposition 9 of [7]). *The μ-projected codes are the duals of μ-subspace subcodes. More precisely, the dual of the μ-block subcode $SS_V(C)$ relative to a generator matrix M_V is the μ projected code $\Psi(C^\perp)$, where the $m \times \mu$ projection matrix of π is M_V^T.*

Projected codes and subspace subcodes can be interpreted in term of punctured and shortened codes (see [7] for more details).

The following Corollary gives some bounds on the parameters of a subspace subcode. In addition, we notice that if a code C possesses a decoding algorithm up to t errors, this algorithm can be applied to the μ-block code $SS_V(C)$. Using this approach, it is possible to derive some bounds on the pseudo-dimension and the μ-block minimum distance of a subspace subcode.

Corollary 1 (Corollary 2 of [7]). *Let C be a binary m-block code of parameters $[n, k, d]_{2^m}$. Let $C_\mu = SS_V(C)$ be a μ-subspace subcode of C of parameters $[n, k', d']_{2^\mu}$. We have $k' \geq (km - n(m - \mu))/\mu$ and $d' \geq d$.*

Note that, in this Corollary, k' and d' are related to the μ-block structure of C_μ.

The notion of subspace subcodes of m-block codes can be naturally extended to linear codes over $GF(2)$ by constructing first a binary image of these codes.

2.4 Generalized Subspace Subcodes

The notion of subspace subcodes was generalized in [7]. Let $\overline{V} = \prod_{i=1}^{n} V_i$ be the cartesian product of n $GF(2)$-subspaces of $GF(2^m)$ of dimension μ. The generalized subspace subcode of C relative to $\overline{V}$ is $C_{\overline{V}}(C) = C \cap \overline{V}$.

Its μ-block representation $\mathrm{GSS}_{\overline{V}}(C)$ is constructed as previously by choosing a basis for each μ-subspace V_i.

An equivalent definition for the μ-subspace subcodes of a code C is the following: the generalized subspace subcodes of an m-block code C are the μ-subspace subcodes of the m-block codes that are multiplier equivalent to C. Note that, if our code C is a binary image of a $GF(2^m)$-linear code, the notion of multiplier equivalence at m-block level is more general than those at $GF(2^m)$ level.

The main advantages of the use of generalized subspace subcodes in a cryptographic context are the following:

- Bounds on dimension and minimum distance given in Corollary 1 hold for GSS codes.
- If there exists a decoding algorithm of m-block error correcting capability t, it can be used to decode up to μ-block errors in the GSS code.
- The projection from $GF(2)^m$ to $GF(2)^\mu$ destroys any $GF(2^m)$ underlying structure. In particular it avoids some classical attacks such as the square attack or the Sidel'nikov Shestakov attack.

In addition, if we consider a code $C_{\overline{V}}(C)$, its dual contains a lot of words of weight μ corresponding to parity-check matrices of the subspaces V_i. However, if we take a μ-subspace subcode $\mathrm{GSS}_{\overline{V}}(C)$, these parity-check equations related to the V_i's are removed, since the redundancy was suppressed.

3 Quasi-cyclic Codes in the Context of GSS Codes

A quasi-cyclic permutation σ of order ℓ and index s is a permutation acting on a set of $n = s\ell$ elements whose decomposition into orbits is constituted of s orbits of same length ℓ. A quasi-cyclic code is then a code of length n invariant under a quasi-cyclic permutation.

There are basically two ways to order the support of a quasi-cyclic code of order ℓ and index s. The first one consists in taking the decomposition under the orbits of σ. The quasi-cyclic permutation acts as a simultaneous application of the shift permutation on each block of size ℓ. In that case, we say that the quasi-cyclic code is in orbit order. The second one is to consider the quasi-cyclic permutation as the s-th power of the shift permutation, and we say that the quasi-cyclic code is in shift order.

The classical notion of block circulant matrix remains ambiguous, since it may mean that each block is circulant or the circulant permutation is applied to the blocks. In the sequel, we denote by P_s the matrix permutation which transforms a quasi-cyclic code in orbit order into a quasi-cyclic code in shift order.

For quasi-cyclic m-block codes, it is better to use the representation in shift order. We index codewords from 0 to $n-1$. If $x \in \mathbb{E}^n$, we set $x = (\overline{x}_0, ..., \overline{x}_{n-1})$ and $\overline{x}_i = (x_{i,0}, ..., x_{i,m-1}) \in \mathbb{E}$. In addition, we set $x_{im+j} = x_{i,j}$ in such a way that $x = (x_0, ..., x_{nm-1}) \in GF(2)^{nm}$.

Let $\overline{\sigma} \in \mathrm{Sym}(n)$ be the cyclic shift on m-blocks: $\overline{\sigma}(i) = i + 1 \pmod{n}$. As usual, $\overline{\sigma}$ acts on $\mathbb{E}^n$ as follows:

$$\overline{\sigma}(x) = (\overline{x}_{\overline{\sigma}^{-1}(0)}, ..., \overline{x}_{\overline{\sigma}^{-1}(n-1)}) = (\overline{x}_{n-1}, \overline{x}_0, ..., \overline{x}_{n-2}).$$

In a similar way, if $\sigma \in \mathrm{Sym}(nm)$ is the cyclic shift at bit level, *i.e.* $\sigma(i) = i+1 \pmod{nm}$, then $\sigma^m(x) = \overline{\sigma}(x)$.

Definition 3. *Let ℓ be a divisor of n and s such that $n = s\ell$. An m-block C is quasi-cyclic of order ℓ and index s if it holds that $\overline{\sigma}^s(C) = C$.*

Clearly, at bit level, such a code is quasi-cyclic of order ℓ and index ms.

The following proposition characterizes the multipliers which preserve the quasi-cyclicity of an m-block code.

Proposition 2. *Let C be an m-block quasi-cyclic code of order ℓ and index s. Let $\Lambda = (\Lambda_0, ..., \Lambda_{n-1}) \in GL_2(m)^n$ be a multiplier. If $\Lambda_i = \Lambda_j$ for all i, j such that $(i - j) = 0 \pmod{s}$, then $C' = \Lambda(C)$ is quasi-cyclic of order ℓ and index s.*

Proof. The condition $\Lambda_i = \Lambda_j$ for all i, j such that $(i - j) = 0 \pmod{s}$ means that the multipliers must be constant on each orbit under $\overline{\sigma}^s$. The proof is a straightforward verification.

If the projection $\Psi = (M_0, ..., M_{n-1})$ is well-chosen the quasi-cyclicity of a parent code is preserved by the GSS construction. This property was studied in [7] Section C.

Proposition 3 (Proposition 14 of [7]). *Let C be an m-block quasi-cyclic code of order ℓ and index s. Let $\Psi = (M_0, ..., M_{n-1})$ be a projection map such that each matrix M_i^T is a generator matrix of a subspace V_i of dimension μ of E. If $M_i = M_j$ for all i, j such that $(i - j) = 0 \pmod{s}$, then $C'^{\perp} = \Psi(C^{\perp})$ and $GSS_{\overline{V}}(C)$ are μ-block quasi-cyclic codes of order ℓ and index s.*

Note that, at binary level, these codes are quasi-cyclic of order ℓ and index $s\mu$.

If the quasi-cyclic property of a GSS code is inherited from the parent code, we say that it is an induced quasi-cyclic code.

3.1 Induced Quasi-cyclic GSS Codes of GRS Codes

In our construction, we use quasi-cyclic Generalized Reed Solomon codes in order to construct a quasi-cyclic public binary code.

We use the GRS codes in our construction for the following reasons: these codes are MDS, *i.e.* have the optimal minimum distance for a length n and a dimension k. They have an efficient decoding algorithm that can be applied to the public code.

The quasi-cyclicity of our public code reduces the size of the public key. When the length n divides $2^m - 1$ it is easy to construct a lot of quasi-cyclic GRS code. It was done for instance in [4]. More details on the construction of random quasi-cyclic GRS code can be found in [7].

3.2   Permutations Preserving the Quasi-cyclicity

In this Section, we consider a binary quasi-cyclic code of order ℓ and index ν. In order to facilitate the presentation of permutation preserving the quasi-cyclicity, we use a representation of codewords under a union of orbits derived from the representation of block codes:

$$c = (\bar{c}_1, ..., \bar{c}_\nu) \in GF(2)^{\nu\ell} \text{ and } \bar{c}_i = (c_{i,0}, ..., c_{i,\ell-1}) \in GF(2)^\ell.$$

If σ is the cyclic shift acting on $GF(2)^\ell$, the quasi-cyclic permutation is then $\Sigma(c) = (\sigma(\bar{c}_1), ..., \sigma(\bar{c}_\nu))$.

There are two types of permutations which preserve the quasi-cyclicity:

- Type 1 permutations are permutations of orbits of c:
  If $\pi \in \mathrm{Sym}(\nu)$, then $\Pi_1(c) = (\bar{c}_{\pi^{-1}(1)}, ..., \bar{c}_{\pi^{-1}(\nu)})$.
- Type 2 permutations need to choose ν independent powers of the circular shift $\sigma \in \mathrm{Sym}(\ell)$ and apply them to the ν orbits of c:
  If $\Pi_2 = (\sigma^{i_1}, ..., \sigma^{i_\nu}) \in \mathrm{Sym}(\ell)^\nu$ then $\Pi_2(c) = (\sigma^{i_1}(\bar{c}_1), ..., \sigma^{i_\nu}(\bar{c}_\nu))$.

Clearly, these permutations preserve the quasi-cyclicity. One can verify that, in general, there is no more permutation which preserves the quasi-cyclicity.

4   PKE Based on Subspace Subcodes of QC-GRS Codes

One crucial point in public key code-based cryptography is not only to hide the structure of the secret code, but also to reduce the size of the public key. The use of generalized subspace subcodes of Reed-Solomon codes not only allows us to mask the original code, but also, under certain conditions, to use quasi-cyclic codes in order to reduce the key size.

Our proposition is based on cyclic or quasi-cyclic generalized subspace subcodes of Reed-Solomon codes. There have been several attempts to use quasi-cyclic or quasi-dyadic GRS codes to design a Public Key Cryptosystem (eg. [4,24]), most of them have been severely attacked [2,12,14,15]. Our approach is new and radically different from previous ones in many ways.

- The first point is the fact that the underlying Reed-Solomon code is not secret, so the classical attacks such as folding attacks or square code attacks have no more purpose.
- Secondly, in our construction, we do not use subfield subcodes, but subspace subcodes. Moreover, we remove the m-block structure by performing permutations at bit level.
- The last point is the fact that the public code is a small subcode which is a mono-generator quasi-cyclic code. In particular, such a code has parameters similar to those of random mono-generator quasi-cyclic codes of same length and order.

Remark 1. The authors of the paper [19] use only the notion of *extended codes* that we call *q-ary image* in our paper. In addition, we use a block-puncturing technique which is a block projections on vector subspaces. In the case of subspace subcode the projections are same for all blocks and in the case of Generalized subspace subcode the projections are different. These allow us to make the underlying structure of the Reed-Solomon code not secret. This is not used in the paper [19].

4.1   Key Derivation Algorithm

Some public parameters must be published in the specification of our algorithm before the application of the key derivation algorithms:

These parameters are constituted of

- The degree of extension m of the fields $GF(2^m)$. An order ℓ and an index s of a quasi-cyclic code of length $n = \ell s$ over $GF(2^m)$. Note that ℓ must be a divisor of $2^m - 1$.
- The integer $\mu \leq m$, which is the dimension projection subspaces V_i.
- A public quasi-cyclic GRS code GRS(k) of order ℓ and length n over $GF(2^m)$. The dimension k, the minimum distance $d = n - k + 1$ and the error correcting capability $t = \lfloor (d-1)/2 \rfloor$ are also public. We set $\nu = s\mu$.
- A public generator matrix $\mathcal{G}$ of a binary image C of GRS(k). The basis $\mathcal{B}$ used for computing this binary image is not a secret and could be published. quasi-cyclic m-block code C of order ℓ, index s, length $n = s\ell$ and pseudo-dimension k over $\mathbb{E} = GF(2)^m$.

From a practical point of wiev, the quasi-cyclic code GRS(k) is in shift order so that, at binary level, C is quasi-cyclic code of length $N = nm$, order ℓ and index sm in shift order.

Secret Key. The secret key is constituted of two parts:

1. A set of s random binary matrices M_i of size $m \times \mu$ and full rank μ.
2. The two random permutation matrices Π_1 and Π_2 respectively of Type 1 and Type 2.

From the matrices $M_i's$, we can compute a projection matrix $\Psi = \text{Diag}(M_1, ..., M_1, M_2, ..., M_2, ..., M_s, ..., M_s)$ corresponding to a random projection matrix which is constant on each orbit. This projection preserves the μ-block quasi-cyclicity in orbit order. So, we have to use the corresponding projection matrix in shift order $\Phi = P_s^{-1} \Psi P_s$.

The size of the secret key is the following:

- the s matrices M_i: $sm\mu$ bits,
- the permutation matrix Π_1: $s\mu \log_2(s\mu)$ bits.
- the permutation matrix Π_2: $s\mu \log_2(\ell)$ bits.

The full size of the secret key is then $sm\mu + s\mu \log_2(s\mu) + s\mu \log_2(\ell)$ bits.

Key Derivation Algorithm

1. Compute a parity check matrix (at binary level) $\mathcal{H}$ of C.
2. Compute a generator matrix G of the code C having as a parity-check matrix $\mathcal{H}\Phi$.
3. Compute $G' = G \times \Pi_2$ be a generator matrix of the image C' of C by the permutations Π_1 and Π_2.
4. Choose randomly an element c of C'. Let denote $< c >$ the code generated by the ℓ images of c under the quasi-cyclic permutation. If $\dim(< c >) = \ell$, then $C_{\text{pub}} =< c >$, else, choose another c.

Public Key. The public key is the codeword c. The public code is then $C_{\text{pub}} =< c >$, and we fix for generator matrix the matrix G_{pub} obtained by applying recursively the quasi-cyclic permutation σ_ℓ to c. We will see in the encryption algorithm that it is not necessary to compute G_{pub}, it can be done on the fly.

In addition, we propose at the end of Sect. 5 an improvement on the choice of v which reduce the size of the public key to $(\nu - 1)\ell$ bits.

4.2   Decoding Algorithm

In this Section, we will explain how we will correct up to $t = \lfloor (d-1)/2 \rfloor$ on a codeword of C_{pub} by means of a decoder of the GRS code.

Before describing this algorithm, it is necessary to introduce a matrix allowing to invert the operation of shortening in the construction of the subspace subcode.

Using the classical the theorem of incomplete basis, we complete each matrix M_i of size $m\mu$ and rank μ into a square invertible matrix M_i' such that the restriction of M_i' to its μ first columns is M_i. The choice of M_i' has no influence on the decoding procedure.

By analogy with the key derivation algorithm, we set
$\Psi' = \text{Diag}(M_1', ..., M_1', M_2', ..., M_2', ..., M_s', ..., M_s')$ and $\Phi' = P_s^{-1}\Psi'P_s$.

Let c be a codeword of C_{pub} and e be a binary vector of length $N = n\mu$ and weight t. The decoding algorithm is the following

- Input: the vector $y = c + e$.
- Compute $y' = y\Pi^{-1}$.
- Compute $y'' = y'P_s$ (in order to reconstruct the μ-block structure)
- To each μ-block of coordinates of y'', add $m - \mu$ 0 (Inversion of the shortening procedure). We obtain an m-block codeword z of length n.
- Compute $z' = z\Psi'^{-1}$.
- From z', using the basis $\mathcal{B}$ of $GF(2^m)$ over $GF(2)$ reconstruct a word $x \in GF(2^m)^n$.
- Under our hypothesis, x is of the form $x = c' + e'$, $c' \in GRS(k)$ and $e' \in GF(2^m)^n$ is of weight t.
- Using the GRS decode, we can recover c' and e'.
- Output the codeword c and the error vector e obtained by inverting all the previous steps.

4.3 Encryption and Decryption Algorithm

As soon as we have a binary code C_{pub} with a secret decoder up to t errors, we can construct a McEliece like (or a Nieddereiter like) PKE.

The only improvement comes from the fact that our public code is quasi-cyclic and generated by a single generator vector v.

Encryption Algorithm

- Known parameters $v \in GF(2)^N$ and the error correction capability t
- Input: a message $m = (m_0, ..., m_{\ell-1}) \in GF(2)^\ell$.
- Coding part:
    $c := m_0 v \in GF(2)^N$
    for $i := 1$ to $\ell - 1$, $v := \sigma(v)$, $c := c + m_i v$.
    end for.
- Choose randomly an error vector $e \in GF(2)^N$ of weight t.
- Output the cipher text $y = c + e$.

Decryption Algorithm

- Input: $y = c + e$.
- Recover the error e and the codeword c with the decoding algorithm.
- Recover the message m from c and G_{pub}.

5 Design Rationale

Publishing the GRS Code: We think that the resistance of our cryptosystem is not based on the knowledge of a particular choice of a quasi-cyclic GRS code. This is why we propose to publish it. In consequence, we do not have to worry about folding attacks [2,14,15] and square code attacks [12].

The Generalized Subspace Subcode Operation: The projection map Ψ used at Step 2 destroys the underlying $GF(2^m)$ structure of GRS codes. It was shown in [7] Section IV that GSS codes can be constructed by taking an m-block multiplier equivalent code, and by shortening the $m - \mu$ last coordinates of each block. For the dual code, it consists in puncturing these last coordinates. So our problem is close to the problem of reconstructing a permuted punctured code, which is proven to be NP-complete in [31].

Importance of Permutations: The permutation equivalence of binary codes is a well studied problem. In general, this problem can be solved by the support splitting algorithm in polynomial time [28]. However, there are some difficult instances such as codes with large permutation group, or large Hull (intersection of a code with its dual).

Then such a permutation is combined with the choice of a subcode, this problem becomes NP-complete, as proven in [6].

In addition, a secret permutation is always difficult to take into account in a problem modelisation in order to design an algebraic attack, *i.e.* when we try to solve a problem under the form of a polynomial system.

The Subcode Strategy: In a first approach, it seems that restricting our code to a mono-generator subcode is a disadvantage since it decreases the size of our code without increasing the correction capacity of our algorithm.

In fact, there is much advantage in the use of such a code.

For practical instantiations, even if our algorithm is far from optimality in term of error correction capability, it is sufficient to ensure a good resistance against brute force decoding.

The use of a single generator decreases the size of the public key, since we just have to publish a single codeword of length $N = \nu\ell$.

The association of a permutation and the choice of a subcode is an efficient strategy to mask the structure of a code. The following problem was proven to be NP-complete in [6]:

Problem 1 (ES Equivalence Subcode Problem). Given two linear codes $\mathcal{C}$ and $\mathcal{D}$ of length of n and respective dimension k' and k, $k' \leq k$, over the finite field $GF(2)$, does a permutation σ of the support such that $\sigma(\mathcal{C})$ is a subcode of $\mathcal{D}$ exist?

The subcode strategy produces a public code that we we think to be random-like.

Improving the Subcode Strategy: Let us consider the public code C_{pub} in orbit order. If we look at the restriction of C_{pub} it is easy to see that we obtain either the null code, or the repetition code or the full space. The first two cases correspond to degenerate situations that it is preferable to avoid. So, we restrict our public key to codes that admit a systematic generator matrix. Such codes are in bijection with the first row of this systematic generator matrix.

So, without loss of generality, it is possible to limit the random choice of the mono-generators to this kind of codewords.

Note that it is sufficient to publish the $(\nu - 1)\ell$ last coordinates of c, so the size of the public key is reduced to $K_{size} = (\nu - 1)\ell$.

6   Security Analysis

There are essentially two families of attacks against this kind of PKE: the structural attacks, where the attacker tries to find the secret key or an equivalent secret key, and the decoding brute force attacks, which use decoding techniques for random codes.

In this Section, we present some security arguments. We discuss about the randomness of our public code. Then we are examining depleted versions of our cryptosystem to better understand what its security is about.

Number of Secret Parameters: The secret key is composed of the two following parts:

1. The s secret projection matrices M_i, $1 \leq i \leq s$.
2. The secret permutations Π_1 and Π_2.

The number of projection matrices of size $m \times \mu$ is $N_{m,\mu} = \prod_{i=0}^{\nu}(2^m - 2^i)$. So, the number of choice for Ψ is $N_{proj} = N_{m,\mu}^s$.

The number of Type 1 and Type 2 permutations are respectively $\nu!$ and ℓ^ν. Taking in account some equivalences, the number of permutations Π for an exhaustive search can be reduced to $N_{per} = \nu!\ell^{\nu-1}/\mu!$.

Randomness of C_{pub}: Our security assumption is the fact that, even if the initial GRS code is known, the code C_{pub} is indistinguishable of a random mono-generator quasi-cyclic code of same order and index.

In our PKE, we use an error correcting algorithm which is far from the true error correcting capability of our codes. Indeed, since we take a small subcode in our subcode strategy, we do not know the true minimum distance of C_{pub}. Using the Magma algebra system [11], we performed some experimental test on their true minimum distance up to the computational limit. We observed that our public codes have the same minimum distance distribution as random mono-generator codes of same parameters. In addition, as expected, these values are over the binary Gilbert-Varshamov bound.

We want to underline that the duals of our codes have a small minimum distance, due to the great dimension of these codes. However, as noticed in Section 2.4, the codewords with small weight are not related to the μ-block structure.

Analysis if the Projection Ψ is Known: Suppose that the projection Ψ is known. In that situation, we know the generator matrix G of GSS code C. The problem to solve is the following:

Problem 2. Given the codes C and C_{pub}, find Type 1 and Type 2 permutations Π_1 and Π_2 such that C_{pub} is a subcode of $\Pi_2 \circ \Pi_1(C)$.

This problem is a particular case of the equivalence-subcode problem 1. We do not know any algorithm other than the exhaustive search on permutations to solve this problem. The complexity of such algorithm is $\mathcal{O}(Num_{per})$ where $Num_{per} = \nu!\ell^{\nu-1}/\mu$ is the number of permutations up to equivalence.

Analysis if the Permutations Π_1 and Π_2 are Known: Suppose that the permutations Π_1 and Π_2 are known. We consider the coefficients of the s matrices M_I as unknowns. Let $\mathcal{H}$ be the parity-check matrix of the code C as described on the part on the construction of the public key. We denote by G_{pub}

a generator matrix of the public code C_{pub} and by Ψ the projection matrix. This matrix contains the unknown coefficients of the M_i's.

We obtain a linear system $\mathcal{H} \times \Psi \times G_{pub}^T = 0$. In fact, since the code is ℓ-quasi-cyclic, this system can be reduced to $\mathcal{H} \times \Psi \times c^T = 0$ where c is the generator of the public code C_{pub}.

The number of rows of $\mathcal{H}$ is $nm - km$, so it is a system constituted of $nm - km$ equations with $(m\mu)^s$ unknowns. For the parameters given in Sect. 7, it is possible to solve this system. So, clearly, the resistance of our proposal is based on the hardness of the reconstruction of the secret permutations. This part only increases the exhaustive search of a factor $\mathcal{O}((n-k)m)^3)$, which is greater than $\mathcal{O}(2^{3m})$ in our examples.

Decoding Random Binary Codes: There are a lot of papers about random decoding (see e.g. [10,13]), however all the known algorithms are based on the search of an information set without error. So we choose as the lower bound of the security of this attack the probability of finding randomly an information set without error: $RD(n,k,t) = \binom{n-t}{k}/\binom{n}{k}$.

Clearly, this lower bound is far from the true complexity of this kind of attack, since it neglects the cost of the search and verification algorithm.

7   Instantiation

Our examples of parameters are constructed as follows:

- First, we fix the parameters of a cyclic GRS code $[n,k,d = n+1-k]$ over $GF(2^m)$. In practice, we set $n = 2^m - 1$.
  In order to limit the size of our public key, in our examples $\ell = n$ and $s = 1$. We choose also the size of the projection, *i.e.* the integer $\mu < m$.
- We choose randomly a public cyclic or quasi-cyclic GRS code GRS_k of parameters $[n,k,d]$. We compute a public binary image C_{bin} of GRS_k.
- The parameter k must be sufficiently large so that the public code C_{pub} has dimension ℓ (cf. subcode strategy). This step must be experimentally validated.

In the following table we give four proposals parameters (Table 1).

The first three proposals are based on a cyclic GRS code. They claim a security of 128 bit. One can notice that, since in this situation $\mu = \nu$, it is not necessary to apply Type 1 permutations. Indeed, they are already integrated in choice of projection matrices M_i. The last proposal is based on a quasi-cyclic GRS code of index 2. Clearly, this choice causes an increase in the size of the public key, but makes it possible to reach a security of 192 bits.

Table 1. Proposal parameters.

Parameters	Proposal 1	Proposal 2	Proposal 3	Proposal 4
m	12	12	13	14
ℓ	4095	4095	8191	5461
s	1	1	1	2
μ	9	10	11	11
k	2300	2300	6000	5000
t	897	897	1095	296
GRS code	[4095,2300,1796]	[4095,2300,1796]	[8191,6000,2192]	[10922,5000,5923]
GSS code	[36855,15315]	[40950,40950]	[90101,61618]	[120142,37234]
C_{pub}	[36855,4095]	[40950,4095]	[90101,8191]	[120142,5461]
N_{proj}	2^{107}	2^{119}	2^{142}	2^{307}
N_{per}	2^{559}	2^{638}	2^{797}	2^{1029}
Decoding	2^{154}	2^{137}	2^{151}	2^{201}
Key size	4 Kbytes	4.5 Kbytes	10 Kbytes	14 Kbytes
Security	128 bits	128 bits	128 bits	192 bits

8   Comparison with Other Proposals

For the same level of security (at least 128 bits), in the original McEliece cryptosystem one can use a binary Goppa code of parameters [1750,1520,91] which leads to a public key of size 42,6 KB.

The following Tables compare the size of our proposal with those of the main unbroken code-based PKEs with a level of security of 128 bits and those of NIST's second round [27] (Table 2).

Table 2. Comparison with other McEliece like PKE, security:128 bits

McEliece like PKE	public key	secrete key
Proposal 1	4 Kbytes	0.31 Kbytes
Proposal 2	4.5 Kbytes	0.34 Kbytes
Proposal 3	10 Kbytes	0.41 Kbytes
Gabidulin matrix codes [5]	11.3 Kbytes	–
Hiden Gabidulin codes [20]	11.5 Kbytes	–
Classical Goppa Codes [21]	42.6 Kbytes	–
HQC-128-1 [27]	6.17 Kbytes	0.25 Kbytes
RQC-I [27]	8.53 Kbytes	0,04 Kbytes
LEDAcrypt1 [27]	4.49 Kbytes	0.47 Kbytes
LEDAcrypt2 [27]	6.52 Kbytes	0.47 Kbytes
BIKE-2 [27]	2.54 Kbytes	0.25 Kbytes

In the Post Quantum Cryptography Competition [26], most of code based submissions are in fact Key Encapsulation mechanisms (KEM) and not PKE. To our knowledge, the only remaining candidate is LEDAcrypt PKE [1], which is based on LDPC (Low Density Parity Check Codes). On that kind of code, there is always a non-negligible probability of non-decryption. The initial submitted parameters with the same level of security lead to a size of public key between 3.4 Kbyte and 6.2 Kbyte, which is close to our proposals.

Conclusion

In this paper, we designed code-based cryptosystems using a subcode of generalized subspace subcodes of structured RS codes. We give some instantiations using subspace subcodes of structured GRS codes. In our scheme the underlying Reed-Solomon code is not secret, so the classical attacks such as square code attacks have no more purpose. Our cryptosystems are secure against some specific attacks as: folding attack and the reconstruction of induced quasi-cyclic GSS codes. In addition, the security of our system relies on the *Equivalence Subcodes Problem* which is proved NP-complete in [6]. Furthermore for well chosen parameters our cryptosystems seem to be secure against the recent best known algebraic attacks. By using the subspace technique, they have a good key size with security parameters compared to other code-based PKEs.

References

1. Baldi, M., Barenghi, A., Chiaraluce, F., Pelosi, G., Santini, P.: LEDAcrypt: Low-density parity-check code-based cryptographic systems, 30 March 2019. https://www.ledacrypt.org/LEDAcrypt/
2. Barelli, E.: On the security of some compact keys for McEliece scheme. CoRR abs/1803.05289 (2018). http://arxiv.org/abs/1803.05289
3. Berger, T.P., El Amrani, N.: Codes over $\mathcal{L}(GF(2)^m, GF(2)^m)$, MDS diffusion matrices and cryptographic applications. In: El Hajji, S., Nitaj, A., Carlet, C., Souidi, E.M. (eds.) C2SI 2015. LNCS, vol. 9084, pp. 197–214. Springer, Cham (2015). https://doi.org/10.1007/978-3-319-18681-8_16
4. Berger, T.P., Cayrel, P.-L., Gaborit, P., Otmani, A.: Reducing key length of the McEliece cryptosystem. In: Preneel, B. (ed.) AFRICACRYPT 2009. LNCS, vol. 5580, pp. 77–97. Springer, Heidelberg (2009). https://doi.org/10.1007/978-3-642-02384-2_6
5. Berger, T.P., Gaborit, P., Ruatta, O.: Gabidulin matrix codes and their application to small ciphertext size cryptosystems. In: Patra, A., Smart, N.P. (eds.) INDOCRYPT 2017. LNCS, vol. 10698, pp. 247–266. Springer, Cham (2017). https://doi.org/10.1007/978-3-319-71667-1_13
6. Berger, T.P., Gueye, C.T., Klamti, J.B.: A NP-complete problem in coding theory with application to code based cryptography. In: El Hajji, S., Nitaj, A., Souidi, E.M. (eds.) C2SI 2017. LNCS, vol. 10194, pp. 230–237. Springer, Cham (2017). https://doi.org/10.1007/978-3-319-55589-8_15

7. Berger, T.P., Gueye, C.T., Klamti, J.B.: Generalized subspace subcodes with application in cryptology. to appear in IEEE Transactions on Information Theory, Online ISSN: 1557–9654, pp. 1–17 (2019). https://doi.org/10.1109/TIT.2019.2909872

8. Berlekamp, E.R., McEliece, R.J., van Tilborg, H.C.A.: On the inherent intractability of certain coding problems. IEEE Trans. Inf. Theory **24**(3), 384–386 (1978)

9. Bernstein, D.J., Buchmann, J., Dahmen, E.: Post Quantum Cryptography. Springer-Verlag, Heidleberg (2008). https://doi.org/10.1007/978-3-540-88702-7

10. Bernstein, D.J., Lange, T., Peters, C.: Attacking and defending the McEliece cryptosystem. In: Buchmann, J., Ding, J. (eds.) PQCrypto 2008. LNCS, vol. 5299, pp. 31–46. Springer, Heidelberg (2008). https://doi.org/10.1007/978-3-540-88403-3_3

11. Bosma, W., Cannon, J., Playoust, C.: The Magma algebra system. I. The user language. J. Symbolic Comput. **24**, 235–265 (1997). Computational algebra and number theory

12. Couvreur, A., Gaborit, P., Gauthier-Umaña, V., Otmani, A., Tillich, J.P.: Distinguisher-based attacks on public-key cryptosystems using Reed-Solomon codes. Des. Codes Crypt. **73**(2), 641–666 (2014)

13. Finiasz, M., Sendrier, N.: Security bounds for the design of code-based cryptosystems. In: Matsui, M. (ed.) ASIACRYPT 2009. LNCS, vol. 5912, pp. 88–105. Springer, Heidelberg (2009). https://doi.org/10.1007/978-3-642-10366-7_6

14. Faugère, J., Otmani, A., Perret, L., de Portzamparc, F., Tillich, J.: Structural cryptanalysis of McEliece schemes with compact keys. Des. Codes Crypt. **79**(1), 87–112 (2016)

15. Faugère, J.-C., Otmani, A., Perret, L., Tillich, J.-P.: Algebraic cryptanalysis of McEliece variants with compact keys. In: Gilbert, H. (ed.) EUROCRYPT 2010. LNCS, vol. 6110, pp. 279–298. Springer, Heidelberg (2010). https://doi.org/10.1007/978-3-642-13190-5_14

16. Gabidulin, E.M., Paramonov, A.V., Tretjakov, O.V.: Ideals over a noncommutative ring and their application in cryptology. In: Davies, D.W. (ed.) EUROCRYPT 1991. LNCS, vol. 547, pp. 482–489. Springer, Heidelberg (1991). https://doi.org/10.1007/3-540-46416-6_41

17. Hattori, M., McEliece, R.J., Lin, W.: Subspace subcodes of Reed-Solomon codes. In: Proceedings of IEEE International Symposium on Information Theory 1994, p. 430. IEEE (1994)

18. Huffman, W.C.: Groups and codes. In: Pless, V.S., Huffman, W.C. (eds.) Handbook of Coding Theory. Elsevier, Amsterdam (1998)

19. Karan K. and Rosenthal J. and Weger V.: Encryption scheme based on expanded reed-solomon codes. arXiv:1906.00745

20. Loidreau, P.: A new rank metric codes based encryption scheme. In: Lange, T., Takagi, T. (eds.) PQCrypto 2017. LNCS, vol. 10346, pp. 3–17. Springer, Cham (2017). https://doi.org/10.1007/978-3-319-59879-6_1

21. McEliece, R.: A public-key cryptosystem based on algebraic coding theory. DSN Prog. Rep., Jet Prop. Lab., California Institute of Technology, Pasadena, CA, Rep. 44, pp. 114–116, January 1978

22. May, A., Meurer, A., Thomae, E.: Decoding random linear codes in $\tilde{\mathcal{O}}(2^{0.054n})$. In: Lee, D.H., Wang, X. (eds.) ASIACRYPT 2011. LNCS, vol. 7073, pp. 107–124. Springer, Heidelberg (2011). https://doi.org/10.1007/978-3-642-25385-0_6

23. May, A., Ozerov, I.: On computing nearest neighbors with applications to decoding of binary linear codes. In: Oswald, E., Fischlin, M. (eds.) EUROCRYPT 2015. LNCS, vol. 9056, pp. 203–228. Springer, Heidelberg (2015). https://doi.org/10.1007/978-3-662-46800-5_9

24. Misoczki, R., Barreto, P.S.L.M.: Compact McEliece keys from Goppa codes. In: Jacobson, M.J., Rijmen, V., Safavi-Naini, R. (eds.) SAC 2009. LNCS, vol. 5867, pp. 376–392. Springer, Heidelberg (2009). https://doi.org/10.1007/978-3-642-05445-7_24
25. Niederreiter, H.: Knapsack-type cryptosystems and algebraic coding theory. Prob. Control Inf. Theory **15**, 159–166 (1986)
26. NIST: Post-quantum crypto project, December 2016. http://csrc.nist.gov/groups/ST/post-quantum-crypto/
27. NIST: Round 2 Submissions. https://csrc.nist.gov/projects/post-quantum-cryptography/round-2-submissions, 14 August 2019
28. Sendrier, N.: Finding the permutation between equivalent codes: the support splitting algorithm. IEEE Trans. Inf. Theory **46**(4), 1193–1203 (2000)
29. Sendrier, N.: QC-MDPC-McEliece: a public-key code-based encryption scheme based on quasi-cyclic moderate density parity check codes. In: Workshop "Post-Quantum Cryptography: Recent Results and Trends". Fukuoka, Japan, November 2014
30. Shor, P.W.: Algorithms for quantum computation: discrete logarithms and factoring. In: 35th Annual Symposium on Foundations of Computer Science, Santa Fe, New Mexico, USA, 20–22 November 1994, pp. 124–134. IEEE Computer Society (1994)
31. Wieschebrink, C.: Two NP-complete problems in coding theory with an application in code based cryptography. In: Proceedings of IEEE ISIT 06, pp. 1733–1737. IEEE, Seattle (2006)

Quantum Resistant Public Key Encryption Scheme polarRLCE

Jingang Liu[1(✉)], Yongge Wang[2], Zongxinag Yi[1], and Dingyi Pei[1]

[1] Guangzhou University, Guangzhou, China
xy07liu@126.com
[2] UNC Charlotte, Charlotte, USA
yonwang@uncc.edu

Abstract. In order to reduce the key size of the original McEliece cryptosystem, several variants exploited the particular structure in the public key. Unfortunately, most of these variants are vulnerable to structural attacks because of the algebraic structure of the underlying codes. In this work, we propose the first efficient secure scheme based on polar codes (i.e., *polarRLCE*), which is inspired by RLCE scheme, a candidate for the NIST post-quantum cryptography standardization. We show that, with the proper choice of parameters, using polar codes, it is possible to design an encryption scheme to achieve the intended security level while keeping a reasonably small key size. In addition, possible attacks are outlined and the key size of several choices of parameters is compared to those of known schemes with the same security level. It is shown that our proposal has the apparent advantage to decrease the key size, especially on the high-security level.

Keywords: McEliece cryptosystem · Polar codes · Code-based cryptography · Post-quantum cryptography

1 Introduction

It is well known that several computation intensive tasks may be significantly accelerated through algorithms running on quantum computer, like Shor [39] and Grover [19] algorithm. While current cryptographic protocols, such as RSA and Diffie-Hellman are proven to be vulnerable under the quantum algorithms. This fact pushed cryptographic research to focus on post-quantum solutions, i.e., finding new primitives based on more well suited mathematical problems that may still be difficult to solve for a quantum computer. With this in mind, the National Institute of Standards and Technology (NIST) is now beginning to prepare for the transition into Post-Quantum Cryptography (PQC) and had launched a call for PQC standardization project [31], and this ongoing standardization has moved on to 2nd round thus far. Due to its inherent resistance

Part of this work was done when the first author visited the UNC Charlotte. Yongge Wang was supported in part by Qatar Foundation Grant NPRP8-2158-1-423.

© Springer Nature Switzerland AG 2019
C. T. Gueye et al. (Eds.): A2C 2019, CCIS 1133, pp. 114–128, 2019.
https://doi.org/10.1007/978-3-030-36237-9_7

to attacks by quantum computers, code-based cryptography is one of the main candidates for the PQC standardization call, alongside multivariate and lattice-based schemes.

Code-based cryptography is accepted as quantum computing resistant based on a hard coding theory problem: decoding a random linear code in some metric. Historically, the conservative and well-understood choices for code-based cryptography are the McEliece cryptosystem [25] and its dual variant by Niederreiter [30] using binary Goppa codes. However, they suffer from the disadvantage of having large public key sizes, in spite of the fast encryption and decryption operations. It is therefore of utmost importance to seek ways to reduce the key sizes for code-based cryptosystems while keeping their security level. After the original proposal of code based encryption scheme by McEliece [25] which was based on binary Goppa codes, several variants have been proposed using different codes that allow for smaller keys or more efficient encoding and decoding algorithms, e.g., algebraic geometric (AG) codes [24], Generalized Reed-Solomon (GRS) codes [6,44], low density parity check (LDPC) codes [26], Reed-Muller (RM) codes [40], Low-rank parity check (LRPC) codes [20], among others. Although the original McEliece cryptosystem remains secure, most of these variants have been successfully cryptanalyzed [4,11,12,27,38]. Despite their promising features, the alternative codes need to be handled carefully due to the too much structural.

It is noteworthy that Wang [45,46] proposed a random linear code based quantum resistant public key encryption scheme, refer to it as RLCE scheme. The idea of RLCE scheme is to use a distortion matrix that mixes some random columns with the structured ones. The advantage of the RLCE scheme is that its security does not depend on any specific structure of underlying linear codes, instead, it based on the $\mathcal{NP}$-hardness of decoding random linear codes. In such a manner, previous attacks regarding GRS codes based on the technique of filtration distinguisher are no longer work.

Polar codes, introduced by Arikan in [1], have received much attention since they are the first class of error-correcting codes that provably achieve capacity for any symmetric binary discrete memoryless channel (B-DMC) with very efficient encoding and decoding algorithm, whose time complexity scales as $\mathcal{O}(nlogn)$, where n is the length of the code. Because of the good performance and low complexity, polar codes have been adopted for use in future wireless communication systems (e.g., 5G cellular systems).

Related Work. Within this thread of research, there are two heuristic variants [23,42] of the McEliece cryptosystem based on polar codes. The first one [42] was broken by Bardet et al. [7] using the structure of the minimum weight codewords. They manage to solve the code equivalence problem for polar codes and thus completely break the scheme [42] based on polar codes. The second variant was presented by Hooshmand et al. [23] which suggested using the subcode of polar codes. However, we found that the proposal in [23] is useless in practice since 80% ciphertexts could not be decrypted, as we will discuss in Sect. 2.4.

Our Contribution. In this paper, we combine the idea of RLCE scheme by inserting random columns, then propose the first efficient secure scheme based on polar codes (i.e., *polarRLCE*), which can avoid the attack of [7]. Furthermore, possible attacks are outlined and the key size of several choices of parameters is compared to those of known schemes with the same security level. We show that the existing attacks on the proposal scheme do not seem to be effective. More importantly, our proposal has an apparent advantage to decrease the key size, especially on the high-security level. It allows us to reconsider polar codes as a good candidate for using in code-based cryptography.

The rest of this paper is organized as follows. Some necessary preliminaries such as notation and definitions are given in the next Section. In Sect. 3, we present the precise description of the construction of the polarRLCE scheme. Section 4 discusses the known cryptanalytic attacks against our proposal. Furthermore, we give the choice of suggested parameters and key size for the achievable security level. Finally, some concluding remarks are made in Sect. 5.

2   Preliminaries

In this section, we introduce some of the basic background information necessary to follow this paper. Throughout the paper, we will denote vectors by lower-case bold letters, e.g., $\mathbf{m}$. And denote matrices by upper-case bold letters, e.g., $\mathbf{A}$.

2.1   Coding Theory

Definition 1 (Linear codes). *An $[n, k]$ linear code $\mathcal{C}$ over a finite field $\mathbb{F}_q$ is a k-dimension linear subspace of $\mathbb{F}_q^n$.*

Definition 2 (Generator matrix). *A $k \times n$ matrix $\mathbf{G}$ with entries from $\mathbb{F}_q$ having row-span $\mathcal{C}$ is a generator matrix for the $[n, k]$ linear code $\mathcal{C}$.*

One can specify a linear code $\mathcal{C}$ via a generator matrix $\mathbf{G} \in \mathbb{F}_q^{k \times n}$ or a parity check matrix $\mathbf{H} \in \mathbb{F}_q^{(n-k) \times n}$ via

$$\mathcal{C} := \{\mathbf{xG} \in \mathbb{F}_q^n \mid \mathbf{x} \in \mathbb{F}_q^k\} \quad \text{or} \quad \mathcal{C} := \{\mathbf{c} \in \mathbb{F}_q^n \mid \mathbf{Hc}^T = 0\}$$

If $\mathbf{G} \in \mathbb{F}_q^{k \times n}$ or $\mathbf{H} \in \mathbb{F}_q^{(n-k) \times n}$, i.e., each matrix entry is chosen uniformly at random from $\mathbb{F}_q$, then we call $\mathcal{C}$ a random linear code.

We now only consider binary codes, i.e., $q = 2$. The hamming weight $w_H(\mathbf{x})$ of a binary vector in $\mathbf{x} \in \mathbb{F}_2^n$ is the number of nonzero entries in the vector. And the minimum hamming distance of the code $\mathcal{C}$ is defined as $d = min \{w_H(\mathbf{x-y})\}$.

Definition 3 (Punctured and shortened codes). *Given a* $[n, k]$ *linear code* C, *let subset* $I \subset \{1, \cdots, n\}$. *Then we define the punctured code* $\mathcal{P}_I(C)$ *and the shortened code* $\mathcal{S}_I(C)$ *as*

$$\mathcal{P}_I(C) = \{(\mathbf{c}_i)_{i \notin I} \mid \mathbf{c} \in C\},$$
$$\mathcal{S}_I(C) = \{(\mathbf{c}_i)_{i \notin I} \mid \exists\ \mathbf{c} \in C,\ s.t.\ \forall i \in I, \mathbf{c}_i = 0\}.$$

Given a subset I of the set of coordinates of a vector $\mathbf{x}$, we denote by $\mathcal{P}_I(\mathbf{x})$ the vector $\mathbf{x}$ punctured at I, that is to say, whose i-th entry has been deleted for any $i \in I$.

Lemma 1. *Let* C *be a code of dimension* k *and generator matrix* $\mathbf{G}$. *Then the matrix* $\mathbf{G}_\mathcal{P}$ *obtained from* $\mathbf{G}$ *by deleting the columns whose index in* I *is a generator matrix for* $\mathcal{P}_I(C)$.

2.2 McEliece's Public-Key Cryptosystem

The McEliece's scheme [25] can be described as follows. The private key consists of three matrices: $k \times n$ generator matrix $\mathbf{G}$ of a linear code which can correct up to t errors, a random $k \times k$ invertible matrix $\mathbf{S}$ and $n \times n$ random permutation matrix $\mathbf{P}$. The public key is $\mathbf{G}' = \mathbf{SGP}$. To encrypt the message $\mathbf{m} \in \mathbb{F}_2^k$, the sender chooses a random error $\mathbf{e} \in \mathbb{F}_2^n$ with $w_H(\mathbf{e}) \leq t$, then the corresponding ciphertext is computed as $\mathbf{c} = \mathbf{m}\mathbf{G}' + \mathbf{e}$. Finally, the legitimate receiver then recovers the message by applying efficient decoding algorithm $\mathcal{D}$ on $\mathbf{c}$.

2.3 Polar Codes Construction

We first recall the basic facts about polar codes. As shown in the seminal work by Arikan [1], for any B-DMC, there exists a polar code of block length $n = 2^m$ which is characterized by the information bit set $\mathcal{A}$ with exponentially small word-error rate under successive cancellation (SC) decoder. A polar code may be specified completely by $(n, k, \mathcal{F})$, where n is the length of a codeword in bits, k is the number of information bits encoded per codeword, and $\mathcal{F}$ is a set of $n - k$ integer indices called frozen bit locations from $\{0, 1, \cdots, n - 1\}$. The k more reliable subchannels (based on the polarization phenomenon) with indices in set $\mathcal{A}$ carry information bits and the rest subchannels included in the complementary set $\mathcal{A}^c$ (i.e., the set $\mathcal{F}$) can be set to fixed bit values, such as all zeros. Generally, the challenge is to select the information bits set $\mathcal{A}$, or more precisely, the methods that are proposed for finding the indices of good polarized channels.

For a binary polar code of length $n = 2^m$ bits, the polar encoding of an input vector is done by the polarization transformation matrix $\mathbf{G}_n = \mathbf{F}^{\otimes m}$ which is the m-th Kronecker power of the 2×2 kernel matrix

$$\mathbf{F} = \begin{pmatrix} 1 & 0 \\ 1 & 1 \end{pmatrix}.$$

For a given noise channel, the generator matrix $\mathbf{G}$ of a $[n, k]$ polar code is defined as the submatrix of $\mathbf{G}_n$ consisting of k rows with indices corresponding to information set $\mathcal{A} = \{i_1, i_2, \cdots, i_k\}$. Roughly speaking these rows are chosen in such a way that it gives good performances for the successive cancellation (SC) decoder. These codes come equipped with SC decoder whose decoding complexity scales as $\mathcal{O}(nlogn)$, for more details please refer to Arikan's seminal paper [1].

The idea of exploiting polar codes in cryptography came in a natural way since polar codes benefit of various interesting properties like: can achieve Shannon capacity for the class of binary discrete memoryless channels, attain better performance (lower decoding errors) because of the channel polarization along with the increased block length, posses efficient encoding and decoding procedures, etc. Even though polar codes are closely related to RM codes, the techniques used for the cryptanalysis of RM codes do not work on polar codes.

2.4   The Proposal by Hooshmand et al. [23]

The error-correcting capacity of polar codes [1] does not only depend on the code length but also on other factors like the code rate and the designed channel. However, the error-correcting capacity was merely set to be a fixed value of $t = 2\sqrt{n} - 1$ in the proposal by Hooshmand et al. [23], they didn't consider the error probability of decoding. For instance, they claimed that one can use [2048, 1750]-polar code with $t = 89$, which followed *Theorem 8* in [28]. Actually, the *Theorem 8* from [28] is only suitable for the concatenated polar codes with respect to the length of burst-errors as stated in [28]. Nevertheless, the proposal in [23] used random errors through the encryption process. With these parameters in [23], we performed numerical simulation using MATLAB R2018a where 10^5 decoding trails are exploited under SC decoder [1], the experiment result indicates that the decoding error probability is nearly 0.8, i.e., 80% ciphertexts could not be decrypted and cannot be employed in a practical environment. With respect to our proposal, the error probability is approximately 2^{-14} (see next section).

3   Our Proposed Scheme of polarRLCE

More precisely, the procedures of our polarRLCE are shown as follows,

Key generation: According to the construction of polar code in Sect. 2.3,
- Choose a $[n, k]$ polar code with the generator matrix $\mathbf{G}$ of length n and the dimension k.
- Generate w random column vectors $\mathbf{r}_1, \mathbf{r}_2, \cdots, \mathbf{r}_w$, and let

$$\mathbf{G}_1 = \left(\mathbf{g}_1, \cdots, \mathbf{g}_{n-w}, \mathbf{g}_{n-w+1}, \mathbf{r}_1, \cdots, \mathbf{g}_n, \mathbf{r}_w\right)$$

be the $k \times (n + w)$ matrix obtained by inserting w random $k \times 1$ column vectors $\mathbf{r}_i$ into matrix $\mathbf{G}$.

- To mix the columns, choose w random non-singular binary 2×2 matrices $\mathbf{A}_1, \mathbf{A}_2, \cdots, \mathbf{A}_w$. Denote $\mathbf{A}$ the $(n + w) \times (n + w)$ block-diagonal matrix

$$
\mathbf{A} = \begin{pmatrix} \mathbf{I}_{n-w} & & & (0) \\ & \mathbf{A}_1 & & \\ & & \ddots & \\ (0) & & & \mathbf{A}_w \end{pmatrix}
$$

- Let $\mathbf{S}$ be a randomly chosen $k \times k$ non-singular matrix and $\mathbf{P}$ be the $(n + w) \times (n + w)$ permutation matrix.
- The public key $k \times (n + w)$ matrix is defined as

$$
\mathbf{G}_{pub} = \mathbf{S}\mathbf{G}_1\mathbf{A}\mathbf{P}
$$

Then the public key and private key are given respectively by

$$
\mathbf{G}_{pub} \text{ and } (\mathbf{G}, \mathbf{S}, \mathbf{P}, \mathbf{A}).
$$

Encryption: Let $\mathbf{m} \in \mathbb{F}_2^k$ be the message to be encrypted, we randomly generate error vector $\mathbf{e} \in \mathbb{F}_2^{n+w}$ such that the hamming weight $w_H(\mathbf{e}) \leq t$. Compute the corresponding ciphertext

$$
\mathbf{c} = \mathbf{m}\mathbf{G}_{pub} + \mathbf{e}.
$$

Decryption: To decrypt the received ciphertext $\mathbf{c}$,
- Calculate $\mathbf{c}\mathbf{P}^{-1}\mathbf{A}^{-1} = (c_1', c_2', \cdots, c_{n+w}')$.
- Delete w entries at the $\mathbf{r}_i$ position of row vector $(c_1', c_2', \cdots, c_{n+w}')$. We denote the obtained n-length vector by

$$
\mathbf{c}' = \left(c_1', c_2', \cdots, c_{n-w+1}', c_{n-w+3}', c_{n-w+5}', \cdots, c_{n+w-1}' \right).
$$

- It is easy to check that $\mathbf{c}' = \mathbf{m}\mathbf{S}\mathbf{G} + \mathbf{e}'$ for some error vector $\mathbf{e}' \in \mathbb{F}_2^n$, where $w_H(\mathbf{e}') \leq t$. Then using the efficient decoding algorithm, one can recover the corresponding message $\mathbf{m}$.

For the purpose of constructing polar code used in our proposed variant scheme, we consider here the binary symmetric channel (BSC) with crossover probability $\epsilon = 0.05$. For instance, to achieve 128-bit security, for reliable decoding and keeping reasonably small key size with enough security level, we will set the choice of parameters such that $n = 2^{11}$, $k = 500$ and $w = 50$. Following the method of [15], validated through exhaustive simulation, we can choose the error weight of $t = 285$ with the reasonable error probability is approximately 2^{-14}.

Remark 1. Please note that our scheme allows occasional decryption failures for valid ciphertexts (similar to some NIST PQC submissions), which is inherited from the decoding algorithm. However, for the good performance of polar codes, one can easily resolve this issue through repeated encryption as presented by Eaton et al [17] that can reduce the decryption failure rate to a level negligible in the security parameter, without altering the whole parameters.

4   Security Analysis

In this section we will discuss several best known possible attacks against our proposed polarRLCE scheme in Sect. 3. There are two main attacks to thwart, i.e. key structural attack and decoding attack.

Furthermore, if the code $\mathcal{C}$, whose generator matrix is used as a part of the public key, could be distinguished, then an adversary could exploit the structure of $\mathcal{C}$, and this would also possibly allow the adversary to develop faster attacking algorithms. Indeed, most of these variations of McEliece system are vulnerable to structural attacks because of the algebraic structure of underlying codes.

4.1   Brute Force Attack

A brute force attack is a trial-and-error method used to obtain the correct keys. For our proposed scheme, recall that the private key $(\mathbf{G}, \mathbf{S}, \mathbf{P}, \mathbf{A})$ is obtained randomly. Moreover, the number of candidates invertible scrambling matrices $\mathbf{S}_{k \times k}$ over $\mathbb{F}_2$ is

$$\mathcal{N} = \prod_{i=1}^{k} (2^k - 2^{i-1}) = 2^{k^2} \prod_{i=1}^{k} (1 - 2^i) > 2^{k^2 - 2}.$$

By putting in the suggested parameters (with 128-bit security) $n = 2048$, $k = 500$ and $w = 50$, it turns out that it is as well infeasible to retrieve the other three elements building the private key just by guessing, since there exist $(n + w)! = 2048! \gg 2^{128}$ different matrices $\mathbf{P}$ and nearly $\mathcal{N} = 2^{500^2} \gg 2^{128}$ choices for $\mathbf{S}$. Moreover, the candidate of block-diagonal matrix $\mathbf{A}$ is $6^w = 2^{129.25}$. Hence, the complexity of the exhaustive search attack against our scheme has an exponential time, which indicates this attack is impractical.

4.2   Square Attack

In this section, we study the square attack on our polarRLCE. There has been an increased interest in the square (Schur product) of linear codes in the last years (cf. [13]). Another and more recent application of the Schur product concerns cryptanalysis of code-based public key cryptosystems. In this context, the Schur product is a very powerful operation which can help to distinguish secret codes from random ones.

In fact, the method of inserting random columns or rows in the secret matrix have indeed proposed [21,44] to avoid structural attacks on similar versions of the McEliece cryptosystem. Although this proposal had effectively avoided the original attack, recent studies [11,33] have shown that in the case of GRS codes or RM codes, the random columns can be found through considerations of the dimension of the Square code.

Definition 4 (Schur product).   *Let* $\mathbf{x}, \mathbf{y} \in \mathbb{F}_2^n$, *the Schur product of two vectors is denoted by*

$$\mathbf{x} * \mathbf{y} = (x_1 y_1, x_2 y_2, \cdots, x_n y_n)$$

Definition 5 (Square code). *Let $\mathcal{A}$, $\mathcal{B}$ be codes of length n. The Schur product of two codes is the vector space spanned by all products $\mathbf{a} * \mathbf{b}$ with $\mathbf{a} \in \mathcal{A}$ and $\mathbf{b} \in \mathcal{B}$ respectively.*

$$\langle \mathcal{A} * \mathcal{B} \rangle = \langle \{ \mathbf{a} * \mathbf{b} \mid \mathbf{a} \in \mathcal{A} \text{ and } \mathbf{b} \in \mathcal{B} \} \rangle$$

If $\mathcal{A} = \mathcal{B}$, then we call $\langle \mathcal{A} * \mathcal{A} \rangle$ the square code of $\mathcal{A}$ and denote it by $\langle \mathcal{A}^2 \rangle$. The impact of square code on code-based cryptosystem becomes clear when we study the dimension of these constructions.

Definition 6 (Schur matrix). *Let $\mathbf{G}$ be a $k \times n$ matrix, with rows $(g_i)_{1 \leq i \leq k}$. The Schur matrix of $\mathbf{G}$, denoted by $S(\mathbf{G})$ consists of the rows $g_i * g_j$ for $1 \leq i \leq j \leq k$.*

we observe that if $\mathbf{G}$ is a generator matrix of a code $\mathcal{C}$ then its Schur matrix $S(\mathbf{G})$ is a generator matrix(at least subjected to $S(\mathbf{G})$) of the square code of $\mathcal{C}$. For the $k \times n$ matrix $\mathbf{G}$, we note that $S(\mathbf{G})$ at most has the size $\binom{k+1}{2} \times n$.

It is well-known that the square of a linear code $\mathcal{C}[n, k]$ has the dimension

$$dim(\mathcal{C}^2) \leq min\{n, \frac{1}{2}k(k+1)\} \tag{1}$$

and a random linear code attains this upper bound with high probability.

Looking at the definition of the square code, we observe that it is generated by all possible products of two elements of a basis of the code. Therefore, it is natural to expect that the dimension of the square code is "as large as possible". In other words, for a randomly chosen linear code $\mathcal{R}$, we expect that the inequality (1) is actually an equality with very high probability.

Let us consider the recent work [16] which reported that it might possibly exist a heuristic distinguisher, if given two specific weakly decreasing sets. However, in the case of our polarRLCE scheme, such sets couldn't be found easily because of the extended public codes by inserting random columns.

To illustrate the square attack, we performed simulation by generating 10000 random sets of the public key matrix. Our experiment result shows that, as in the case of the proposed polarRLCE scheme, the square code of the public code can always reach the maximal dimension bound. Considering the choice of parameters (with 128-bit security) such that $n = 2048$, $k = 500$ and $w = 50$. So we can obtain the $k \times (n + w)$ public key matrix $\mathbf{G_{pub}}$. Denote the extended public code as $\mathcal{C}_{pub}$. Hence from inequality (1), we have

$$dim(\mathcal{C}_{pub}^2) = dim\big(S(\mathbf{G_{pub}})\big) \leq min\{n + w, \frac{1}{2}k(k+1)\}.$$

For the proposed parameters, we observed experimentally that the dimension of the public matrix by square product always reach maximum, that is to say,

$$dim\big(S(\mathbf{G_{pub}})\big) = n + w = 2098.$$

Furthermore, after conducted random puncturing operations, $\mathcal{P}_I(\mathbf{G}_{pub})$, alternatively, we can obtain

$$dim\big(S(\mathcal{P}_I(\mathbf{G}_{pub}))\big) = n + w - |I|.$$

On the basis of the observations made as stated above, we claim that the technique of square attack regarding our polarRLCE could not be used to distinguish from random codes.

4.3   Key-Recovery Attack

The key-recovery attack is one of the important ways of structural attack, consists in recovering the private key from the public key. In this case, the methods are specific to the code family. In order to compute the private key of a given public key, it is often reduced to solve the code equivalence problem:

Definition 7. *Let* $\mathbf{G}$ *and* $\mathbf{G}^*$ *be the generating matrices for two* $[n, k]$ *binary linear codes. Given* $\mathbf{G}$ *and* $\mathbf{G}^*$ *does there exist a* $k \times k$ *binary invertible matrix* $\mathbf{S}$ *and* $n \times n$ *permutation matrix* $\mathbf{P}$ *such that* $\mathbf{G}^* = \mathbf{SGP}$?

This problem was first studied by Petrank et al. over the binary field. And the most common algorithm used to solve this problem is the Support Splitting Algorithm (SSA) [41]. SSA is very efficient in the random case, but cannot be used in the case of codes with large hulls or codes with large permutation group such as Goppa codes, polar codes, etc.

However, a very effective structural attack on the variant [42] using polar code was introduced by Bardet et al. in [7]. Firstly, they managed to determine exactly the structure of the minimum weight codewords of the original polar codes. Then solved the code equivalence problem for polar codes with respect to decreasing monomial codes. Notice that this attack is very specific to the simple usage of polar codes in [42].

Regarding our proposal, there is a really effective way of protecting the scheme since the structure of the private code is someway shattered by inserting random elements. Thus, even though one can find enough low weight codewords, while they are not subject to the original polar code because of these public codes hold extended length $n + w$ which generated by the public key $\mathbf{G}_{pub}$. The natural way is to perform puncturing operations, but an exponential number of codewords since there are

$$\binom{n + w}{w} = \binom{2098}{50} = 2^{336.68}.$$

On the other hand, the adversary cannot identify the inserted positions by distinguishing attack or square attack as stated on Sect. 4.2. So the code equivalence problem becomes even more complicated to solve in this case. Therefore, the attack by [7,16] does not apply directly to our proposed polarRLCE scheme.

Finally, we notice that very recently, Couvreur et al. [14] presented a key-recovery attack on RLCE scheme (only partially broken). They showed that it is possible to distinguish some keys from random codes by computing the square of some shortened public code. The set of positions $\{1, \cdots, n + w\}$ can be splitted into four parts, based on the fact that the entry of any GRS codeword satisfies a specific expression formalization, and then recognize the twin positions. While polar codes out of this situation because of the different structure between polar codes and GRS codes.

According to the aforementioned analysis, and the fact that we found no other distinguishing methods for our proposal, we claim that it is indeed able to avoid the key-recovery attack.

4.4 Message Decoding Attack

Message decoding attack is an important issue in code based cryptography. One possibility attack to recover the message is information set decoding (ISD), which means to decode a random linear codes without exploiting any structural property of the code. The ISD algorithm searches for an information set such that the error positions are all out of the information set. The work factor of ISD clearly increases with the number of errors added in the encryption process. Thus, when choosing parameters, we will focus mainly on defeating attacks of the ISD family.

This technique was first introduced by Prange [34]. Hereafter, numerous different algorithmic techniques have been explored to improve complexity of ISD algorithm. Among several variants [18,29,37,43] and generalizations, it is noteworthy that most modern ISD algorithms are based on Stern's [37] algorithm, which incorporates collision search methods to speed-up decoding.

So we will move on to analyze the complexity of Stern's algorithm. Similarly, they try to find a t-weight codeword in a $[n, k]$ linear code $\mathcal{C}$ generated by $\mathbb{F}_2^{k \times n}$. More precisely, apart from the generator matrix $\mathbf{G}$, the algorithm takes as input additional integer parameters p, l such that $0 \leq p \leq t$ and $0 \leq l \leq n - k$. We refer to the improved version of ISD attack algorithm in [5,35], which is a generalization of Stern's algorithm [37]. And they pointed out that for small fields (e.g. in our case, $\mathbb{F}_2$), choosing the new algorithm parameters c and r ($1 < r \leq c$) should be taken into account, which can yield good speedups on the Gaussian elimination cost of each iteration.

For a practical evaluation of the ISD running times and corresponding security level, similarly, most of the NIST PQCrypto code-based submissions exploited this complexity computation tool [35] to determine the security level of their proposals, we also use this tool to indicate the security level of our implementation. Note that the ciphertext length should be $n + w$ instead of n in our case. According to this computation tool, we test different input parameters to classify expected bit security level $\kappa := 128, 192, 256$ respectively (See Sect. 4.5 for more details).

4.5 Suggested Parameters for PolarRLCE

In this section, we give the suggested parameters in Table 1 for our polarRLCE scheme, with the standard security levels, 128-bit, 192-bit and 256-bit. Besides, a comparison of the public key size for our suggested parameters with RLCE [46] (the secure second group parameters) and the original McEliece [5] scheme is given in Table 2, together with the state-of-the-art NTS-KEM [2] and Classic-McEliece [9] which are moving on to the 2nd round of the NIST PQC Standardization process.

For convenience, we introduce the following notations of each column list in the tables:

- κ : security level.
- w : the number of inserted columns.
- $[n, k, t]$: code length n, code dimension k and t is the error correcting ability.
- $\mathcal{PK}$: public-key size in KBytes.
- $\mathcal{W}_{ISD}$: the work factor of ISD attack algorithm.

Table 1. Set of parameters for polarRLCE scheme

κ	$[n, k, t]$	w	$\mathcal{PK}$	$\mathcal{W}_{ISD}$
128	$[2^{11}, 500, 285]$	50	97.53	130.62
192	$[2^{12}, 585, 760]$	75	256.08	193.84
256	$[2^{12}, 960, 610]$	100	379.22	257.35

Table 2. Public-key size comparison (in KBytes)

κ	Our	McEliece	RLCE	NTS-KEM	Classic-McEliece
128	97.53	187.69	187.53	312	255
192	256.08	489.4	446.36	907.97	511.88
256	379.22	936.02	1212.86	1386.43	1326

We remark that the parameters given in Table 1 may be vulnerable to the attack under quantum random oracle model ($QROM$). Here we present the parameters solely to illustrate the rationality our construction which, to our best knowledge, are secure against current known attacks.

From Table 2, we can see that our scheme reduce the public key size of the original McEliece scheme by at least 52%. It has the apparent advantage to decrease the key size, especially on the high-security level. However, compared to the candidates based on Hamming (Rank) quasi-cyclic (QC) codes, the public key size of our proposal is inferior to them. Nevertheless, a new type of statistical analysis, called reaction attacks [22,32], are threatening these scheme with specific QC structure of the underlying codes [3,10]. As a final remark, it would be required to consider the impact of reaction attacks even without QC structure in our proposal.

5   Conclusion

To summarize, we have proposed a new variant of code-based encryption scheme by exploring polar codes and benefits the lower encoding and decoding complexity. We show that, for the proper choice of parameters together with the state-of-the-art cryptanalysis, it is still possible to design secure schemes to achieve the intended security level while keeping a reasonably small key size, using polar code.

Despite the promising features, the semantically secure is not considered in this work yet; i.e., it does not resistant to adaptively chosen-ciphertext attacks (*IND-CCA2*). For instance, encryption of the same message twice outputs two different ciphertexts which can be compared to find out the original message since it is highly unlikely that errors were added in the same positions. However, it is possible to produce *IND-CCA2* security via the technique of key encapsulation mechanism (KEM). Very recently, [8,36] present some generic construction to obtain *IND-CCA2* security, it's presumably exploited in our scheme. We leave this line of inquiry for future research since it falls outside the scope of this paper.

Acknowledgments. We would like to thank Dr. Vlad Dragoi for insightful discussions. We are also grateful to the anonymous reviewers of A2C 2019 for their valuable feedback.

References

1. Arikan, E.: Channel polarization: a method for constructing capacity achieving codes for symmetric binary-input memoryless channels. IEEE Trans. Inf. Theory **55**(7), 3051–3073 (2009)
2. Albrecht, M., Cid, C., Paterson, K.G., Tjhai, C.J., Tomlinson, M.: NTS-KEM. https://nts-kem.io/. Accessed Aug 2019
3. Aragon, N., Barreto, P., Bettaieb, S., Bidoux, L., Blazy, O., Deneuville, J.C., et al.: BIKE: bit flipping key encapsulation. https://bikesuite.org/. Accessed Aug 2019
4. Baldi, M., Chiaraluce, G.F.: Cryptanalysis of a new instance of McEliece cryptosystem based on QC-LDPC codes. In: IEEE International Symposium on Information Theory - ISIT 2007, pp. 2591–2595. Nice, France, March 2007
5. Bernstein, D.J., Lange, T., Peters, C.: Attacking and defending the McEliece cryptosystem. In: Buchmann, J., Ding, J. (eds.) PQCrypto 2008. LNCS, vol. 5299, pp. 31–46. Springer, Heidelberg (2008). https://doi.org/10.1007/978-3-540-88403-3_3
6. Baldi, M., Bianchi, M., Chiaraluce, F., Rosenthal, J., Schipani, D.: Enhanced public key security for the McEliece cryptosystem. J. Cryptology **29**(1), 1–27 (2016)
7. Bardet, M., Chaulet, J., Dragoi, V., Otmani, A., Tillich, J.-P.: Cryptanalysis of the McEliece public key cryptosystem based on polar codes. In: Takagi, T. (ed.) PQCrypto 2016. LNCS, vol. 9606, pp. 118–143. Springer, Cham (2016). https://doi.org/10.1007/978-3-319-29360-8_9
8. Bernstein, D.J., Persichetti, E.: Towards KEM unification. IACR Cryptology ePrint Archive, Report 2018/526 (2018)
9. Bernstein, D.J., Chou, T., Lange, T., von Maurich, I., Misoczki, R., et al.: Classic McEliece. https://classic.mceliece.org/. Accessed Aug 2019

10. Baldi M., Barenghi A., Chiaraluce F., Pelosi G., Santini P.: LEDAcrypt. https://www.ledacrypt.org/. Accessed Aug 2019
11. Couvreur, A., Gaborit, P., Gauthier-Umaa, V., Otmani, A., Tillich, J.P.: Distinguisher based attacks on public-key cryptosystems using Reed-Solomon codes. Des. Codes Crypt. **73**(2), 641–666 (2014)
12. Couvreur, A., Marquez, C.I., Pellikaan, R.: A polynomial time attack against algebraic geometry code based public key cryptosystems. In: Proceedings of IEEE International Symposium on Information Theory - ISIT, vol. 2014, pp. 1446–1450 (2014)
13. Cascudo, I., Cramer, R., Mirandola, D., Zmor, G.: Squares of random linear codes. IEEE Trans. Inf. Theory **61**(3), 1159–1173 (2015)
14. Couvreur, A., Lequesne, M., Tillich, J.-P.: Recovering short secret keys of RLCE in polynomial time. In: Ding, J., Steinwandt, R. (eds.) PQCrypto 2019. LNCS, vol. 11505, pp. 133–152. Springer, Cham (2019). https://doi.org/10.1007/978-3-030-25510-7_8
15. Dragoi, V.: Algebraic approach for the study of algorithmic problems coming from cryptography and the theory of error correcting codes. Ph.D. thesis, University of Rouen, France, July 2017
16. Drăgoi, V., Beiu, V., Bucerzan, D.: Vulnerabilities of the McEliece variants based on polar codes. In: Lanet, J.-L., Toma, C. (eds.) SECITC 2018. LNCS, vol. 11359, pp. 376–390. Springer, Cham (2019). https://doi.org/10.1007/978-3-030-12942-2_29
17. Eaton, E., Lequesne, M., Parent, A., Sendrier, N.: QC-MDPC: a timing attack and a CCA2 KEM. In: Lange, T., Steinwandt, R. (eds.) PQCrypto 2018. LNCS, vol. 10786, pp. 47–76. Springer, Cham (2018). https://doi.org/10.1007/978-3-319-79063-3_3
18. Finiasz, M., Sendrier, N.: Security bounds for the design of code-based cryptosystems. In: Matsui, M. (ed.) ASIACRYPT 2009. LNCS, vol. 5912, pp. 88–105. Springer, Heidelberg (2009). https://doi.org/10.1007/978-3-642-10366-7_6
19. Grover, L.K.: A fast quantum mechanical algorithm for database search. In: Proceedings of the 28th Annual ACM Symposium on Theory of Computing, pp. 212–219. ACM press, May 1996
20. Gaborit, P., Murat, G., Ruatta, O., Zemor, G.: Low rank parity check codes and their application in cryptography. In: The Proceedings of Workshop on Coding and Cryptography, WCC 2013, Borgen, Norway, pp. 167–179 (2013)
21. Gueye, C.T., Mboup, E.H.M.: Secure cryptographic scheme based on modified Reed Muller codes. Int. J. Secur. Appl. **7**(3), 55–64 (2013)
22. Guo, Q., Johansson, T., Stankovski, P.: A key recovery attack on MDPC with CCA security using decoding errors. In: Cheon, J.H., Takagi, T. (eds.) ASIACRYPT 2016. LNCS, vol. 10031, pp. 789–815. Springer, Heidelberg (2016). https://doi.org/10.1007/978-3-662-53887-6_29
23. Hooshmand, R., Shooshtari, M.K., Eghlidos, T., Aref, M.R.: Reducing the key length of McEliece cryptosystem using polar codes. In: 2014 11th International ISC Conference on Information Security and Cryptology (ISCISC), pp. 104–108. IEEE (2014)
24. Janwa, H., Moreno, O.: McEliece public key cryptosystems using algebraic-geometric codes. Des. Codes Crypt. **8**(3), 293–307 (1996)
25. McEliece, R.J.: A public-key cryptosystem based on algebraic coding theory. Jet Propulsion Laboratory DSN Progress Report, 42–44, pp. 114–116 (1978)

26. Monico, C., Rosenthal, J., Shokrollahi, A.: Using low density parity check codes in the McEliece cryptosystem. In: Proceedings of IEEE International Symposium on Information Theory - ISIT 2000, Sorrento, Italy, p. 215, June 2000
27. Minder, L., Shokrollahi, A.: Cryptanalysis of the Sidelnikov cryptosystem. In: Naor, M. (ed.) EUROCRYPT 2007. LNCS, vol. 4515, pp. 347–360. Springer, Heidelberg (2007). https://doi.org/10.1007/978-3-540-72540-4_20
28. Mahdavifar, H., El-Khamy, M., Lee, J., Kang, I.: Performance limits and practical decoding of interleaved Reed-Solomon polar concatenated codes. IEEE Trans. Commun. $62(5)$, 1406–1417 (2014)
29. May, A., Ozerov, I.: On computing nearest neighbors with applications to decoding of binary linear codes. In: Oswald, E., Fischlin, M. (eds.) EUROCRYPT 2015. LNCS, vol. 9056, pp. 203–228. Springer, Heidelberg (2015). https://doi.org/10.1007/978-3-662-46800-5_9
30. Niederreiten, H.: Knapsack-type cryptosystems and algebraic coding theory. Prob. Control Inform. Theory $15(2)$, 159–166 (1986)
31. NIST: Post quantum crypto project (2017). http://csrc.nist.gov/groups/ST/post-quantum-crypto/. Accessed 19 May 2017
32. Nilsson, A., Johansson, T., Wagner, P.S.: Error amplification in code-based cryptography. IACR Trans. Cryptographic Hardware Embed. Syst. $2019(1)$, 238–258 (2019)
33. Otmani, A., Kalachi, H.T.: Square code attack on a modified Sidelnikov cryptosystem. In: El Hajji, S., Nitaj, A., Carlet, C., Souidi, E.M. (eds.) C2SI 2015. LNCS, vol. 9084, pp. 173–183. Springer, Cham (2015). https://doi.org/10.1007/978-3-319-18681-8_14
34. Prange, E.: The use of information sets in decoding cyclic codes. IRE Trans. Inf. Theory $8(5)$, 5–9 (1962)
35. Peters, C.: Information-set decoding for linear codes over $\mathbb{F}_q$. In: Sendrier, N. (ed.) PQCrypto 2010. LNCS, vol. 6061, pp. 81–94. Springer, Heidelberg (2010). https://doi.org/10.1007/978-3-642-12929-2_7
36. Persichetti, E.: On the CCA2 security of McEliece in the standard model. In: Baek, J., Susilo, W., Kim, J. (eds.) ProvSec 2018. LNCS, vol. 11192, pp. 165–181. Springer, Cham (2018). https://doi.org/10.1007/978-3-030-01446-9_10
37. Stern, J.: A method for finding codewords of small weight. In: Cohen, G., Wolfmann, J. (eds.) Coding Theory 1988. LNCS, vol. 388, pp. 106–113. Springer, Heidelberg (1989). https://doi.org/10.1007/BFb0019850
38. Sidelnikov, V.M., Shestakov, S.O.: On insecurity of cryptosystems based on generalized Reed-Solomon codes. Discrete Math. Appl. $2(4)$, 439–444 (1992)
39. Shor, P.W.: Polynomial time algorithms for discrete logarithms and factoring on a quantum computer. In: Adleman, L.M., Huang, M.-D. (eds.) ANTS 1994. LNCS, vol. 877, pp. 289–289. Springer, Heidelberg (1994). https://doi.org/10.1007/3-540-58691-1_68
40. Sidelnikov, V.M.: A public-key cryptosystem based on binary Reed-Muller codes. Discrete Math. Appl. $4(3)$, 191–208 (1994)
41. Sendrier, N.: Finding the permutation between equivalent linear codes: the support splitting algorithm. IEEE Trans. Inf. Theory $46(4)$, 1193–1203 (2000)
42. Shrestha, S.R., Kim, Y.S.: New McEliece cryptosystem based on polar codes as a candidate for post-quantum cryptography. In: 14th International Symposium on Communications and Information Technologies (ISCIT), pp. 368–372. IEEE (2014)
43. Canto Torres, R., Sendrier, N.: Analysis of information set decoding for a sub-linear error weight. In: Takagi, T. (ed.) PQCrypto 2016. LNCS, vol. 9606, pp. 144–161. Springer, Cham (2016). https://doi.org/10.1007/978-3-319-29360-8_10

44. Wieschebrink, C.: Two NP-complete problems in coding theory with an application in code based cryptography. In: IEEE International Symposium on Information Theory - ISIT 2006, Seattle, USA, pp. 1733–1737. IEEE, Los Alamitos (2006)
45. Wang, Y.: Quantum resistant random linear code based public key encryption scheme RLCE. In: IEEE International Symposium on Information Theory - ISIT 2006, pp. 2519–2523. IEEE, Barcelona (2016)
46. Wang, Y.: RLCE-KEM, December 2017. https://csrc.nist.gov/Projects/PostQuantumCryptography/Round1Submission

Security Evaluation Against Side-Channel Analysis at Compilation Time

Nicolas Bruneau[1,2], Charles Christen[3], Jean-Luc Danger[1,4],
Adrien Facon[1,5], and Sylvain Guilley[1,4,5(✉)]

[1] Secure-IC S.A.S, 35510 Cesson-Sévigné, France
`sylvain.guilley@secure-ic.com`
[2] STMicroelectronics, 13790 Rousset, France
[3] Direction Générale de l'Armement, 35170 Bruz, France
[4] Télécom-Paris, Institut Polytechnique de Paris, 91400 Saclay, France
[5] Département d'informatique de l'ENS, CNRS, PSL University, 75005 Paris, France

Abstract. Masking countermeasure is implemented to thwart side-channel attacks. The maturity of high-order masking schemes has reached the level where the concepts are sound and proven. For instance, Rivain and Prouff proposed a full-fledged AES at CHES 2010. Some non-trivial fixes regarding refresh functions were needed though. Now, industry is adopting such solutions, and for the sake of both quality and certification requirements, masked cryptographic code shall be checked for correctness using the same model as that of the theoretical protection rationale (for instance the probing leakage model).

Seminal work has been initiated by Barthe et al. at EUROCRYPT 2015 for automated verification at higher orders on concrete implementations.

In this paper, we build on this work to actually perform verification from within a compiler, so as to enable timely feedback to the developer. Precisely, our methodology enables to provide the actual security order of the code at the intermediate representation (IR) level, thereby identifying possible flaws (owing either to source code errors or to compiler optimizations). Second, our methodology allows for an exploitability analysis of the analysed IR code. In this respect, we formally handle all the symbolic expressions in the static single assignment (SSA) representation to build the optimal distinguisher function. This enables to evaluate the most powerful attack, which is not only function of the masking order d, but also on the number of leaking samples and of the expressions (e.g., linear vs non-linear leakages).

This scheme allows to evaluate the correctness of a masked cryptographic code, and also its actual security in terms of number of traces in a given deployment context.

Keywords: Cryptographic code · Compilation · Intermediate representation (IR) · Static single assignment (SSA) · Side-channel analysis · Masking protection · Compositional countermeasure · Formal

N. Bruneau—Work done while at Secure-IC S.A.S.

C. T. Gueye et al. (Eds.): A2C 2019, CCIS 1133, pp. 129–148, 2019.
https://doi.org/10.1007/978-3-030-36237-9_8

analysis · Optimal side-channel attacks · Taylor expansion of distinguishers

1 Introduction

Context. With the massive deployment of Internet of Things (IoT), many devices are placed in-the-field which handle sensitive information. Typically, they must authenticate themselves, hence protect the integrity of public keys, and they handle private information, hence must ensure the confidentiality of secret encryption keys. The IoT devices are programmed in software, and their cryptographic stack deserves special attention.

The industry has put forward methodologies to ensure the protection of keys. A survey on this topic is freely available as a Technical Report from the ETSI [21]. For instance, cryptographic keys derivation, storage, and usage are confined in so called Trusted Execution Environments (TEEs). It is customary to study the security of digital assets according to their usage: data *at rest*, *in transit*, and *in computation*. The TEE takes care of keys at rest; cryptographic mechanisms, such as key establishment and key wrapping, allow to protect keys in transit; the hard problem is that of protecting keys in computation.

Indeed, software cryptographic code is vulnerable to side-channel attacks. They come in two flavors: first those which exploit conditional control-flow and/or table lookups (such as substitution boxes, or sboxes S), which fall into the class of so-called cache-timing attacks. Software techniques exist to make control-flow and table lookups uniform. In such situation, implementations are still vulnerable to a second kind of side-channel, which consists in physical spying of manipulated values, thanks to power or electromagnetic analysis.

State-of-the-Art. Several approaches have been put forward. One of them consists in the use of whitebox cryptography (WBC). The idea is that even if an attacker has access to the code (including keys) in source code, there is no way for her to extract the key. The technique is based on data obfuscation in the code. WBC implementations aim an linearizing the control flow and hiding data into obscure (random-looking) tables. However, WBC has little (practical) theoretical foundation, and without surprise, several structural attacks have been demonstrated [10]. Other approaches revolve around "signal-to-noise" minimization. One approach is to cancel the signal leakage, through balancing. This suits leakage in the control-flow: typically, algorithms such as AES have a data-independent control flow, therefore this kind of leakage is easy to plug perfectly. Now, so-called vertical leakage (that is, leakage of values, cf. ISO/IEC 20085-1 [23]) is harder to cancel. Balancing approaches are possible [26], albeit result in unpredictable security level and complex coding style. Another approach is the "masking" countermeasure, which aims at introducing some noise in the implementation. It consists in randomly sharing sensitive variables so that a side-channel attacker collect many meaningless leakages, since information is dissolved into several shares (conventionally, the number of shares is denoted by d). Therefore, such protections consume a lot of randomness (which must be uniform, independent, etc.). Today, many masking protections are constructively designed to be dth-order secure. The parameter $d > 0$ is a design security metric whereby each tuple of strictly less than d shares is independent from the clear sensitive variables.

For masking schemes such as Ishai-Sahai-Wagner (CRYPTO 2003 [22]) or Rivain-Prouff (CHES 2010 [27]) the security proof is based on composability: it is possible to design widgets for basic operations (e.g., field addition and multiplication) which form a universal set. Subsequently, combining them allows to build arbitrary computations. Reuse of variables shall be dealt with cautiously. In practice, reused variables usually benefit from refresh. This kind of masking is mature from a theoretical standpoint, and therefore is diffusing in the industry.

The problem is therefore to attest of the actual security, not of the principle of masking, but on the very implementation under consideration. There is one field of research which consists in checking a complete implementation (Barthe et al., EUROCRYPT 2015 [2]—known as MaskVerif). Since the combinatorics of verification is large, the proof employs heuristics taylored for the masking scheme.

Now, in practice, the attacker must perform a dth-order attack. But the attacker will maximize her advantage, and so she is not expected to be satisfied by one combination of d shares. Here we face a paradox: the larger the order d, the more possible combinations, hence it is relevant to study whether in practice, the security level in terms of data complexity (number of traces to recover the key) is still increasing with parameter d.

In this paper, we show how to compute optimal attacks, with tradeoffs regarding data and computational complexities (as in Bruneau et al., J. of Cryptology 2018 [16]). We aim at analyzing real-world implementations, irrespective of the source code language they are written in. Moreover, we want to consider optimized code. For this reason, we analyze the intermediate representation generated by a compiler, and generate the formula for the multivariate high-order distinguisher after having simplified the leaking terms.

Results show that monovariate high-order attacks are underestimating the security level by orders of magnitude, especially for high noise levels.

Contributions. The contributions in this paper are as follows:

- Application of dth-order masking correctness verification on an implementation extracted from within a compiler (at the IR level, after all optimization passes);
- Use of automated proof tools paradigm to generate optimal distinguishers for side-channel attacks;
- Trade-off regarding attack computation and attack efficiency, based on a truncated Taylor expansion of the side-channel distinguisher;

– Cautionary note that increasing d^1 does not increase exponentially the number of traces to recover the key.

Outline. The rest of this paper is structured as follows. An introduction on masking schemes and their need for formal verification has already be given in the present Sect. 1. Next the state-of-the-art automation method for the verification of a full-fledged implementations is recalled in Sect. 2. This methodology and its coding is leveraged to build automatically optimal distinguishers to attack at best the implementation. This allows to derive the most realistic security level (as in practice, the attacker will face more harsh attacking conditions, such as a degraded leakage model). This is the topic of Sect. 3, where we also highlight compromise between data and computational efficiency. Eventually, conclusion and perspectives are provided in Sect. 4.

2 Reminder About Automated Proof of Masking Schemes

2.1 Multi-variate and High-Order Side-Channel Attacks

State-of-the-art attacks against masked software consist in manual construction of leakage models. For instance, evaluation is done in [1, §5.2] on a first-order masking scheme, by using an absolute difference between two samples of the leakage. In [7], a combination is used albeit with unchosen coefficients, since those are coefficients from discrete Fourier or Hartley transforms. In [13], the dimensionality is reduced by an additive combination of samples, weighted by a profile obtained in a characterization phase.

It shall be noticed that software execution of software implementing masking countermeasure might leak a variable multiple times. Typically, the variable can be popped from the stack, then processed, and finally pushed back to the stack. It can also be copied at different places for different usages. Additionally, masking schemes themselves might involve multiple random variables. Typically, table recomputation countermeasure needs to address all entries of a lookup table, and process them with the same mask. Therefore, many samples in time leak the mask, and it is possible to combine them all in order to build the most efficient attack.

A methodology to build an attack of masked and shuffled table recomputation countermeasure is described in [16]. The method is empirical, because the *optimal distinguisher*, described in [14], is computationally too complex. The reason is that the optimal distinguisher shall be averaged over all masks m (if the countermeasure is of order d, m consists in a tuple of $(d-1)$ independent random variables). An approach is to develop the expression of the optimal distinguisher

---

[1] In this paper, we use the same letter d for the number of shares necessary to recover information on sensitive variables (designer's perspective) and the smallest attack order (evaluator's perspective). Actually, those values match in practice, assuming that the implementation is not flawed.

using a Taylor expansion, as described in [15]. This *systematic* approach allows for automation of attacks, by enabling a trade-off between accuracy of the distinguisher (Taylor expansion at high order) and computational complexity (Taylor expansion at low order). Namely, the mathematical formula for optimal attacks is explained below:

- Optimal attack consists in guess the key $\hat{k}$ according to maximum likelihood approach [14]:

$$\hat{k} = \operatorname*{argmax}_{k} \sum_{q=1}^{Q} \log \sum_{m} \exp -\frac{1}{2\sigma^2}(x_q - f(t_q, k, m))^2, \tag{1}$$

where:
  - q is the traces index,
  - t_q are the known texts, e.g., plaintexts,
  - m are the masks (whose distribution does not depend on q),
  - f is the leakage model, e.g.,

$$f(t_q, k, m) = w_H(S(t_q \oplus k) \oplus m) \tag{2}$$

  here assumed, but obtained by profiling in real attacks,
  - x_q are the leakages $x_q = f(t_q, k^*, m) + n_q$, k^* being the correct key,
  - σ^2 is the (centered) noise variance (n_q is a sample of this noise, that is

$p_N(n_q) = \frac{1}{\sqrt{2\pi\sigma^2}} \exp -\frac{n_q^2}{2\sigma^2}$).
- For attack simplification, the following Taylor expansion:

$$\log \mathbb{E} \exp(tX) = \sum_{n=1}^{\infty} \kappa_n \frac{t^n}{n!} \tag{3}$$

is leveraged
  - for an expectation $\mathbb{E}$ over X, and
  - where κ_n is the cumulant of order n of random variable X,
  starting at order $n = d$ and stopping strictly before ∞ for tractability reasons[2].

In this article, we combine all the relevant tuples which leak information, up to a predetermined order chosen by the attacker. If this order is less than the protection order, then (provided the masking is perfect [11]), our algorithm finds no terms. This is consistent with the targeted security order: no attack shall be feasible hence the distinguisher is constant for all key hypotheses; the "argmax" in Eq. 1 returns the full keyspace. Such result attests of countermeasure "correctness", meaning the formal verification that the countermeasure is "correctly" implemented (i.e., without flaws). But if we specify an order greater than that of the countermeasure, then our algorithm lists all the leaking terms, one shall consider to achieve the optimal attack rounded at a given order, as explicited theoretically in [15]. Besides, we assume that the code is constant-time (control flow does not depend on the data), hence traces are well aligned and can readily be used for performing a vertical high-order attack.

---

[2] At infinite order, the expansion of Eq. (3) is not considered, rather the original expression of Eq. (1) is used.

2.2   Analysis of Code at Intermediate Representation

Applying countermeasures at the LLVM Intermediate Representation (IR) level, as produced by clang/LLVM's middle-end, has already been hinted in [5,6]. But this approach has limitations, as shown in [20], because [5] actually assumes as "masked data" any intermediate variable which even depends in a non-uniform way of masking material.

Like in "Side-Channel Robustness Analysis of Masked Assembly Codes using a Symbolic Approach" (PROOFS 2017, JCEN 2019 [8]), the control-flow of the program to analyse must be statically known; in particular, there must be no indirect jump and the number of loop iterations must be known at compile time. Fortunately, cryptographic algorithms usually fulfill this requirement, as well as their algorithms implementation. Indeed, otherwise, there is the possibility of a cache-timing attack (see for instance review in [17]).

2.3   Probing Leakage Model

In order to capture security correctness, the "probing leakage model" (stemming from initial work by Blömer et al. [11]) has been put forward. A design is dth-order secure if any tuple of strictly less than d intermediate variables carries no information on sensitive variables. Gilles Barthe and coauthors have been automating the verification of such property, recognized as that of "non-interference". It uses simplification heuristics, for instance to show that $M \oplus E$, where M is a randomly distributed mask and E an expression which does not depend on M, is simply distributed as another uniformly distributed random mask M'. M' cannot be distinguished from M, hence *sensitive* expression E is not exposed. It is able to attest of the soundness of a masking scheme, or to find explicit counter-examples. On top of these interesting results, we also provide the design of the best possible attack (namely the *optimal attack*).

3   Contribution: Automated Attack Construction

3.1   Framework Concept

The analyses we conduct on the cryptographic code are sketched in Fig. 1. The nominal compilation flow is represented in the leftmost column of the figure (on the running example of LLVM). The rest of the figure represents the security analysis. The column in the middle represents the symbolic analyses, based on expressions for each SSA. The rightmost column represents the concrete analysis, whereby variables are assigned values, as per a series of (say ≈ 100) attack simulations.

The compilation is conducted in a nominal way until intermediate representation is reached. Here, we also let optimization passes be executed. The outcome is a list of static single assignments (SSA). Altogether, these expressions are the inputs of our analysis.

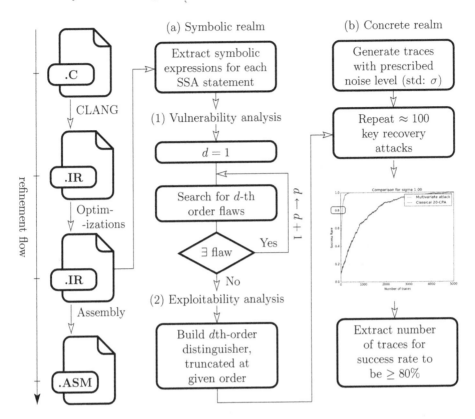

Fig. 1. Big picture of the two pipelined analyses conducted in the optimized intermediate representation code

(a) They are turned into symbolic expressions, named terms. They play the role of "intermediate variables" in side-channel papers such as [11].

    (a)-(1) Algorithm of MaskVerif is applied. The size d of the tuples is incremented (starting from $d = 1$) until we find a dependency into at least one tuple of expressions.

    * If this size is strictly inferior to the intended masking order, then a flaw has been found. Recall that as the analysis is at optimized IR level, the flaw can be structural (i.e., already present in the source code, as happened in the past for some countermeasures such as [18, 29]) or caused by an optimization pass which either removes some randomness or swaps operations (resulting in countermeasure break, see [28]).

    * Otherwise there is no flaw (result obtained by *formal proof*) and the masking order is exactly verified on the optimized IR representation. This however only indicates the absence of vulnerability from a design-for-security perspective. This means that the design matches its security specification.

(a)-(2) The optimal distinguisher at order d (see generic formula in (1)) is derived. Its expression is rounded at a given order and simplified, for evaluation efficiency.

(b) Concrete security shall be attested by actual exploitation of the leaking terms. This complementary step is mandatory to have a concrete idea about the attacker effort to extract the key. Differences might show up owing to the multiplicity of leakages, the confusion in the terms (linear terms induce leakages harder to exploit, compared to sboxes), the variety of terms, etc. This is achieved based on simulated traces, computed by a leakage model (typically the classical Hamming weight model, cf. Eq. (2)), and assuming a certain level of additive noise (i.i.d. for each expression and normally distributed as $\mathcal{N}(0, \sigma^2)$). The optimal distinguisher from step (a)-(2) is evaluated under q traces, and the indicator of attack success ($\hat{k} \stackrel{?}{=} k^*$) is computed. This process is reproduced ≈ 100 times, and averaged indicator yield the success rate. This curve is (globally–after smoothing) increasing. The number of traces for which success rate is equal to 80% is returned.

3.2   Framework Implementation

Analysis Parameters. The parameters of our analysis are:

- the order d of the masking in the input C files; they are immune to side-channel flaws if this order is equal to the one found in the vulnerability analysis (1) of Fig. 1.
- the optimal attack distinguisher Taylor expansion order ($\geq d$ if no flaw). In practice, an expansion at minimum degree (sufficient to distinguish) performs good results in terms of attack distinguishing power (meaning that the optimal attack extracts keys with similar number of traces). Higher degrees can sometimes help or sometimes not—this is difficult to estimate, and up to today, the exact degree where to cut the distinguisher off is considered an open problem.
- the leakage function associated with each share (i.e., each expression in the SSA extracted upon compilation), and
- the simulated measurement noise added to the leakage of each share (since masking works only as a countermeasure in the presence of some noise, hence evaluation must take place in this condition).

Source Code to Analyze. From a syntactic point of view, arbitrary source code can be inputted, since only optimized IR is analyzed. The countermeasures shall be applied at source-code level. Some constraints can nonetheless be enforced, such as `__attribute__((optnone))` in Listing 1.1 of Sect. A.1. The tool must nonetheless be informed about the name of the top-most function to analyze and the mask names. In the exemplar Listing 1.1, those are respectively "sbox" and "r" (short for random) at line 111.

Analysis Toolset. The tools used in the analysis are described here-after:

(a) symbolic expressions are generated by an LLVM plugin called SAW, contributed by Galois Inc. on GitHub.

  (a)-(1) The expressions happen to be huge: they are simplified using SAGE. Then, the are parsed and loaded into Julia. A script into Julia performs the Non-Interference analysis for each tuple of d expressions, exactly as explained in MaskVerif [2].

  (a)-(2) The optimal distinguisher is computed (and simplified to the best extend, for instance by regrouping identical leakages).

(b) It is then applied on these traces. In practice, since the attack is computationally intensive, it is first translated in C language which is compiled with maximum amount of optimizations. Further, this C code itself contains many precomputations before the attack proper is launched.

3.3 Why Check at the IR Level?

The choice of the validation level (refer to vertical axis in Fig. 1) results from a compromise:

– it shall be as low as possible, for the validation to be as close as possible to the final product (notice that we do not expect to go as down as the concrete evaluation, since this will be the task of a third-party certification laboratory), while

– it shall be possible to conduct a formal analysis, so as to prove the security order (or to formally detect flaws) and to devise the best possible attack.

From this tradeoff, we position the analysis after the optimization passes, hence we analyze the IR code which is the closest to the actual assembly (i.e., machine code) which will be executed. This means that we detect all faults potentially caused by the compiler[3], which could break the countermeasure.

Moreover, from a practical point of view, the method consisting in outputting proof elements from the compiler allows to streamline the evaluation: the user codes the countermeasure in the language of its own choice, and then an automated verdict is provided[4].

3.4 Application

Codes applying a masking strategy as that put forward in [11] can be analyzed. A classical example is the Ishai-Sahai-Wagner computations [22], extending protection of bits (elements of $\mathbb{F}_2$) to words (elements of $\mathbb{F}_{2^l}$) [27]. A concrete code which can be analysed is listed in Appendix A.1.

---

[3] The LLVM IR is lowered to machine instructions, and some optimizations can still be performed on this representation. In particular, some memory accesses can be gathered, some peephole optimizations may remove some useless computations, selection of some instructions may disrupt the intended control flow, instruction scheduling may reorder computations and register allocation can introduce flaws.

[4] This is the way all Secure-IC pre-silicon tools, namely Virtualyzr® and Catalyzr®, work.

In this article, we take as a running example the operations in a Galois field of characteristic two. Those are suitable for the computation of block ciphers, such as AES and even PRESENT [12]. AES fits naturally in $\mathbb{F}_{2^8}$ whereas PRESENT is nibble-oriented, hence can be represented in $\mathbb{F}_{2^4}$. In both AES and PRESENT, the only complex part regarding masking is the sbox, because it is non-linear. The expression of the PRESENT sbox (see table at page 453 of [12]) is obtained from Lagrange polynomial interpolation. We use MAGMA [31] to compute the extrapolation in $\mathbb{F}_{16}$ represented as $\mathbb{F}_2[x]/\langle x^4 + x + 1 \rangle$. The element of this field $(\mathbb{F}_{16}, +, \cdot)$ are denoted according to the convention below:

- $[0, 1, 2, 3, 4, \ldots, 15]$ .... (decimal)
- $[\text{0x0}, \text{0x1}, \text{0x2}, \text{0x3}, \text{0x4}, \ldots, \text{0xf}]$ .... (hexadecimal)
- $[0, 1, x, x+1, x^2, \ldots, x^3 + x^2 + x + 1]$ .... (polynomial)

We get for the sbox of PRESENT the expression:

$$
\begin{aligned}
sbox(A) = {} & \text{0xc} + \text{0x7} \cdot A^2 + \text{0x7} \cdot A^3 + \text{0xe} \cdot A^4 + \text{0xa} \cdot A^5 + \text{0xc} \cdot A^6 \\
& + \text{0x4} \cdot A^7 + \text{0x7} \cdot A^8 + \text{0x9} \cdot A^9 + \text{0x9} \cdot A^{10} + \text{0xe} \cdot A^{11} \\
& + \text{0xc} \cdot A^{12} + \text{0xd} \cdot A^{13} + \text{0xd} \cdot A^{14},
\end{aligned}
$$

which is efficiently computed using Horner's method:

$$
\begin{aligned}
sbox(A) = {} & \text{0xc} + A^2 \cdot (\text{0x7} + A \cdot (\text{0xe} + A \cdot (\text{0xa} + A \cdot (\text{0xc} + A \cdot (\\
& \text{0x4} + A \cdot (\text{0x7} + A \cdot (\text{0x9} + A \cdot (\text{0x9} + A \cdot (\text{0xe} + A \cdot (\\
& \text{0xc} + A \cdot (\text{0xd} + A \cdot (\text{0xd} + A)))))))))))).
\end{aligned}
$$

This expression is implemented in C language in function sbox in the Listing 1.1 of Appendix A.1. The attacked function is actually the composition of AddRoundKey and sBoxLayer, namely $A \mapsto sbox(A \oplus k^*)$, as done customarily in side-channel analysis.

Also, for the sake of tractability, we provide examples on masking of order $d = 1$. Indeed, our framework (described in previous Sect. 3.1) is, as of today, limited in the "concrete realm" of Fig. 1. The attack part is slow, though it has been translated from Julia to C for more efficiency. We leave the case $d > 1$ as a venue for further work, and as we shall see, results for $d = 1$ are already very rich.

One shall beware of the way the multiplication is performed, indeed, it is known that lookup-table based multiplications in characteristic two Galois fields do leak information [19]. In our implementation, the multiplication is constant-time (see function mult at line 19 of Listing 1.1).

Using the tool, we prove the correctness of code in Listing 1.1 of Appendix A.1. Notice that slight changes, such as asserting macro at line 50, would have the tool detect a first-order flaw. On the same code, we compare the traditional second-order (bi-variate) attack with our attack (multi-variate analysis, denoted MVA, extracted from the compiler). For these attacks, we use a Taylor expansion at order 2. The results for a few selected noise levels (in terms

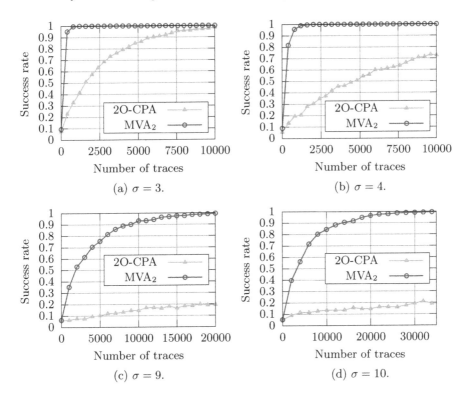

Fig. 2. Comparison of success rate for classical bi-variate attack and our multi-variate attack

of noise standard deviation σ) are represented in Fig. 2. It clearly appears that the optimal attack, even though rounded at order 2 (i.e., the first order at which there is a leakage, since the implementation is first-order secure), is significantly more efficient than the customary 2O-CPA.

The multivariate attack is performing better and better relative to the classical bi-variate attack as the noise level increases. This was already remarked in some papers, such as those analysing substitution table masking with recomputation [16, 30][5].

The speed of the attack can be summarized as the number of traces to reach 80% of success rate. We represent the number of traces to reach 80% of success rate for bivariate and multivariate attacks. It can be seen that the number of traces is indeed increasing exponentially, although not at the same rate. We represent in Fig. 3 the number of traces to recover the key (with success probability

---

[5] Battistello et al. [4] also notice that great multiplicity helps attacks, albeit in the different context of low-noise implementations (e.g., software running on top of a CPU). Anyway, such results highlight well that high dimensionality significantly favors the attackers, and that this aspect is often overlooked when simply analysing the security of a masking scheme only in terms of its degree (i.e., number of shares).

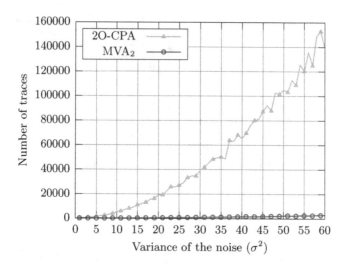

Fig. 3. Comparison between the number of traces to recover the key (with success probability of 80%) between the classical bi-variate attack and our multivariate attack (MVA$_2$, meaning that it is rounded at degree 2)

of 80%) between the classical bi-variate attack and our optimal distinguisher (termed "multivariate attack"). Again, we see that the difference increases with the noise level (quantified by the noise standard deviation σ). Moreover, we see here that this difference increases faster than linearly with σ. Therefore, we really insist that the security of a masking scheme cannot only be considered by considering the number of shares, but also the dimensionality of the leakages.

We clearly see that our method allows to build really powerful attacks (compared with naïve state-of-the-art attack for which dimensionality = order). The combination of multiple leaking points makes the computation of the attack non-trivial, but the result is that, with equal amount of traces, the multivariate attack (and actually not the best one, simply an approximation with a Taylor expansion) surpasses the schoolbook attack by orders of magnitudes. However, we underline that the goal of our study is not to design stronger attacks. Actually, we aim at providing after compilation the clearest possible image of the real security-level of the produced code. Therefore, the intent is to help the developer decide whether his implementation is secure enough vis-à-vis a security objective. In this respect, we had to devise a setup in which the strongest possible concrete attack is deployed.

The method of Taylor expansion is akin the soft analytical side-channel analysis. Actually, in the case of low-noise, such method is realistic, whereas our method is not. The Table 1 compares the state-of-art-art approaches in terms of attack depending on the noise level σ (for a normalized leakage).

The impact of the Taylor expansion order is negligible in our case-study, as illustrated in Fig. 4, which compares MVA rounded at order 2nd and 3rd. The difference is very limited, and in practice, the MVA$_2$ actually performs a

Table 1. Optimal attack methods in highly multivariate contexts

No noise	Low noise	High noise
Soft analytical [32]	SAT- or Pseudo-Boolean-solvers [25]	Taylor expansion [15]

little better than the MVA_3. Therefore, in order to save computation time, we recommend to apply the software verification scheme (Fig. 1) only at minimum order. The exact number of terms (which quantifies the complexity of the attack) is:

- MVA_2: 595 terms,
- MVA_3: 49011 terms.

In addition, the MVA_3 attack is more complex to perform, since the terms depend on σ, hence for the two values of noise in Fig. 4, two different distinguisher values shall be computed.

3.5   Limitations

Our tool checks the IR hence is unaware of registers and memory allocation. In particular, low-order leakage might arise owing to resources reuse, which, obviously, our tool cannot predict.

Besides, our method, like former methods [2,3], remains tailored for perfect sharings [11], but not custom maskings, e.g., first order masking schemes where masks are reused or so-called low-entropy masking schemes, e.g., masking schemes where the masks are not uniformly distributed [9,24]. For instance, the code segment in Appendix A.2 is perfectly masked at order one but cannot be analyzed by our method (which uses same heuristics as [2]).

4   Conclusion and Perspectives

In this article, we tackled the issue of analyzing the security level of a software cryptographic code, during its compilation. For the analysis to be complete, we leverage both a vulnerability analysis (already put forward by Barthe et al. at EUROCRYPT 2015) and an exploitability analysis (based on a simulation of the optimal attack, computed within the compiler and tailored for the inputted code).

The first phase (vulnerability analysis) allows to verify that there is no flaw in the source code down to the optimized IR code. The second phase (exploitability analysis) investigates in which respect the implementation is robust in a given context (characterized by a noise level).

The two phases provide a clear view of the suitability of the compiled code for its execution in an operational environment.

As a byproduct of this work, we show that it is possible to automate crafted distinguishers which perform significantly better than simple theoretical attacks (e.g., d-valued d-th order CPA). The improvement results from the signal-to-noise increase by aggregating multiple high-order univariate analyses. The resulting multivariate high-order attacks can be extremely efficient. The reason is that there is an exponential way to combine leakages, and that this exponential advantage is comparable to the exponential complexity increase of the sharing order. We show concrete examples where security metrics (number of shares d) do not trivially reflect the practical security level in terms of attacks.

As a perspective, we intend to optimize the attacks (third column of Fig. 1). Indeed, using parallelism (vectorization) and precomputations, those could be drastically speeded-up. Moreover, we also intend to study the vulnerability and exploitability of attacks not only at the IR level, but also directly on assembly code. Eventually, it could be beneficial to extend the methodology (today geared from Barthe et al. method) to masking schemes which are not full-entropy or that reuse masks.

Acknowledgments. This work has been partly financed via TEAMPLAY, a project from European Union's Horizon20202 research and innovation program, under grand agreement N° 779882 (https://teamplay-h2020.eu/).

A   Example of Input Codes For Analysis

A.1   Codes Which can be Analyzed in Our Framework

Two examples of codes which can be analyzed are provided here-after in Listing 1.1. The selection between the two codes is achieved by defining macro SBOX_TYPE to either cube or present at line 117.

```
1   /*
2    * Regarding the refresh algorithm, there is no need at order 1, but beware of higher-order!
3    * https://eprint.iacr.org/2015/359.pdf
4    */
5
6   #include <stdio.h>
7   #include <stdlib.h>
8   #include <stdint.h>
9
10  static unsigned int    gnr = 0;
11
12  /* This function must be optimized, otherwise it will include a test! */
13  uint8_t multx(uint8_t x)
14  {
15      uint8_t y = (x & 0x08) ? ((0x0f & (x << 1)) ^ 0x03) : 0x0f & (x << 1);
16      return y;
17  }
18
19  uint8_t mult(uint8_t x, uint8_t y)
20  {
21  #if 0 // Not to use with SAW plugin of LLVM
22      uint8_t z = 0;
23      uint8_t b = 0x8;
24      for ( int i=3; i>=0; i--)
25      {
26          z = multx(z);
27          z ^= x & (-!!(y&b));       // Constant-time multiplication
28          b >>= 1;
29      }
30      return z;
31  #else
32      return x*y;  // To simplify the analysis on an abstracted code
33  #endif
34  }
35
36  uint8_t __attribute__((optnone)) rnd(uint8_t r[])
37  {
38      return r[gnr++];
39  }
40
41  void refresh_masks(uint8_t x[2], uint8_t r[])
42  {
43      uint8_t new_r = rnd(r);
44      x[0] ^= new_r;
45      x[1] ^= new_r;
46  }
47
48  void __attribute__((optnone)) secmult(uint8_t x[2], uint8_t y[2], uint8_t z[2], uint8_t r[])
49  {
50  #if 0 // Security bug: the new_r self-demasks [To use for testing]
51      uint8_t new_r = rnd(r);
52      z[0] = (mult(x[0],y[0]) ^ new_r) ^ (mult(x[0],y[1]) ^ new_r) ^ mult(x[1],y[0]);
53      z[1] = mult(x[1],y[1]);
54  #else
55      uint8_t r01 = rnd(r);
56      uint8_t r10 = (r01 ^ mult(x[0],y[1]));
57      r10 ^= mult(x[1],y[0]);
58
59      z[0] = mult(x[0],y[0]);
60      z[0] ^= r01;
61      z[1] = mult(x[1],y[1]);
62      z[1] ^= r10;
63  #endif
64  /* Another option from \cite{2003-11949} would be:
65      c_1 = r                                                        (1)
66      c_2 = (((a_1 b_1 + r) + a_1 b_2 ) + a_2 b_1 ) + a_2 b_2 . (2)
67  */
68  }
69  void secsquare(uint8_t x[2], uint8_t z[2], uint8_t r[])
70  {
71      z[0] = mult(x[0],x[0]);
72      z[1] = mult(x[1],x[1]);
73  }
74
75  const uint8_t a[15] = { 12, 0, 7, 7, 14, 10, 12, 4, 7, 9, 9, 14, 12, 13, 13 };
76
77  uint8_t l1(uint8_t x)
78  {
```

```
79          uint8_t x_2 = mult(x,x);
80          uint8_t x_4 = mult(x_2,x_2);
81          uint8_t x_8 = mult(x_4,x_4);
82          return   mult(a[1],x) ^ mult(a[2],x_2) ^ mult(a[4],x_4) ^ mult(a[8],x_8);
83      }
84
85      uint8_t l3(uint8_t x)
86      {
87          uint8_t x_2 = mult(x,x);
88          uint8_t x_4 = mult(x_2,x_2);
89          uint8_t x_8 = mult(x_4,x_4);
90          return   mult(a[3],x) ^ mult(a[6],x_2) ^ mult(a[12],x_4) ^ mult(a[9],x_8);
91      }
92
93      uint8_t l5(uint8_t x)
94      {
95          uint8_t x_2 = mult(x,x);
96          return   mult(a[5],x) ^ mult(a[10],x_2);
97      }
98
99      uint8_t l7(uint8_t x)
100     {
101         uint8_t x_2 = mult(x,x);
102         uint8_t x_4 = mult(x_2,x_2);
103         uint8_t x_8 = mult(x_4,x_4);
104         return   mult(a[7],x) ^ mult(a[14],x_2) ^ mult(a[13],x_4) ^ mult(a[11],x_8);
105     }
106     /*
107      * The function which be symbolically interpreted by SAW
108      * All the shares are named 'm', and the masks 'r'
109      */
110     //                    DATA IN        DATA OUT        RANDOMS
111     void  sbox(         uint8_t m[2], uint8_t y[2], uint8_t r[])
112     {
113         uint8_t x_2[2];
114         uint8_t x_3[2];
115         uint8_t x_5[2];
116         uint8_t x_7[2];
117     #if 1 // DEBUG, SBOX_TYPE=cube
118         secsquare( m, x_2, r );
119         refresh_masks(m,r);
120         secmult( m, x_2, y, r );
121     #else // SBOX_TYPE=present
122
123         x_2[0] = mult(m[0],m[0]);
124         x_2[1] = mult(m[1],m[1]);
125         secmult( m, m, x_2, r );
126         secsquare( m, x_2, r );
127         refresh_masks(m,r);
128         refresh_masks(x_2,r);
129         secmult( m, x_2, y, r );
130
131         x_3[0] = mult(x_2[0],m[0]);
132         x_3[1] = mult(x_2[1],m[1]);
133         y[0] = x_2[0];
134         y[1] = x_2[1];
135         secmult( m, x_2, x_3, r );
136         refresh_masks(m,r);
137         refresh_masks(x_2,r);
138         refresh_masks(x_3,r);
139
140         x_5[0] = mult(x_3[0],x_2[0]);
141         x_5[1] = mult(x_3[1],x_2[1]);
142         secmult( x_2, x_3, x_5, r );
143         refresh_masks(x_2,r);
144         refresh_masks(x_3,r);
145
146         refresh_masks(x_5,r);
147         x_7[0] = mult(x_5[0],x_2[0]);
148         x_7[1] = mult(x_5[1],x_2[1]);
149
150         secmult( x_2, x_5, x_7, r );
151         refresh_masks(x_2,r);
152         refresh_masks(x_5,r);
153         refresh_masks(x_7,r);
154         y[0] = a[0] ^ l1(m[0]) ^ l3(x_3[0]) ^ l5(x_5[0]) ^ l7(x_7[0]);
155         y[1] =          l1(m[1]) ^ l3(x_3[1]) ^ l5(x_5[1]) ^ l7(x_7[1]);
156     #endif
```

```
157     }
158
159     int   main()
160     {
161         uint8_t  r[100];
162         for ( int   i=0;  i<100;  i++){
163             r[i]  =  rand();
164         }
165
166         uint8_t  x[2];
167         uint8_t  y[2];
168
169         const   uint8_t  sbox_ref[16]  = {  12,  5,  6,  11,  9,  0,  10,  13,  3,  14,  15,  8,  4,  7,  1,  2 };
170         for ( uint8_t  i=0;  i<16;  ++i )
171         {
172             uint8_t  mask  =  rand();
173             x[0]  =  i^mask;
174             x[1]  =       mask;
175             sbox(x,y,r);
176             printf(  "%d  --> %d  (ref  = %d)\n"   ,  x[0]^x[1],  y[0]^y[1],  sbox_ref[i]);
177         }
178         return   0;
179     }
```

Listing 1.1: C code representing polynomial computations

A.2   Code Which Cannot be Analyzed

In this section, we present one example of code which cannot be analyzed (*automatically*) since simplifications as per Barthe [2] do not apply. Indeed, the masks are not used as in ISW [22]:

-   in ISW: masks are added (XORed) and subsequently subtracted (XORed), whereas
-   in Alg. 1.2: the masks are involved in computation as selection variable in a choice.

The listing 1.2 presents both a straightforward multiplexor and a multiplexor protected at first-order.

```
1     /* Unprotected function, computing a selection (= multiplexor) */
2     uint8_t MUX( uint8_t a, uint8_t b, uint8_t c )
3     {
4         // return c?b:a; // At bit-level
5         return (c&b)|(~c&a); // At word-level
6     }
7
8     /* Function whose 1st-order security can be demonstrated */
9     uint8_t MUX_masked(uint8_t am, uint8_t bm, uint8_t cm, uint8_t r[] /* m, m2 */ )
10    {
11        uint8_t m  = r[0];
12        uint8_t m2 = r[1];
13        return (((am &  ~m2) ^ (bm &  m2)) & (~(cm ^ (m^m2)))) ^ \
14               (((am &  m2) ^ (bm & ~m2)) &  (cm ^ (m^m2)));
15    }
```

Listing 1.2: Function which conditionally returns one of the two arguments, protected against side-channel attacks at order $d = 1$,

B   Multi-variate Attack at Degrees Two and Three

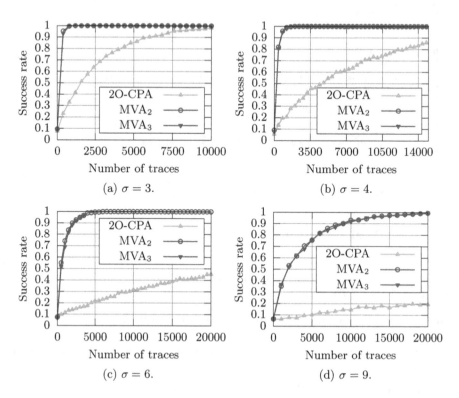

Fig. 4. Comparison of success rate for classical bi-variate attack and our multi-variate attack at degrees two and three

References

1. Balasch, Josep, Gierlichs, Benedikt, Reparaz, Oscar, Verbauwhede, Ingrid: DPA, bitslicing and masking at 1 GHz. In: Güneysu, Tim, Handschuh, Helena (eds.) CHES 2015. LNCS, vol. 9293, pp. 599–619. Springer, Heidelberg (2015). https://doi.org/10.1007/978-3-662-48324-4_30
2. Barthe, G., Belaïd, S., Dupressoir, F., Fouque, P.-A., Grégoire, B., Strub, P.-Y.: Verified proofs of higher-order masking. In: Oswald, E., Fischlin, M. (eds.) EUROCRYPT 2015. LNCS, vol. 9056, pp. 457–485. Springer, Heidelberg (2015). https://doi.org/10.1007/978-3-662-46800-5_18
3. Barthe, G., Dupressoir, F., Faust, S., Grégoire, B., Standaert, F.-X., Strub, P.-Y.: Parallel implementations of masking schemes and the bounded moment leakage model. In: Coron, J.-S., Nielsen, J.B. (eds.) EUROCRYPT 2017. LNCS, vol. 10210, pp. 535–566. Springer, Cham (2017). https://doi.org/10.1007/978-3-319-56620-7_19

4. Battistello, A., Coron, J.-S., Prouff, E., Zeitoun, R.: Horizontal side-channel attacks and countermeasures on the ISW masking scheme. In: Gierlichs, B., Poschmann, A.Y. (eds.) CHES 2016. LNCS, vol. 9813, pp. 23–39. Springer, Heidelberg (2016). https://doi.org/10.1007/978-3-662-53140-2_2

5. Bayrak, A.G., Regazzoni, F., Novo, D., Brisk, P., Standaert, F.-X., Ienne, P.: Automatic application of power analysis countermeasures. IEEE Trans. Comput. **64**(2), 329–341 (2015)

6. Bayrak, A.G., Regazzoni, F., Novo, D., Ienne, P.: Sleuth: automated verification of software power analysis countermeasures. In: Bertoni, G., Coron, J.-S. (eds.) CHES 2013. LNCS, vol. 8086, pp. 293–310. Springer, Heidelberg (2013). https://doi.org/10.1007/978-3-642-40349-1_17

7. Belgarric, P., et al.: Time-frequency analysis for second-order attacks. In: Francillon, A., Rohatgi, P. (eds.) CARDIS 2013. LNCS, vol. 8419, pp. 108–122. Springer, Cham (2014). https://doi.org/10.1007/978-3-319-08302-5_8

8. El Ouahma, I.B., Meunier, Q., Heydemann, K., Encrenaz, E.: Side-channel robustness analysis of masked assembly codes using a symbolic approach. J. Cryptographic Eng. 1–12 (2019). https://doi.org/10.1007/s13389-019-00205-7.

9. Bhasin, S., Danger, J.-L., Guilley, S., Najm, Z.: A low-entropy first-degree secure provable masking scheme for resource-constrained devices. In: Proceedings of the Workshop on Embedded Systems Security, WESS 2013, Montreal, Quebec, Canada, 29 September–4 October 2013, pp. 7:1–7:10. ACM (2013)

10. Billet, O., Gilbert, H., Ech-Chatbi, C.: Cryptanalysis of a white box AES implementation. In: Selected Areas in Cryptography, pp. 227–240 (2004)

11. Blömer, J., Guajardo, J., Krummel, V.: Provably secure masking of AES. In: Handschuh, H., Hasan, M.A. (eds.) SAC 2004. LNCS, vol. 3357, pp. 69–83. Springer, Heidelberg (2004). https://doi.org/10.1007/978-3-540-30564-4_5

12. Bogdanov, A., et al.: PRESENT: an ultra-lightweight block cipher. In: Paillier, P., Verbauwhede, I. (eds.) CHES 2007. LNCS, vol. 4727, pp. 450–466. Springer, Heidelberg (2007). https://doi.org/10.1007/978-3-540-74735-2_31

13. Bruneau, N., Danger, J.-L., Guilley, S., Heuser, A., Teglia, Y.: Boosting higher-order correlation attacks by dimensionality reduction. In: Chakraborty, R.S., Matyas, V., Schaumont, P. (eds.) SPACE 2014. LNCS, vol. 8804, pp. 183–200. Springer, Cham (2014). https://doi.org/10.1007/978-3-319-12060-7_13

14. Bruneau, N., Guilley, S., Heuser, A., Rioul, O.: Masks will fall off. In: Sarkar, P., Iwata, T. (eds.) ASIACRYPT 2014. LNCS, vol. 8874, pp. 344–365. Springer, Heidelberg (2014). https://doi.org/10.1007/978-3-662-45608-8_19

15. Bruneau, N., Guilley, S., Heuser, A., Rioul, O., Standaert, F.-X., Teglia, Y.: Taylor expansion of maximum likelihood attacks for masked and shuffled implementations. In: Cheon, J.H., Takagi, T. (eds.) ASIACRYPT 2016. LNCS, vol. 10031, pp. 573–601. Springer, Heidelberg (2016). https://doi.org/10.1007/978-3-662-53887-6_21

16. Bruneau, N., Guilley, S., Najm, Z., Teglia, Y.: Multivariate high-order attacks of shuffled tables recomputation. J. Cryptol. **31**(2), 351–393 (2018)

17. Carré, S., Facon, A., Guilley, S., Takarabt, S., Schaub, A., Souissi, Y.: Cache-timing attack detection and prevention. In: Polian, I., Stöttinger, M. (eds.) COSADE 2019. LNCS, vol. 11421, pp. 13–21. Springer, Cham (2019). https://doi.org/10.1007/978-3-030-16350-1_2

18. Coron, J.-S., Prouff, E., Rivain, M.: Side channel cryptanalysis of a higher order masking scheme. In: Paillier, P., Verbauwhede, I. (eds.) CHES 2007. LNCS, vol. 4727, pp. 28–44. Springer, Heidelberg (2007). https://doi.org/10.1007/978-3-540-74735-2_3

19. Danger, J.-L., et al.: On the performance and security of multiplication in $GF(2^N)$. Cryptography **2**(3), 25 (2018)
20. Eldib, H., Wang, C., Schaumont, P.: Formal verification of software countermeasures against side-channel attacks. ACM Trans. Softw. Eng. Methodol. **24**(2), 11:1–11:24 (2014)
21. ETSI/TC CYBER. Security techniques for protecting software in a white box model. ETSI TR 103 642 V1.1.1, October 2018
22. Ishai, Y., Sahai, A., Wagner, D.: Private circuits: securing hardware against probing attacks. In: Boneh, D. (ed.) CRYPTO 2003. LNCS, vol. 2729, pp. 463–481. Springer, Heidelberg (2003). https://doi.org/10.1007/978-3-540-45146-4_27
23. ISO/IEC JTC 1/SC 27/WG 3. ISO/IEC CD 20085–1:2017 (E). Information technology - Security techniques – Test tool requirements and test tool calibration methods for use in testing non-invasive attack mitigation techniques in cryptographic modules – Part 1: Test tools and techniques, 25 January 2017
24. Nassar, M., Souissi, Y., Guilley, S., Danger, J.-L.: RSM: a small and fast countermeasure for AES, secure against first- and second-order zero-offset SCAs. In: DATE, pp. 1173–1178. IEEE Computer Society, Dresden, Germany, 12–16 March 2012. (TRACK A: "Application Design", TOPIC A5: "Secure Systems")
25. Oren, Y., Weisse, O., Wool, A.: A new framework for constraint-based probabilistic template side channel attacks. In: Batina, L., Robshaw, M. (eds.) CHES 2014. LNCS, vol. 8731, pp. 17–34. Springer, Heidelberg (2014). https://doi.org/10.1007/978-3-662-44709-3_2
26. Rauzy, P., Guilley, S., Najm, Z.: Formally proved security of assembly code against power analysis - a case study on balanced logic. J. Cryptogr. Eng. **6**(3), 201–216 (2016)
27. Rivain, M., Prouff, E.: Provably secure higher-order masking of AES. In: Mangard, S., Standaert, F.-X. (eds.) CHES 2010. LNCS, vol. 6225, pp. 413–427. Springer, Heidelberg (2010). https://doi.org/10.1007/978-3-642-15031-9_28
28. Roy, D.B., Bhasin, S., Guilley, S., Danger, J.-L., Mukhopadhyay, D.: From theory to practice of private circuit: a cautionary note. In: 33rd IEEE International Conference on Computer Design, ICCD 2015, New York City, NY, USA, 18–21 October 2015, pp. 296–303. IEEE Computer Society (2015)
29. Schramm, K., Paar, C.: Higher order masking of the AES. In: Pointcheval, D. (ed.) CT-RSA 2006. LNCS, vol. 3860, pp. 208–225. Springer, Heidelberg (2006). https://doi.org/10.1007/11605805_14
30. Tunstall, M., Whitnall, C., Oswald, E.: Masking tables—an underestimated security risk. In: Moriai, S. (ed.) FSE 2013. LNCS, vol. 8424, pp. 425–444. Springer, Heidelberg (2014). https://doi.org/10.1007/978-3-662-43933-3_22
31. University of Sydney (Australia). Magma Computational Algebra System. http://magma.maths.usyd.edu.au/magma/. Accessed 22 Aug 2014
32. Veyrat-Charvillon, N., Gérard, B., Standaert, F.-X.: Soft analytical side-channel attacks. In: Sarkar, P., Iwata, T. (eds.) ASIACRYPT 2014. LNCS, vol. 8873, pp. 282–296. Springer, Heidelberg (2014). https://doi.org/10.1007/978-3-662-45611-8_15

Subliminal Hash Channels

George Teşeleanu[1,2(✉)] [iD]

[1] Simion Stoilow Institute of Mathematics of the Romanian Academy,
21 Calea Grivitei, Bucharest, Romania
[2] Advanced Technologies Institute, 10 Dinu Vintilă, Bucharest, Romania
tgeorge@dcti.ro

Abstract. Due to their nature, subliminal channels are mostly regarded as being malicious, but due to recent legislation efforts users' perception might change. Such channels can be used to subvert digital signature protocols without degrading the security of the underlying primitive. Thus, it is natural to find countermeasures and devise subliminal-free signatures. In this paper we discuss state-of-the-art countermeasures and introduce a generic method to bypass them.

1 Introduction

As more and more countries require individuals and providers to hand over passwords and decryption keys [7], we might observe an increase in the usage of *subliminal channels*. Subliminal channels are secondary channels of communication hidden inside a potentially compromised communication channel. The concept was introduced by Simmons [39–41] as a solution to the *prisoners' problem*. In the prisoners' problem *Alice* and *Bob* are incarcerated and wish to communicate confidentially and undetected by their guard *Walter* who imposes to read all their communication. Note that *Alice* and *Bob* can exchange a secret key before being incarcerated.

A special case of subliminal channels are secretly embedded trapdoor with universal protection (SETUP) attacks. By combining subliminal channels with public key cryptography Young and Yung devised a plethora of mechanisms [45–48] to leak a user's private key or message. Although the authors assume a black-box environment[1] in [15] is pointed out that these mechanism can also be implemented in open source software due to the code's sheer complexity and the small number of experts who review it. SETUP attacks are meant to capture the situation in which the manufacturer of a black-box device is also an adversary or employed by an adversary.

According to the classified documents leaked by Snowden [12,34] the NSA made efforts for subverting cryptographic standards. More precisely, there are strong indications of the existence of a backdoor in the Dual-EC generator [13]. This backdoor is a direct application of Young and Yung's work.

---

[1] A black-box is a device, process or system, whose inputs and outputs are known, but its internal structure or working is not known or accessible to the user (*e.g.* tamper proof devices, closed source software).

© Springer Nature Switzerland AG 2019
C. T. Gueye et al. (Eds.): A2C 2019, CCIS 1133, pp. 149–165, 2019.
https://doi.org/10.1007/978-3-030-36237-9_9

Snowden's revelations rekindled the study of backdoors. Thus, more examples of backdoor embedding methods were found [14,22,24,42,43], methods for protecting users against them were developed [11,18,21,26,36,37] and implementations were exploited in the wild [19,20].

Most subliminal channels or SETUP attacks use random numbers to convey information undetected. In consequence, all the proposed countermeasures focus on sanitizing the random numbers used by a system. In the case of digital signatures, a different but laborious method for inserting a subliminal channel in a system is presented in [44]. Instead of using random numbers as information carriers, *Alice* uses the hash of the message to convey the message for *Bob*. In order to achieve this, *Alice* makes small changes to the message until the hash has the desired properties. Note that the method presented in [44] bypasses all the countermeasures mentioned so far.

This paper studies a generic method that allows the prisoners to communicate through the subliminal-free signatures found in [11,18,21,26,36,37]. To achieve our goal we work in a scenario where all messages are time-stamped before signing. Note that we do not break any of the assumptions made by the subversion-free proposals. This work is motivated by the fact that most end-users to do not verify the claims made by manufacturers[2]. Moreover, users often do not know which should be the outputs of a device [30]. A notable incident in which users where not aware of the correct outputs and trusted the developers is the Debian incident [17].

Structure of the Paper. We introduce notations and definitions in Sect. 2. By adapting and improving the mechanism from [44] we introduce new hash channels in Sect. 3. A series of experiments is conducted in Sect. 4. Applications are provided in Sect. 5. We conclude in Sect. 6. Additional definitions are given in Appendix A.

2   Preliminaries

Notations. Throughout the paper λ and κ will denote security parameters. We let ROM denote the random oracle model. The number of bits of an element x is denoted by $|x|$ and $x\|y$ represents the concatenation of the strings x and y. The set $\{0,1\}^{\ell}$ consists of bit strings of length ℓ.

The action of selecting a random element x from a sample space X is represented by $x \xleftarrow{\$} X$. We also denote by $x \leftarrow y$ the assignment of value y to variable x. The encryption of a message $m \in \{0,1\}$ using one-time pad is denoted by $\omega \leftarrow m \oplus b$, where b is a random bit used only once.

2.1   Diffie-Hellman Assumptions

Definition 1 (Hash Diffie-Hellman - HDH). *Let $\mathbb{G}$ be a cyclic group of order q, g a generator of $\mathbb{G}$, $\mathbb{G}_m$ a set and $H : \mathbb{G} \rightarrow \mathbb{G}_m$ a hash function. Let A be*

---

[2] Manufacturers might implement subversion-free signatures just for marketing purposes, while still backdooring some of the devices produced.

a probabilistic polynomial-time algorithm (PPT algorithm) which returns 1 on
input (g^x, g^y, z) if $H(g^{xy}) = z$. We define the advantage

$$ADV_{\mathbb{G},g,H}^{\text{HDH}}(A) = |Pr[A(g^x, g^y, H(g^{xy})) = 1 | x, y \xleftarrow{\$} \mathbb{Z}_q^*]$$
$$- Pr[A(g^x, g^y, z) = 1 | x, y \xleftarrow{\$} \mathbb{Z}_q^*, z \xleftarrow{\$} \mathbb{G}_m]|.$$

If $ADV_{\mathbb{G},g,H}^{\text{HDH}}(A)$ is negligible for any PPT algorithm A, we say that the Hash
Diffie-Hellman problem is hard in $\mathbb{G}$.

Hashed Diffie-Hellman Key Exchange (HKE). Based on the HDH assumption we describe a key exchange protocol[3] in Fig. 1. A formal analysis of this
design can be found in [8,9,23].

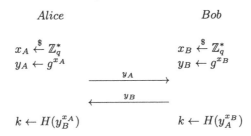

Fig. 1. The Hashed Diffie-Hellman key exchange protocol.

2.2   Digital Signatures

Definition 2 (Signature Scheme). *A Signature Scheme consists of four PPT
algorithms: ParamGen, KeyGen, Sign and Verification. The first one takes as
input a security parameter and outputs the system's parameters. Using these
parameters, the second algorithm generates the public key and the matching secret
key. The secret key together with the Sign algorithm are used to generate a signature σ for a message m. Using the public key, the last algorithm verifies if a
signature σ for a message m is generated using the matching secret key.*

Remark 1. For simplicity, public parameters will further be considered implicit
when describing an algorithm.

Schnorr Signature. In [38], Schnorr introduces a digital signature based on
the discrete logarithm problem. Later on, the scheme was proven secure in the
ROM by Stern and Pointcheval [35]. We further recall the Schnorr signature.

ParamGen (λ): Generate two large prime numbers p, q, such that $q \geq 2^\lambda$ and
$q|p - 1$. Select a cyclic group $\mathbb{G}$ of order p and let $g \in \mathbb{G}$ be an element of

---

[3] A high level description of the IKE protocols [27,28].

order q. Let $h : \{0,1\}^* \rightarrow \mathbb{Z}_q^*$ be a hash function. Output the public parameters $pp = (p, q, g, \mathbb{G}, h)$.

KeyGen (pp): Choose $x \xleftarrow{\$} \mathbb{Z}_q^*$ and compute $y \leftarrow g^x$. Output the public key $pk = y$. The secret key is $sk = x$.

Sign (m, sk): To sign a message $m \in \{0,1\}^*$, first generate a random number $k \xleftarrow{\$} \mathbb{Z}_q^*$. Then compute the values $r \leftarrow g^k$, $e \leftarrow h(r\|m)$ and $s \leftarrow k - xe \bmod q$. Output the signature (e, s).

Verification (m, e, s, pk): To verify the signature (e, s) of message m, compute $r \leftarrow g^s y^e$ and $u \leftarrow h(r\|m)$. Output true if and only if $u = e$. Otherwise, output false.

2.3   Subliminal Channels and SETUP Attacks

Covert channels [31] have the capability of transporting information through system parameters apparently not intended for information transfer. Subliminal channels and SETUP attacks are special cases of covert channels and achieve information transfer by modifying the original specifications of cryptographic primitives[4]. We further restrict covert channels to two sub-cases: subliminal channels and SETUP attacks.

Definition 3 (Subliminal channel). *A Subliminal channel is an algorithm that can be inserted in a system such that it allows the system's owner to communicate[5] with a recipient without their communication being detected by a third party[6]. It is assumed that the prisoners' communication is encrypted using a secret/public key encryption scheme and the decryption function is accessible to the recipient.*

Definition 4 (Secretly Embedded Trapdoor with Universal Protection - SETUP). *A Secretly Embedded Trapdoor with Universal Protection (SETUP) is an algorithm that can be inserted in a system such that it leaks encrypted private key information to an attacker through the system's outputs. Encryption of the private key is performed using a public key encryption scheme. It is assumed that the decryption function is accessible only to the attacker.*

Remark 2. Note that SETUP mechanisms are special cases of subliminal channels. In the SETUP case, the sender is the system, the recipient is the attacker, while the third party is the owner of the system.

Definition 5 (Covert channel indistinguishability - IND-COVERT). *Let C_0 be a black-box system that uses a secret key sk. Let $\mathcal{E}$ be the encryption scheme used by a covert channel as defined above, in Definitions 3 and 4. We consider*

---

[4] For example, by modifying the way random numbers are generated.

[5] Through the system's outputs.

[6] The sender and receiver will further be called prisoners and the third party warden.

C_1 an altered version of C_0 that contains a covert channel based on $\mathcal{E}$. Let A be a PPT algorithm which returns 1 if it detects that C_0 is altered. We define the advantage

$$ADV_{\mathcal{E},C_0,C_1}^{\text{IND-COVERT}}(A) = |Pr[A^{C_1(sk,\cdot)}(\lambda) = 1] - Pr[A^{C_0(sk,\cdot)}(\lambda) = 1]|.$$

If $ADV_{\mathcal{E},C_0,C_1}^{\text{IND-COVERT}}(A)$ is negligible for any PPT algorithm A, we say that C_0 and C_1 are polynomially indistinguishable.

Remark 3. In some cases, if sk is known, the covert channel can be detected by using its description and parameters. Thus, depending on the context we will specify if A has access to sk or not. If $\mathcal{E}$ is a public key encryption scheme we always assume that A has access to the public key[7].

We consider that the covert channels presented from now on are implemented in a device D that digitally signs messages. In the case of subliminal channels, the prisoners are denoted as *Alice* (sender) and *Bob* (receiver), while *Walter* is the guard. In the case of SETUP attacks, the owner of the device is referred to as *Charlie* and the attacker is usually *Mallory*. When the secret key sk is not known to the PPT algorithm A we assume that sk is stored only in D's volatile memory. Note that *Walter* and *Charlie* believe that D signs messages using the original specifications of the signature scheme implemented in D. When one of the original signature's algorithm is not modified by the covert channel, the algorithm will be omitted when presenting the respective channel.

Throughout the paper, when presenting covert channels, we make use of the following additional algorithms:

- *Subliminal/Malicious ParamGen* − used by the prisoners/attacker to generate their (his) parameters;
- *Subliminal/Malicious KeyGen* − used by the prisoners/attacker to generate their (his) keys;
- *Extract* − used by the recipient to extract the secret message;
- *Recovering* − used by the attacker to recover *Charlie*'s secret key.

The algorithms above are not implemented in D. For simplicity, covert parameters will further be implicit when describing an algorithm.

Two classes of covert channels can be found in literature: the trivial channel and the Young-Yung SETUP attack. We refer the reader to Appendix A.2 for their description.

3   Hash Channels

In order to be valid, legal documents need a timestamp appended to them before being digitally signed [10, 25]. According to [10] the timestamp must include seconds. Note that if the timestamp module is independent from the *Sign* module,

---

[7] Found by means of reverse engineering the system, for example.

then *Walter* or *Charlie* can inject false timestamps into the signing module. Thus, we assume that the timestamp module is integrated in the signing module. Using this framework we achieve a subliminal channel by adapting and simplifying the idea from [44].

Let *lim* be an upper limit for the number of trials and u_t the smallest time unit used by the time stamping algorithm (*e.g.* seconds, milliseconds). We further present our proposed subliminal channel.

Time Stamp (u_t): Output the current time τ including u_t.

Subliminal Sign (m, ω, sk): Generate a random number $k \overset{\$}{\leftarrow} \mathbb{Z}_q^*$ and compute $r \leftarrow g^k$. Let *counter* = 1. Generate τ using the *Time Stamp* algorithm and compute $e \leftarrow h(r\|m\|\tau)$ and *counter* = *counter* + 1, until $e \equiv \omega \bmod 2$ or *counter* = *lim*. Compute $s \leftarrow k - xe \bmod q$. Output the signature (e, s).

Extract (m, e, s, pk) : To extract the embedded message compute $\omega \leftarrow e \bmod 2$. Remark that the probability of event $e \equiv \omega \bmod 2$ is $1 - 1/2^{lim}$.

The security of the Schnorr signature scheme is preserved, since we are not modifying the scheme itself, but the way messages are processed. Let τ_h be the average time it takes device D to compute $h(r\|m\|\tau)$ for fixed bit-size bit_m messages. To avoid detection by *Walter* or *Charlie* the manufacturer writes in D's specification that for a message of size bit_m it takes $lim \cdot \tau$ to sign bit_m messages. Thus, D remains consistent with the specifications (*i.e.* IND-COVERT secure). The main restriction when choosing lim is users' usability. Due to the hash-rate statistics reported for SHA-256 in [2,3] we can assume $\tau_h < 1$ second. Thus, the bottleneck becomes the time stamp (*i.e.* D can not output a signature for time t at time $t-1$). This can be mitigated by including finer time units into the timestamp (*e.g* milliseconds).

Remark 4. Let $z \leftarrow g^t$ be the public key of *Bob* and b a bit *Alice* wants to send to *Bob*. Then, we can easily transition to a public key subliminal channel by using HKE and computing $\omega \leftarrow b \oplus H(z^k)$, where $\mathbb{G}_m = \{0, 1\}$. Since k is fresh for each signature, *Alice* can continuously leak data to *Bob*. Note that we are using HKE to encrypt the message so, ω is indistinguishable from a random bit. Thus, the scheme is IND-COVERT secure under the HDH assumption. We further denote this public key subliminal channel by hash$_p$.

Remark 5. When dealing with longer messages there is a simpler way to transmit them. Thus, let m_i be the ith bit of m and $\mathbb{G}_m = \{0, 1\}^{|m|}$. The device D can leak m to *Bob* through $|m|$ signing sessions by computing $c \leftarrow m \oplus H(z^{k_0})$ and setting $\omega \leftarrow c_i$ for the ith signing session, where $0 \leq i < |m|$. Note that m is successfully transmitted with a probability of $(1 - 1/2^{lim})^{|m|}$. This channel is further denoted by hash$_\ell$. Remark that if we replace *Bob* with *Mallory* and set $m \leftarrow x$, hash$_\ell$ is transformed into a SETUP attack.

Remark 6. If adversary A has access to x, then he can compute all k numbers. Thus, hash$_p$ and hash$_\ell$ can be detected if x can be retrieved from D, while the regular hash channel it is not. Hence, in the public key setting we assume that x is only stored in D's volatile memory[8].

4   Stochastic Detection

In [29], the authors show that the execution time of the Young-Yung attack can be used to distinguish honest devices from backdoored devices. Using Kucner *et al.* observations as a starting point, we run a series of experiments to see if our proposed methods can be detected by measuring their execution time.

We implemented in C using the GMP library [6] the Schnorr signature (normal), the trivial channel (trivial), the Young-Yung attack (yy), the hash channel (hash), the public key hash channel (hash$_p$) and hash$_p$'s extension to long messages (hash$_\ell$). The programs were run on a CPU Intel i7-4790 4.00 GHz and compiled with GCC with the O3 flag activated. In our experiments for each prime size of 2048, 3072, 4096 and 8192 bits, we ran the algorithms with 100 safe prime numbers from [5]. For each prime we measured the average running time for 128 random 2040 byte messages[9] using the function $omp\_get\_wtime()$ [4]. Before signing each message we added a 8 byte timestamp with the current system time in milliseconds ($clock\_gettime()$). The hash function used internally by the algorithms is either SHA256 or SHA512 [1].

When we implemented the hash channels we took advantage of the Merkle-Damgard structure of SHA256 or SHA512. Thus, we computed and stored the intermediary value h_{it} obtained after processing 1984 (SHA256) or 1920 (SHA512) bytes. Then for each trial we used h_{it} to process the last block of the message. Note that the size of the messages was selected such that after h_{it} the SHA functions must process one full message block and a full padding block (worst case scenario). Also, in our experiments lim tends towards infinity.

The results of our experiments are presented in Figs. 3, 4, 5, 6, 7, 8, 9 and 10. We can see from the plots that the Schnorr signature and the hash channel have similar execution times. We further investigated this by computing the absolute time difference between a normal execution (t_{n_1}) and a hash channel execution (t_h) or another normal execution (t_{n_2}). The results are presented in Table 1. Note that the empirical evidence suggests that the normal and hash channel executions are indistinguishable due to the noise added by the operating system.

When we implemented hash$_\ell$ we distributed the HKE protocol execution over 128 Schnorr signature computations. The downside of this method is that first we need to use 128 Schnorr signatures for masking the HKE and then 128 hash channel signatures for leaking the message. The results presented in this section are only for the first part, since experimental data for hash channels is already presented. Note that the first part of hash$_\ell$ is indistinguishable from the normal execution. As in the case of the hash channel, we further investigated

---

[8] The same assumption is make in Young-Yung's attack, since their mechanism can also be detected when x is known to the attacker.

[9] By choosing 128 messages we simulated the following scenario: the secret key x is generated using a PRNG with a seed of 128 bits and D leaks the seed.

| —— normal —— trivial —— yy |
| —— hash —— hash$_p$ —— hash$_\ell$ |

Fig. 2. Plot legend for Figs. 3 to 10

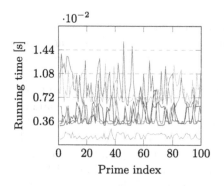

Fig. 3. Prime's size 2048 bits with SHA256

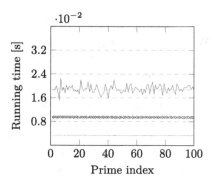

Fig. 4. Prime's size 2048 bits with SHA512

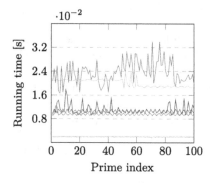

Fig. 5. Prime's size 3072 bits with SHA256

Fig. 6. Prime's size 3072 bits with SHA512

the indistinguishability claim by computing the absolute time difference between a normal execution (t_{n_1}) and a hash$_\ell$ channel execution (t_ℓ). The results are presented in Table 1. Note that the empirical evidence suggests that the normal and hash$_\ell$ channel executions are indistinguishable due to the noise added by the operating system.

Another remark is that the rest of the channels can be easily detected by measuring their execution time. Thus, noise must be added to the Young-Yung attack or to the Schnorr signature in order to make the subliminal channels undetectable. Note that the trivial channel and the public key hash channel have similar execution times. Thus, any technique used to mask the execution time of the trivial channel can also be used for the public key hash channel.

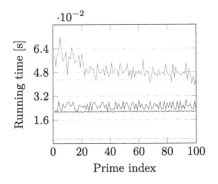

Fig. 7. Prime's size 4096 bits with SHA256

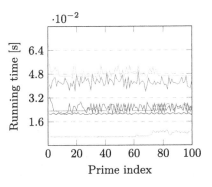

Fig. 8. Prime's size 4096 bits with SHA512

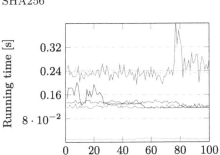

Fig. 9. Prime's size 8192 bits with SHA256

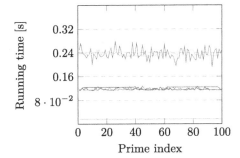

Fig. 10. Prime's size 8192 bits with SHA512

Table 1. Time comparison

Prime's size	SHA	$|t_{n_1} - t_{n_2}|$	$|t_{n_1} - t_h|$	$|t_{n_1} - t_\ell|$
2048	256	0.185621	0.138921	0.149650
	512	0.122050	0.097460	0.101445
3072	256	0.462406	0.105996	0.150646
	512	0.156232	0.156667	0.160375
4096	256	0.523229	0.354953	0.358049
	512	0.118666	0.134868	0.085795
8192	256	1.381028	1.548863	1.586020
	512	0.483661	0.629464	0.468742

Let T be the computation time for one signature. We denote by $E[T]$ and $\sigma[T]$ the expected value and the standard deviation of T. Kucner *et al.* introduce the $R[T] = \sigma[T]/E[T]$ characteristic in order to measure computation time independently of the actual speed of the processor. We computed $R[T]$ for all

Table 2. $R[T]$ characteristic

Prime's size	SHA	Normal	Trivial	yy	hash	hash$_p$	hash$_\ell$
2048	256	0.128196	0.722334	0.322048	0.076789	0.138364	0.147544
	512	0.035308	0.695684	0.131866	0.017010	0.033390	0.094461
3072	256	0.094644	0.751065	0.430531	0.044940	0.063633	0.095584
	512	0.024800	0.718313	0.207609	0.024917	0.044754	0.092278
4096	256	0.131263	0.700937	0.582434	0.043187	0.045583	0.101174
	512	0.051705	0.691449	0.326993	0.125705	0.089238	0.140214
8192	256	0.116920	0.704363	1.156188	0.172762	0.101328	0.132343
	512	0.056456	0.708207	0.594154	0.057846	0.086518	0.121710

channels and the results are presented in Table 2. We can observe from our experiments that the $R[T]$ characteristic fluctuates in practice. Also, from Table 2, it is easy to observe that the $R[T]$ characteristic for the Schnorr signature is always smaller than the one for the trivial channel and the Young-Yung attack. Thus, we can distinguish these two channels from an honest execution. Unfortunately, the results are inconclusive for the rest of the channels.

5 Marketing Backdoors

In this section we provide the reader with state-of-the-art countermeasures used to obtain subliminal-free signatures and show that three proposals are vulnerable to the hash and hash$_\ell$ channels without masking the channels' execution time, while for the rest the channels must be masked. Thus, a manufacturer can market a product as being subliminal free[10], while in reality it is not. Note that our proposed scenario does not violate the assumptions made by the subversion-free protocols.

5.1 Russel *et al.* Subversion-Free Proposal

The authors of [36,37] assume that all the random numbers used by a signature scheme are generated by a malicious RNG (including the key generation step). Based on this assumption, the authors describe and prove secure a generic method for protecting users. Note that both the trivial channel and the Young-Yung attack can be modeled as malicious RNGs. Unfortunately, in the hash channel scenario the security of their method breaks down.

The philosophy behind Russel *et al.* method is to split every generation algorithm into two parts: a random string generation part RG and a deterministic part DG. By extensively testing DG the user can be ensured that the deterministic part is *almost consistent* with the specifications. By using two independent

---

[10] By implementing one of these countermeasures.

RNG modules $Source_1, Source_2$ and hashing their concatenated outputs, any backdoors implemented in the RNGs will not propagate into DG. We further describe an instantiation of [37] using the Schnorr signature scheme.

Random $(Source_1, Source_2)$: Generate $s_1 \overset{\$}{\leftarrow} Source_1$ and $s_2 \overset{\$}{\leftarrow} Source_2$. Output $h(s_1\|s_2)$.

KeyGen (pp): Generate x using the *Random* algorithm and compute $y \leftarrow g^x$. Output the public key $pk = y$. The secret key is $sk = x$.

Sign (m, sk): To sign a message $m \in \{0,1\}^*$, first generate k using the *Random* algorithm. Then, compute the values $r \leftarrow g^k$, $e \leftarrow h(r\|m)$ and $s \leftarrow k - xe \bmod q$. Output the signature (e, s).

5.2 Hanzlik *et al.* Controlled Randomness Proposal

A method for controlling the quality of k is proposed in [26]. In order to do this, the authors use a blinding factor $U \leftarrow g^u$ that is installed by the owner of the device and a counter i. The owner accepts a signature produced by D if and only if *Check* returns **true**. Note that the Young-Yung SETUP attack is not possible due to the blinding factor. We further present their modifications on the *Sign* algorithm.

Sign (m, U, i, sk): To sign a message $m \in \{0,1\}^*$, first generate $k_0 \overset{\$}{\leftarrow} \mathbb{Z}_q^*$, compute $r' \leftarrow g^{k_0}$, $k_1 \leftarrow H(U^{k_0}, i)$ and increment i. Let $k \leftarrow k_0 k_1$. Compute the values $r \leftarrow g^k$, $e \leftarrow h(r\|m)$ and $s \leftarrow k - xe \bmod q$. Output the signature (e, s) and the control data (r', i).

Check (e, s, r', u): Compute $r \leftarrow g^s y^e$ and $\alpha \leftarrow H(r'^u, i)$. Output **true** only if and only if $r = r'^\alpha$. Otherwise, output **false**.

The authors underline that a subliminal channel[11] exists, but due to the limited memory of the signing device, hiding the time needed to implement their proposed channel is difficult. Note that our timestamp method proposed in Sect. 3 is much faster[12] and, thus, in some cases is feasible for bypassing Hanzlik *et al.* mechanism.

5.3 Choi *et al.* Tamper-Evident Digital Signatures

Choi *et al.* [21] introduce the notion of tamper-evidence for digital signatures in order to prevent corrupted nodes to covertly leak secret information. We further provide the tamper-evident Schnorr signature.

ParamGen (λ): Generate two large prime numbers p, q, such that $q \geq 2^\lambda$ and $q | p - 1$. Select a cyclic group $\mathbb{G}$ of order p and let $g \in \mathbb{G}$ be an element of order q. Let $h : \{0, 1\}^* \rightarrow \mathbb{Z}_q^*$ be a hash function and let ℓ be the number of permitted signatures. Output the public parameters $pp = (p, q, g, \mathbb{G}, h, \ell)$.

---

[11] Similar to the trivial channel described in Sect. 2.3.

[12] *i.e.* computing a hash is faster than computing a modular exponentiation.

KeyGen (pp, κ): Choose $x \xleftarrow{\$} \mathbb{Z}_q^*$ and compute $y \leftarrow g^x$. Also, choose $\omega_\ell \xleftarrow{\$} \{0,1\}^\kappa$. For $1 \leq i \leq \ell$, generate $k_i \xleftarrow{\$} \mathbb{Z}_q^*$ and compute $\omega_{i-1} \leftarrow h(g^{k_i} \| \omega_i)$. Output the public key $pk = (y, \omega_0)$. The secret key is $sk = (x, k_1, \ldots, k_\ell, \omega_1, \ldots, \omega_\ell)$.

Sign (m_i, sk): To sign the ith message $m_i \in \{0,1\}^*$, compute the values $r_i \leftarrow g^{k_i}$, $e_i \leftarrow h(r_i \| \omega_i \| m_i)$ and $s_i \leftarrow k_i - xe_i \bmod q$. Output the signature (e_i, s_i, ω_i).

Verification $(m_i, e_i, s_i, \omega_i, pk)$: To verify the signature (e_i, s_i, ω_i) of message m_i, compute $r_i \leftarrow g^{s_i} y^{e_i}$ and $u \leftarrow h(r_i \| \omega_i \| m_i)$. Output true if and only if $u = e_i$ and $\omega_{i-1} \leftarrow h(r_i \| \omega_i)$. Otherwise, output false.

The authors work in the honest key generation model. Thus, the nodes can not manipulate the k_is in any way. Fortunately, our hash channels use messages to leak confidential data. Thus, the nodes can still subliminally transmit data by using our proposed channels.

5.4   Ateniese *et al.* and Bohli *et al.* Subversion-Free Proposals

The authors of [11] propose the usage of re-randomizable signatures[13] and unique signatures (see footnote 13) as countermeasures to backdoors induced by malicious RNGs. These proposals are secure according to their security model [11]. A similar approach can be found in [18], where the authors convert the Digital Signature Algorithm into a deterministic signature. Note that both approaches assume honest key generation.

All these signature schemes work on fixed length messages and internally use a number theoretic hash function[14]. In order to work on variable length messages a standard hash function is used to process the message and the resulting hash is used as input for the Naor-Reingold function. Thus, for each small change in the message we have to recompute the hash $h(m)$, multiply $|h(m)/2|$ integers from $\mathbb{Z}_q^*$ and perform an exponentiation in $\mathbb{G}$. So, our proposed hash channel on average doubles the time necessary to process a message. In this case, the execution time of a hash channel is no longer similar to an honest implementation and, thus, noise must be added to mask the backdoor.

6   Conclusions

In this paper we introduced a scenario in which the security of the subliminal-free methods presented in [21,26,36,37] degrades. We also conducted a series of experiments to show that the hash and hash$_\ell$ channels' executions are indistinguishable from the normal signature executions due to the noise produced by the operating system. Hence, we proved that users must request justifications for every security choice made by the manufacturer and that testing centers must never let the DG modules to modify inputs.

---

[13] See Appendix A.1 for a definition of the concept.
[14] More precisely, the Naor-Reingold pseudo-random function [32,33].

A   Additional Preliminaries

A.1   Definitions

Definition 6. (Computational Diffie-Hellman - CDH). *Let $\mathbb{G}$ be a cyclic group of order q, g a generator of $\mathbb{G}$ and let A be a PPT algorithm that returns an element from $\mathbb{G}$. We define the advantage*

$$ADV_{\mathbb{G},g}^{CDH}(A) = Pr[A(g^x, g^y) = g^{xy} | x, y \xleftarrow{\$} \mathbb{Z}_q^*].$$

If $ADV_{\mathbb{G},g}^{CDH}(A)$ is negligible for any PPT algorithm A, we say that the Computational Diffie-Hellman problem is hard in $\mathbb{G}$.

Remark 7. The CDH assumption is standard and we include it for completeness. The HDH assumption was formally introduced in [8,9], although it was informally described as a composite assumption in [16,49]. According to [16], the HDH assumption is equivalent with the CDH assumption in the ROM. Although an equivalent of the HDH assumption exists in the standard model, in this paper we are working with the Schnorr signature scheme that is secure in the ROM. Thus, the security in the ROM suffices for our purposes.

Definition 7. (Unique Signature Scheme). *Let S be a signature scheme and pk be a public key generated by the KeyGen algorithm of S. We say that S is a Unique Signature Scheme if for any message m and any signatures of m, $\sigma_1 \neq \sigma_2$*

$$Pr[\mathit{Verification}(m, \sigma_1, pk) = \mathit{Verification}(m, \sigma_2, pk) = \mathtt{true}]$$

is negligible.

Definition 8. (Re-Randomizable Signature Scheme). *Let S be a signature scheme and (pk, sk) be a public/secret key pair generated by the KeyGen algorithm of S. We say that S is a Re-Randomizable Signature Scheme if there exists a PPT algorithm ReRand such that for all messages m the output of $ReRand(m, \sigma, pk)$ is statistically indistinguishable from $Sign(m, sk)$.*

A.2   Covert Channels

Trivial Subliminal Channel. The Schnorr signature supports a subliminal channel based on rejection sampling. We further describe the trivial subliminal channel.

Sign (m, sk): Choose $k \xleftarrow{\$} \mathbb{Z}_q^*$ and compute $r \leftarrow g^k$, until $\omega \equiv r \bmod 2$. To sign a message $m \in \{0,1\}^*$ compute the values $e \leftarrow h(r\|m)$ and $s \leftarrow k - xe \bmod q$. Output the signature (e, s).

Extract (e, s) : To extract the embedded message ω compute $\omega \leftarrow g^s y^e \bmod 2$.

Young-Yung SETUP Attack. In [45–48], the authors propose a kleptographic version of Schnorr signatures and prove it IND-COVERT secure in the standard model under the HDH assumption. The algorithms of the SETUP attack are shortly described below. Note that after D signs at least two messages, $Mallory$ can recover $Charlie$'s secret key and, thus, impersonate $Charlie$.

Malicious ParamGen (pp): Let $H : \mathbb{G} \rightarrow \mathbb{Z}_q^*$ be a hash function. Output the public parameter $sp_M = H$. Note that H will be stored in D's volatile memory.

Malicious KeyGen (pp): Choose $x_M \xleftarrow{\$} \mathbb{Z}_q^*$ and compute $y_M \leftarrow g^{x_M}$. Output the public key $pk_M = y_M$. The public key pk_M will be stored in D's volatile memory. The secret key is $sk_M = x_M$; it will only be known by $Mallory$ and will not be stored in the black-box.

Signing Sessions: The possible signing sessions performed by D are described below. Let $i \geq 1$.

Session$_0$(m_0, sk): To sign message $m_0 \in \mathbb{G}$, D does the following

$$k_0 \xleftarrow{\$} \mathbb{Z}_q^*, r_0 \leftarrow g^{k_0}, e_0 \leftarrow h(r_0 \| m_0), s_0 \leftarrow k_0 - xe_0 \bmod q.$$

The value k_0 is stored in D's volatile memory until the end of *Session$_1$*. Output the signature (r_0, s_0).

Session$_i$(m_i, sk, pk_M): To sign message $m_i \in \mathbb{G}$, D does the following

$$z_i \leftarrow y_M^{k_{i-1}}, k_i \leftarrow H(z_i), r_i \leftarrow g^{k_i}, e_i \leftarrow h(r_i \| m_i), s_i \leftarrow k_i - xe_i \bmod q.$$

The value k_i is stored in D's volatile memory until the end of *Session$_{i+1}$*. Output the signature (r_i, s_i).

Recovering $(m_i, e_{i-1}, e_i, s_i, sk_M)$: Compute $r_{i-1} \leftarrow g^{s_{i-1}} y^{e_{i-1}}$, $\alpha \leftarrow r_{i-1}^{x_M}$ and $k_i \leftarrow H(\alpha)$. Recover x by computing $x \leftarrow e_i^{-1}(k_i - s_i) \bmod q$.

References

1. mbed TLS. https://tls.mbed.org
2. Mining Hardware Comparison. https://en.bitcoin.it/wiki/Mining_hardware_compa rison
3. Non-Specialized Hardware Comparison. https://en.bitcoin.it/wiki/Non-specializ ed_hardware_comparison
4. OpenMP. https://www.openmp.org/
5. Safe Prime Database. https://2ton.com.au/safeprimes/
6. The GNU Multiple Precision Arithmetic Library. https://gmplib.org/
7. World Map of Encryption Laws and Policies. https://www.gp-digital.org/world-map-of-encryption/
8. Abdalla, M., Bellare, M., Rogaway, P.: DHAES: An Encryption Scheme Based on the Diffie-Hellman Problem. IACR Cryptology ePrint Archive 1999/7 (1999)

9. Abdalla, M., Bellare, M., Rogaway, P.: The oracle Diffie-Hellman assumptions and an analysis of DHIES. In: Naccache, D. (ed.) CT-RSA 2001. LNCS, vol. 2020, pp. 143–158. Springer, Heidelberg (2001). https://doi.org/10.1007/3-540-45353-9_12

10. Adams, C., Cain, P., Pinkas, D., Zuccherato, R.: RFC 3161: internet X.509 public key infrastructure time-stamp protocol (TSP). Technical report, Internet Engineering Task Force (2001)

11. Ateniese, G., Magri, B., Venturi, D.: Subversion-resilient signature schemes. In: ACM-CCS 2015, pp. 364–375. ACM (2015)

12. Ball, J., Borger, J., Greenwald, G.: Revealed: how US and UK spy agencies defeat internet privacy and security. Guardian **6**, 2–8 (2013)

13. Barker, E., Kelsey, J.: SP 800–90A. Recommendations for Random Number Generation Using Deterministic Random Bit Generators (2012)

14. Bellare, M., Jaeger, J., Kane, D.: Mass-surveillance without the state: strongly undetectable algorithm-substitution attacks. In: ACM-CCS 2015, pp. 1431–1440. ACM (2015)

15. Bellare, M., Paterson, K.G., Rogaway, P.: Security of symmetric encryption against mass surveillance. In: Garay, J.A., Gennaro, R. (eds.) CRYPTO 2014. LNCS, vol. 8616, pp. 1–19. Springer, Heidelberg (2014). https://doi.org/10.1007/978-3-662-44371-2_1

16. Bellare, M., Rogaway, P.: Minimizing the use of random oracles in authenticated encryption schemes. In: Han, Y., Okamoto, T., Qing, S. (eds.) ICICS 1997. LNCS, vol. 1334, pp. 1–16. Springer, Heidelberg (1997). https://doi.org/10.1007/BFb0028457

17. Bello, L.: DSA-1571-1 OpenSSL—Predictable Random Number Generator. https://www.debian.org/security/2008/dsa-1571 (2008)

18. Bohli, J.-M., González Vasco, M.I., Steinwandt, R.: A subliminal-free variant of ECDSA. In: Camenisch, J.L., Collberg, C.S., Johnson, N.F., Sallee, P. (eds.) IH 2006. LNCS, vol. 4437, pp. 375–387. Springer, Heidelberg (2007). https://doi.org/10.1007/978-3-540-74124-4_25

19. Checkoway, S., et al.: A systematic analysis of the juniper dual EC incident. In: ACM-CCS 2016, pp. 468–479. ACM (2016)

20. Checkoway, S., et al.: On the practical exploitability of dual EC in TLS implementations. In: USENIX Security Symposium, pp. 319–335. USENIX Association (2014)

21. Choi, J.Y., Golle, P., Jakobsson, M.: Tamper-evident digital signature protecting certification authorities against malware. In: DASC 2006, pp. 37–44. IEEE (2006)

22. Dodis, Y., Ganesh, C., Golovnev, A., Juels, A., Ristenpart, T.: A formal treatment of backdoored pseudorandom generators. In: Oswald, E., Fischlin, M. (eds.) EUROCRYPT 2015. LNCS, vol. 9056, pp. 101–126. Springer, Heidelberg (2015). https://doi.org/10.1007/978-3-662-46800-5_5

23. Dodis, Y., Gennaro, R., Håstad, J., Krawczyk, H., Rabin, T.: Randomness extraction and key derivation using the CBC, cascade and HMAC modes. In: Franklin, M. (ed.) CRYPTO 2004. LNCS, vol. 3152, pp. 494–510. Springer, Heidelberg (2004). https://doi.org/10.1007/978-3-540-28628-8_30

24. Fried, J., Gaudry, P., Heninger, N., Thomé, E.: A kilobit hidden SNFS discrete logarithm computation. In: Coron, J.-S., Nielsen, J.B. (eds.) EUROCRYPT 2017. LNCS, vol. 10210, pp. 202–231. Springer, Cham (2017). https://doi.org/10.1007/978-3-319-56620-7_8

25. Haber, S., Stornetta, W.S.: How to time-stamp a digital document. In: Menezes, A.J., Vanstone, S.A. (eds.) CRYPTO 1990. LNCS, vol. 537, pp. 437–455. Springer, Heidelberg (1991). https://doi.org/10.1007/3-540-38424-3_32

26. Hanzlik, L., Kluczniak, K., Kutyłowski, M.: Controlled randomness – a defense against backdoors in cryptographic devices. In: Phan, R.C.-W., Yung, M. (eds.) Mycrypt 2016. LNCS, vol. 10311, pp. 215–232. Springer, Cham (2017). https://doi.org/10.1007/978-3-319-61273-7_11

27. Harkins, D., Carrel, D.: RFC 2409: the internet key exchange (IKE). Technical report, Internet Engineering Task Force (1998)

28. Kaufman, C., Hoffman, P., Nir, Y., Eronen, P., Kivinen, T.: RFC7296: internet key exchange protocol version 2 (IKEv2). Technical report, Internet Engineering Task Force (2014)

29. Kucner, D., Kutyłowski, M.: Stochastic kleptography detection. In: Public-Key Cryptography and Computational Number Theory, pp. 137–149 (2001)

30. Kwant, R., Lange, T., Thissen, K.: Lattice klepto. In: Adams, C., Camenisch, J. (eds.) SAC 2017. LNCS, vol. 10719, pp. 336–354. Springer, Cham (2018). https://doi.org/10.1007/978-3-319-72565-9_17

31. Lampson, B.W.: A note on the confinement problem. Commun. ACM **16**(10), 613–615 (1973)

32. Naor, M., Reingold, O.: Number-theoretic constructions of efficient pseudo-random functions. In: FOCS 1997, pp. 458–467. IEEE Computer Society (1997)

33. Naor, M., Reingold, O.: Number-theoretic constructions of efficient pseudo-random functions. J. ACM (JACM) **51**(2), 231–262 (2004)

34. Perlroth, N., Larson, J., Shane, S.: NSA Able to Foil Basic Safeguards of Privacy on Web, vol. 5. The New York Times, New York (2013)

35. Pointcheval, D., Stern, J.: Security proofs for signature schemes. In: Maurer, U. (ed.) EUROCRYPT 1996. LNCS, vol. 1070, pp. 387–398. Springer, Heidelberg (1996). https://doi.org/10.1007/3-540-68339-9_33

36. Russell, A., Tang, Q., Yung, M., Zhou, H.-S.: Cliptography: clipping the power of kleptographic attacks. In: Cheon, J.H., Takagi, T. (eds.) ASIACRYPT 2016. LNCS, vol. 10032, pp. 34–64. Springer, Heidelberg (2016). https://doi.org/10.1007/978-3-662-53890-6_2

37. Russell, A., Tang, Q., Yung, M., Zhou, H.S.: Generic semantic security against a kleptographic adversary. In: ACM-CCS 2017, pp. 907–922. ACM (2017)

38. Schnorr, C.P.: Efficient identification and signatures for smart cards. In: Brassard, G. (ed.) CRYPTO 1989. LNCS, vol. 435, pp. 239–252. Springer, New York (1990). https://doi.org/10.1007/0-387-34805-0_22

39. Simmons, G.J.: The subliminal channel and digital signatures. In: Beth, T., Cot, N., Ingemarsson, I. (eds.) EUROCRYPT 1984. LNCS, vol. 209, pp. 364–378. Springer, Heidelberg (1985). https://doi.org/10.1007/3-540-39757-4_25

40. Simmons, G.J.: Subliminal communication is easy using the DSA. In: Helleseth, T. (ed.) EUROCRYPT 1993. LNCS, vol. 765, pp. 218–232. Springer, Heidelberg (1994). https://doi.org/10.1007/3-540-48285-7_18

41. Simmons, G.J.: Subliminal channels; past and present. Eur. Trans. Telecommun. **5**(4), 459–474 (1994)

42. Teşeleanu, G.: Unifying kleptographic attacks. In: Gruschka, N. (ed.) NordSec 2018. LNCS, vol. 11252, pp. 73–87. Springer, Cham (2018). https://doi.org/10.1007/978-3-030-03638-6_5

43. Teşeleanu, G.: Managing your kleptographic subscription plan. In: Carlet, C., Guilley, S., Nitaj, A., Souidi, E.M. (eds.) C2SI 2019. LNCS, vol. 11445, pp. 452–461. Springer, Cham (2019). https://doi.org/10.1007/978-3-030-16458-4_26

44. Wu, C.K.: Hash channels. Comput. Secur. **24**(8), 653–661 (2005)

45. Young, A., Yung, M.: The dark side of "Black-Box" cryptography or: should we trust capstone? In: Koblitz, N. (ed.) CRYPTO 1996. LNCS, vol. 1109, pp. 89–103. Springer, Heidelberg (1996). https://doi.org/10.1007/3-540-68697-5_8

46. Young, A., Yung, M.: Kleptography: using cryptography against cryptography. In: Fumy, W. (ed.) EUROCRYPT 1997. LNCS, vol. 1233, pp. 62–74. Springer, Heidelberg (1997). https://doi.org/10.1007/3-540-69053-0_6

47. Young, A., Yung, M.: The prevalence of kleptographic attacks on discrete-log based cryptosystems. In: Kaliski, B.S. (ed.) CRYPTO 1997. LNCS, vol. 1294, pp. 264–276. Springer, Heidelberg (1997). https://doi.org/10.1007/BFb0052241

48. Young, A., Yung, M.: Malicious Cryptography: Exposing Cryptovirology. Wiley, Hoboken (2004)

49. Zheng, Y., Seberry, J.: Immunizing public key cryptosystems against chosen ciphertext attacks. IEEE J. Sel. Areas Commun. **11**(5), 715–724 (1993)

Boolean Functions for
Homomorphic-Friendly Stream Ciphers

Claude Carlet[1,2(✉)] and Pierrick Méaux[3]

[1] Department of Informatics, University of Bergen, Bergen, Norway
[2] LAGA, University of Paris 8, Saint-Denis, France
claude.carlet@gmail.com
[3] ICTEAM/ELEN/Crypto Group, Université catholique de Louvain,
Louvain-la-Neuve, Belgium
pierrick.meaux@uclouvain.be

Abstract. The proliferation of small embedded devices having growing but still limited computing and data storage facilities, and the related development of cloud services with extensive storage and computing means, raise nowadays new privacy issues because of the outsourcing of data processing. This has led to a need for symmetric cryptosystems suited for hybrid symmetric-FHE encryption protocols, ensuring the practicability of the FHE solution. Recent ciphers meant for such use have been introduced, such as LowMC, Kreyvium, FLIP, and Rasta. The introduction of stream ciphers devoted to symmetric-FHE frameworks such as FLIP and its recent modification has in its turn posed new problems on the Boolean functions to be used in them as filter functions. We recall the state of the art in this matter and present further studies (without proof).

1 Introduction

The cloud has become nowadays an unavoidable complement to a variety of embedded devices such as mobile phones, smart cards, smart-watches, as these cannot perform all the storage and computing needed by their use. This raises a new privacy concern: it must be impossible to the cloud servers to learn about the data of the users. The first scheme of *fully homomorphic encryption* (*FHE*) realized by Gentry [Gen09] gives a solution to this problem by providing an encryption scheme $\mathbf{C}^H$ preserving both operations of addition and multiplication:

$$\mathbf{C}^H(m + m') = \mathbf{C}^H(m) \boxplus \mathbf{C}^H(m'); \ \mathbf{C}^H(mm') = \mathbf{C}^H(m) \boxdot \mathbf{C}^H(m'). \quad (1)$$

Then, combining these two operations allows to evaluate any polynomial over the algebraic structure where m and m' live, allowing to perform any computation if this structure is a finite field, or even if it is a vector space over a finite field, since we know that any function over such structure is (univariate, resp. multivariate) polynomial (see [TKP68]). Let us recall how such scheme can be used if one

C. T. Gueye et al. (Eds.): A2C 2019, CCIS 1133, pp. 166–182, 2019.
https://doi.org/10.1007/978-3-030-36237-9_10

wants to compute the image of some data by some function, and needs the help of the cloud for that. We first represent the data by elements m_i of a finite field $\mathbb{F}_q$ (or a ring but she shall restrict ourselves to a field), where $i \in I \subseteq \mathbb{N}$; the function, transposed as a function over $\mathbb{F}_q^I$, that we shall denote by $F(m_i, i \in I)$, becomes then a polynomial, according to what we recalled above (or according to the fact that the vector space $\mathbb{F}_q^I$ can be identified with the field $\mathbb{F}_{q^{|I|}}$). If one needs the help of the cloud for the computation, it is sufficient to send $\mathbf{C}^H(m_i)$ for $i \in I$ to the cloud server, which will compute $F(\mathbf{C}^H(m_i), i \in I)$. Thanks to (1), this value will equal $\mathbf{C}^H(F(m_i, i \in I))$ and decryption by the owner of the private decryption key will provide $F(m_i, i \in I)$, and the server will have not learned anything about the m_i nor about $F(m_i, i \in I)$. The computation to perform is transposed as a function F over this field since the homomorphic operations allowed by a FHE scheme are only defined for this field (or ring), and it does not allow to perform other operations using different representation of the data. For example, a FHE scheme for plaintexts from $(\mathbb{F}_{2^n}, +, \times)$ cannot handle the plaintexts as elements from $\mathbb{F}_2^n$, therefore in this case any computation is evaluated based on the univariate representation.

But the theoretical solution we just described is not practical by itself, because the repetitive use of homomorphic encryption (and even in most cases a single use!) requires itself too much computational power and storage capacity than what can offer a device like those listed above. In practice, FHE schemes come from noise-based cryptography such as schemes relying on the learning with errors assumption [Reg05]. Each ciphertext contains an error part, also called noise, hiding the relation with the plaintext and the secret key. When a plaintext is encrypted, the error corresponds to a vector of low norm, and a ciphertext can be decrypted correctly until the amount of noise reaches a fixed bound. When homomorphic operations are performed on ciphertexts, the noise increases. The cornerstone of FHE is a function called bootstrapping, that allows to obtain a ciphertext for m with a low noise from a noisy ciphertext of m. This function corresponds to homomorphically performing the decryption of the FHE scheme and requires an extra key. Bootstrapping is the most costly algorithm of a FHE cryptosystem in terms of computation and storage, and two differ- ent strategies are used to get around this bottleneck. The most spread strategy consists in minimizing the number of bootstrappings during an evaluation; the parameters are taken to allow a bounded number of operations (or levels) on fresh ciphertexts before bootstraping. The best performances are obtained by expressing the functions to evaluate in a way minimizing the noise growth. A more recent strategy initiated by [DM15], later referred as gate-bootstrapping, performs a bootstrapping at each operation, amortizing the cost by combining the two functions (the operation and the bootstrapping) at once. Such strategy can lead to good performances when function F can be evaluated as a circuit with a limited number of gates. For both strategies, some functions F imply a low noise growth (and can be qualified as homomorphic-friendly, but the func- tions qualified this way in the title of this paper are also functions involved in the hybrid symmetric-FHE encryption itself; they need then extra properties, see below); these functions depend on the particular FHE scheme chosen.

The main drawback of FHE constructions is the huge expansion factor, the ratio between the size of the plaintext (in bits) and the size of the corresponding ciphertexts. The expansion factor can be as big as 1.000.000, and it implies the major constraints for small devices. Indeed, doing computations on small devices with these ciphertexts is challenging, and only a limited number of homomorphic ciphertexts can be handled at the same time. A solution to this problem for the user, traditionally called Alice, is to use a *hybrid symmetric-FHE encryption* protocol:

1. Alice sends to the server her public key pk^H associated to the chosen homomorphic encryption protocol and the ciphertext $\mathbf{C}^H(\mathsf{sk}^S)$ corresponding to the homomorphic encryption of her key sk^S associated to a chosen symmetric encryption scheme $\mathbf{C}^S$,
2. she encrypts her data m with $\mathbf{C}^S$, and sends $\mathbf{C}^S(m)$ to the server,
3. the server computes $\mathbf{C}^H(\mathbf{C}^S(m))$ and homomorphically evaluates the decryption of the symmetric scheme on Alice's data; it obtains $\mathbf{C}^H(m)$,
4. the server homomorphically executes polynomial function F on Alice's data, and obtains $\mathbf{C}^H(F(m))$,
5. the server sends $\mathbf{C}^H(F(m))$ to Alice who obtains $F(m)$ by decrypting (whose operation is much less costly than encrypting in FHE).

Such symmetric-FHE framework allows Alice to circumvent the huge costs implied by the expansion factor. In this context she uses homomorphic encryption only in the first step (on a small data, the symmetric key), and then she uses symmetric encryption and homomorphic decryption. Both of these algorithms are operations that can be efficiently performed on limited devices. Consequently, the performance of the whole hybrid framework is mainly determined by the third and fourth steps. Since the homomorphic evaluation of the symmetric decryption algorithm is independent of the applications wanted by Alice, one natural goal consists in making its over-cost as low as possible. It would make correspond the time cost of the framework to the computations of the fourth step only, linking the performances to the complexity of computations delegated. As explained above, for reasons of efficiency, the choice of the symmetric cipher itself $\mathbf{C}^S$ is central in this matter, since its decryption should keep as low as possible the noise of homomorphic ciphertexts. The error-growth given by the basic homomorphic operations is different for each FHE scheme but some general trends can be exhibited, therefore giving guidelines to understand which symmetric schemes are homomorphic-friendly. For the schemes often qualified as second generation such as [BV11, BGV12, FV12], the sum corresponds to adding the noises whereas the error produced in a product is way more important. Each level in the tree representing the multiplication corresponds to a level of noise [HS14], hence the final noise is often well approximated by relating it to the multiplicative depth of the circuit evaluated, or equivalently, $\lceil \log(d)_2 \rceil$ where d is the degree of the (univariate, resp. multivariate) polynomial evaluated (note that there is no ambiguity on the degree since the homomorphic operations are valid over one field or ring only, therefore allowing only one representation). The homomorphic-friendly functions for this generation are the ones with a low multiplicative depth (and a bounded number of additions). For the schemes following the blueprint of [GSW13] such as [KGV14, CGGI16], referred as third generation, the error-growth of a product is asymmetric in the input ciphertexts. This property allows to obtain a small noise for the result of many products, with some limitations. More precisely, products of ciphertexts where always one (of both) has a small noise results in a small noise. This more complex error-growth gives access to other homomorphic-friendly functions, beyond the restrictions of very small degree. Examples of homomorphic-friendly functions for this generation are given by the sums of successive products, or combinations of multiplexers using a fresh variable.

When the decryption function of a symmetric scheme is evaluated as a polynomial in one field rather than combining computations over different representations, such as alternating operations from $\mathbb{F}_2^n$ to $\mathbb{F}_2^n$ and operations from $\mathbb{F}_{2^n}$ to $\mathbb{F}_{2^n}$ on a value from the same register, it often leads to a high degree and many terms. For example, the multiplicative depth of AES is often too large, and its additive depth is still more, to efficiently evaluate it homomorphically. Thus, other symmetric encryption schemes have been proposed in the context of symmetric-FHE frameworks: some block ciphers, like *LowMC* [ARS+16], *Rasta* and *Agrasta* [DEG+18], and stream ciphers such as *Kreyvium* [CCF+16]. These solutions have drawbacks: Kreyvium becomes more and more expensive during the encryption since the noise in the produced ciphertexts increases (or the system has to be reboot often). LowMC provides low noise at each round, but the iteration of rounds makes it unadapted, as almost any other block cipher since the lower bound on the round number for security reverberates on the homomorphic evaluation. We can observe this impact by studying how they can work with HElib [HS14] for instance, where the number of homomorphic levels required is always at least the number of rounds. This is however a minor drawback in this generation (HElib implements the FHE of [BGV12]) for Rasta and Agrasta, since they allow a very small number of rounds. These schemes are also well adapted for *multiparty computation*, but not for all FHE, for example their high number of sums are not well suited for the third generation. In this paper, we focus our study on symmetric encryption schemes that could be tailored for any FHE scheme, and more precisely on the functions used in these homomorphic-friendly constructions.

The *FLIP* cipher is an also very efficient encryption scheme, described in [MJSC16], which tries to minimize the noise involved in homomorphic evaluation. More precisely it intents to optimize the parameters mentioned above, targeting the most homomorphic-friendly functions which are sufficient to ensure security (for example minimizing the multiplicative depth). This scheme is based on a new stream cipher model, called the *filter permutator* (see Fig. 1). It consists in updating at each clock cycle a key register by a permutation of the coordinates. A pseudorandom number generator (PRNG) pilots the choice of the permutation.

The permuted key is then filtered, like in a classical stream cipher, by a Boolean function f whose output provides the *keystream*. Note that the input to f is the whole key register and this is a difference with the classical way of using filter functions. Applying the non-linear filtering function directly on the key bits allows to greatly reduce the noise level in the framework of hybrid symmetric-FHE encryption protocols. More precisely, the noise is given by the evaluation of one function only: the filtering function, rather than by the combination of all the functions used in the decryption algorithm as for other schemes. In theory, there are no big differences between the filter model and the filter permutator: the LFSR is simply replaced by a permutator. Nevertheless, in practice there are huge differences since the filter function has hundreds of input bits instead of about 20.

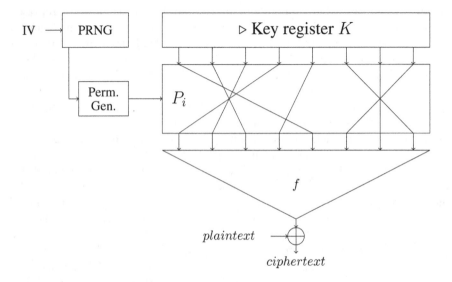

Fig. 1. Filter permutator construction.

In the versions of the cipher proposed in [MJSC16], the function f has $n = n_1 + n_2 + n_3 \geq 500$ variables, where n_2 is even and n_3 equals $\frac{k(k+1)}{2}t$ for some integers k and t. The functions $f(x_0, \ldots, x_{n_1-1}, y_0, \ldots, y_{n_2-1}, z_0, \ldots, z_{n_3-1})$ is defined as:

$$\sum_{i=0}^{n_1-1} x_i + \sum_{i=0}^{n_2/2-1} y_{2i}\, y_{2i+1} + \sum_{j=0}^{t-1} T_k\left(z_{\frac{jk(k+1)}{2}}, z_{\frac{jk(k+1)}{2}+1}, \ldots, z_{\frac{jk(k+1)}{2}+\frac{k(k+1)}{2}-1} \right),$$

where the *triangular function* T_k is defined as:

$$T_k(z_0, \ldots, z_{j-1}) = z_0 + z_1 z_2 + z_3 z_4 z_5 + \cdots + z_{\frac{k(k-1)}{2}} \cdots z_{\frac{k(k+1)}{2}-1}.$$

The filter permutator has been improved in [MCJS19] (see Fig. 2), and some results on this improvement will be published soon in the article at Indocrypt 2019 entitled "Improved Filter Permutators for Efficient FHE: Better Instances and Implementations". There are two modifications. Firstly, at each clock cycle, the function is applied on a part of the key rather than on the whole key register and, secondly, a public vector (called whitening) is added before the computation of f. The subset of the key register used and the whitening are derived from the PRNG's output at each clock cycle, like the permutation. These modifications have no impact on the noise when the cipher decryption is homomorphically evaluated, the final noise is the one given by the evaluation of f only. On the security side, the resulting register extension allows to obtain the same security with simpler functions, and the whitening allows to temper the attacks using guess-and-determine strategies. The combination of both makes possible to study the security more easily, relating it with the Boolean cryptographic criteria of f, and those of the functions obtained by fixing variables in the input to f.

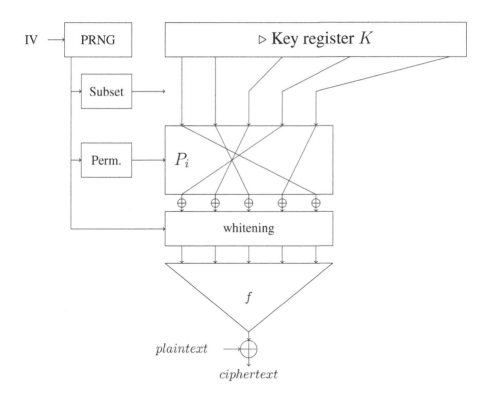

Fig. 2. Improved filter permutator construction.

The attacks which classically apply on the filter model, or slightly modified ones, can apply on the filter permutator or its improved version. Then, the usual Boolean cryptographic criteria have to be taken into consideration for the choice of f. More precisely, for algebraic attacks or variants, the parameters of algebraic immunity, fast algebraic immunity, and number of annihilators of minimal algebraic degree are important. Standard correlation attacks do not apply on these models, but some variations can; this motivates to address the resiliency and nonlinearity of f. Since, unlike the filter model, the key register in the filter permutator is not updated, guessing some key bits importantly simplifies the system of equations given by the keystream. It makes attacks using guess-and-determine strategies more efficient [DLR16] than regular ones. Bounding the complexity of these attacks necessitates to determine the cryptographic parameters of the functions obtained by fixing various variables in f. These security considerations bring us to focus on families of functions with more variables than usually and whose sub-function parameters can be determined or at least be efficiently bounded.

In this article we give without proof the relevant cryptographic parameters for two families of homomorphic-friendly functions. Functions from these families enable to securely instantiate the filter permutator or the improved filter permutator, and allow a very efficient homomorphic evaluation.

2   Preliminaries

For readability we use the notation $+$ instead of $\oplus$ to denote addition in $\mathbb{F}_2$. We denote $\{1, \ldots, n\}$ by $[n]$.

2.1   Boolean Functions and Cryptographic Criteria

Boolean Functions. We recall here some core notions on Boolean functions in cryptography, restricting our study to the single-output Boolean functions.

Definition 1 (Boolean Function). *A Boolean function f in n variables is a function from $\mathbb{F}_2^n$ to $\mathbb{F}_2$. The set of all Boolean functions in n variables is denoted by $\mathcal{B}_n$. We call pseudo-Boolean function a function with input space $\mathbb{F}_2^n$ but output space different from $\mathbb{F}_2$ (e.g. $\mathbb{R}$).*

Definition 2 (Algebraic Normal Form (ANF)). *We call Algebraic Normal Form of a Boolean function f its n-variable polynomial representation over $\mathbb{F}_2$ (i.e. belonging to $\mathbb{F}_2[x_1, \ldots, x_n]/(x_1^2 + x_1, \ldots, x_n^2 + x_n)$):*

$$f(x) = \sum_{I \subseteq [n]} a_I \left(\prod_{i \in I} x_i \right) = \sum_{I \subseteq [n]} a_I x^I,$$

where $a_I \in \mathbb{F}_2$.

Every Boolean functions has a unique ANF. The degree of this unique ANF is called the algebraic degree of the function and denoted by $\deg(f)$.

Boolean Criteria. In this part, we recall the main cryptographic properties of Boolean functions, mostly taken from [Car10]: balancedness, resiliency, nonlinearity, algebraic immunity, fast algebraic immunity and minimal degree annihilator space's dimension.

Definition 3 (Balancedness). *A Boolean function $f \in \mathcal{B}_n$ is said to be balanced if its output is uniformly distributed over $\{0, 1\}$.*

Definition 4 (Resiliency). *A Boolean function $f \in \mathcal{B}_n$ is called m-resilient if any of its restrictions obtained by fixing at most m of its coordinates is balanced. We denote by $\mathsf{res}(f)$ the maximum resiliency (also called resiliency order) of f and set $\mathsf{res}(f) = -1$ if f is unbalanced.*

Definition 5 (Nonlinearity). *The nonlinearity $\mathsf{NL}(f)$ of a Boolean function $f \in \mathcal{B}_n$, where n is a positive integer, is the minimum Hamming distance between f and all the affine functions in $\mathcal{B}_n$:*

$$\mathsf{NL}(f) = \min_{g,\, \deg(g) \leq 1} \{d_H(f, g)\},$$

with $d_H(f, g) = \#\{x \in \mathbb{F}_2^n \mid f(x) \neq g(x)\}$ the Hamming distance between f and g, and $g(x) = a \cdot x + \varepsilon$; $a \in \mathbb{F}_2^n, \varepsilon \in \mathbb{F}_2$ (where $\cdot$ is some inner product in $\mathbb{F}_2^n$; any choice of an inner product will give the same value of $\mathsf{NL}(f)$).

Definition 6 (Algebraic Immunity and Annihilators). *The algebraic immunity of a Boolean function $f \in \mathcal{B}_n$, denoted as $\mathsf{AI}(f)$, is defined as:*

$$\mathsf{AI}(f) = \min_{g \neq 0}\{\deg(g) \mid fg = 0 \text{ or } (f+1)g = 0\},$$

where $\deg(g)$ is the algebraic degree of g. The function g is called an annihilator of f (or $f + 1$).
We additively use the notation $\mathsf{dAN}(f)$ for the minimum algebraic degree of non null annihilator of f:

$$\mathsf{dAN}(f) = \min_{g \neq 0}\{\deg(g) \mid fg = 0\}.$$

We also use the notation $\mathcal{DAN}(f)$ for the dimension of the vector space made of the annihilators of f of degree $\mathsf{AI}(f)$ and the zero function. Note that, for every function f we have $\mathcal{DAN}(f) \leq \binom{n}{\mathsf{AI}(f)}$, because two distinct annihilators of algebraic degree $\mathsf{AI}(f)$ cannot have in their ANF the same part of degree $\mathsf{AI}(f)$ (their difference being itself an annihilator).

Definition 7. (Fast Algebraic Immunity [ACG+06]). *The fast algebraic immunity of a Boolean function $f \in \mathcal{B}_n$, denoted as $\mathsf{FAI}(f)$, is defined as:*

$$\mathsf{FAI}(f) = \min\{2\mathsf{AI}(f), \min_{1 \leq \deg(g) < \mathsf{AI}(f)} \deg(g) + \deg(fg)\}.$$

Families of Boolean Functions. In the very constrained framework in which we are, where we need to maximize the ratio $\frac{security}{complexity}$ of our functions in a much more drastic way than for classical stream ciphers, we highlight three families of functions: direct sum of monomials, threshold functions, and XOR-Threshold functions. We begin by introducing the secondary construction (*i.e.* construction of functions using already defined functions as building blocks) called direct sum, which will lead to the first of these families. This secondary construction is usually considered as unadapted to the design of cryptographic Boolean functions, because the *decomposability* of the functions it provides may be used in attacks. But in our framework, the number of variables is more than 500 (while in classical stream ciphers it is about 20) and this changes the situation. Moreover, the direct sum will lead us to a quite interesting class of functions (direct sums of monomials), well adapted to our framework, and for which we shall be able to determine the cryptographic parameters of all functions in the class. This is the first time that all functions in a whole class of functions can be evaluated (with the exception of the Maiorana-McFarland class, see [Car10], but the functions in this latter class do not have quite good algebraic immunity). Before, the contributions of the papers were to construct functions achieving provably good characteristics; the corresponding classes contained only one or at most a few functions in each number of variables. Concretely, in most cases, these functions had optimal algebraic immunity (this was necessary because of the rather small number of variables). Here we shall have much more flexibility for the choice of functions within the class.

Definition 8 (Direct Sum). *Let f be a Boolean function of n variables and g a Boolean function of m variables, f and g depending on distinct variables, the direct sum h of f and g is defined by:*

$$h(x, y) = f(x) + g(y), \quad \text{where } x \in \mathbb{F}_2^n \text{ and } y \in \mathbb{F}_2^m.$$

The direct sum has been generalized into the indirect sum (see [Car10]) which provides more complex functions, better adapted to classical stream ciphers. It seems that using the indirect sum does not allow to have simple enough functions for our framework nor to determine exactly the cryptographic parameters of all the functions in a class. We focus more precisely on direct sums of monomials, which consist of functions where each variable appears at most once in the ANF.

Definition 9 (Direct Sum of Monomials). *Let f be a non constant Boolean function of n variables, we call f a Direct Sum of Monomials (or DSM) if the following holds for its ANF:*

$$\forall (I, J) \text{ such that } a_I = a_J = 1, \ I \cap J \in \{\emptyset, I \cup J\}.$$

Definition 10. (Direct Sum Vector [MJSC16]). *Let f be a DSM, we define its direct sum vector:*
$$\mathbf{m}_f = [m_1, m_2, \ldots, m_k],$$

of length $k = \deg(f)$, *where* m_i *is the number of monomials of degree* i, $i > 0$, *of* f:

$$m_i = |\{a_I = 1, \ such \ that \ |I| = i\}|.$$

The function f *associated to the direct sum vector* $\mathbf{m}_f = [m_1, m_2, \ldots, m_k]$ *has* $M = \sum_{i=1}^{k} m_i$ *monomials in its ANF and has* $N \geq \sum_{i=1}^{k} i m_i$ *variables.*

As we wrote, we shall be able to determine all parameters of all functions in this class.

We also define the family of threshold functions, which is a super-class of that of majority functions.

Definition 11 (Threshold Function). *For any positive integers* $d \leq n + 1$ *we define the Boolean function* $\mathsf{T}_{d,n}$ *as:*

$$\forall x = (x_1, \ldots, x_n) \in \mathbb{F}_2^n, \quad \mathsf{T}_{d,n}(x) = \begin{cases} 0 & if \ \mathsf{w_H}(x) < d, \\ 1 & otherwise. \end{cases}$$

Definition 12 (Majority Function). *For any positive odd integer* n *we define the Boolean function* MAJ_n *as:*

$$\forall x = (x_1, \ldots, x_n) \in \mathbb{F}_2^n, \quad \mathsf{MAJ}_n(x) = \begin{cases} 0 & if \ \mathsf{w_H}(x) \leq \lfloor \frac{n}{2} \rfloor, \\ 1 & otherwise. \end{cases}$$

Note that threshold functions are symmetric functions (changing the order of the input bits does not change the output), which have been the focus of many studies *e.g.* [Car04, CV05, DMS06, QLF07, SM07, QFLW09]. Note also that $\mathsf{MAJ}_n = \mathsf{T}_{\lceil \frac{n+1}{2} \rceil, n}$. These functions can be described more succinctly through the simplified value vector.

Definition 13 (Simplified Value Vector). *Let* f *be a symmetric function in* n *variables, we define its simplified value vector:*

$$\mathbf{s} = [w_0, w_1, \ldots, w_n]$$

of length n, *where for each* $k \in \{0, \ldots, n\}$, $w_k = f(x)$ *where* $\mathsf{w_H}(x) = k$, *i.e.* w_k *is the value of* f *on all inputs of Hamming weight* k.

Note that for a threshold function, we have $w_k = 0$ for $k < d$ and 1 otherwise, so the simplified value vector of a threshold function $\mathsf{T}_{d,n}$ is the $n + 1$-length vector of d consecutive 0's and $n + 1 - d$ consecutive 1's.

We will also be interested in functions obtained by a direct sum of a linear direct sum of monomials and a threshold function, called XOR-THR (or XOR-MAJ when the threshold function happens to be a majority function).

Definition 14. (XOR-THR Function). *For any positive integers* k, d *and* n *such that* $d \leq n + 1$ *we define* $\mathsf{XOR}_k + \mathsf{T}_{d,n}$ *for all* $z = (x_1, \ldots, x_k, y_1, \ldots, y_n) \in \mathbb{F}_2^{k+n}$ *as:*

$$(\mathsf{XOR}_k + \mathsf{T}_{d,n})(z) = x_1 + \cdots + x_k + \mathsf{T}_{d,n}(y_1, \ldots, y_n) = \mathsf{XOR}_k(x) + \mathsf{T}_{d,n}(y).$$

Boolean Functions and Bit-Fixing. In this part, we give the necessary vocabulary relatively to bit-fixing (as defined e.g. in [AL16]) on Boolean functions, the action consisting in fixing the value of some variables of a Boolean function and then considering the resulting Boolean function. These notions are important when guess-and-determine attacks are investigated.

Definition 15 (Bit-fixing Descendant). *Let f be a Boolean function in n variables (x_i, for $i \in [n]$), let ℓ be an integer such that $0 \leq \ell < n$, let $I \subset [n]$ be of size ℓ (i.e. $I = \{i_1, \ldots, i_\ell\}$ with $i_j < i_{j+1}$ for all $j \in [\ell - 1]$), and let $b \in \mathbb{F}_2^\ell$, we denote as $f_{I,b}$ the ℓ-bit fixing descendant of f on subset I with binary vector b the Boolean function in $n - \ell$ variables:*

$$f_{I,b}(x') = f(x) \mid \forall j \in [\ell], \; x_{i_j} = b_j,$$

where $x' = (x_i)_{i \in [n] \setminus I}$.

Definition 16 (Bit-fixing Stability). *Let $\mathcal{F}$ be a family of Boolean functions, $\mathcal{F}$ is called bit-fixing stable, or stable relatively to guess and determine, if for all functions $f \in \mathcal{F}$ such that f is a n-variable function with $n > 1$, the following holds:*

- *for all number of variables ℓ such that $0 \leq \ell < n$,*
- *for all choice of positions $1 \leq i_1 < i_2 < \cdots < i_\ell \leq n$,*
- *for all value of guess $(b_1, \ldots, b_\ell) \in \mathbb{F}_2^\ell$,*

at least one of these properties is fulfilled: $f_{I,b} \in \mathcal{F}$, or $f_{I,b}+1 \in \mathcal{F}$, or $\deg(f_{I,b}) = 0$.

Remark 1. Both DSM and XOR-THR functions are bit-fixing stable families (as well as the set of threshold functions). Therefore, if the cryptographic parameters of any functions of one of these families are determined then the complexity of all attacks using guess-and-determine strategies on any filtering function of this family can be derived.

3 Parameters of Direct Sums of Monomials

In this section we give the relevant parameters relatively to Boolean cryptographic criteria of DSM functions. The DSM are a generalization of triangular and FLIP functions [MJSC16]. Their very sparse ANF is the reason why they are homomorphic-friendly. Note that such function in n variables can be computed with at most $n - 1$ additions and multiplications. Regarding the satisfaction of the constraints of the second generation of FHE schemes, secure functions can have a multiplicative depth as low as 2 or 3 [MCJS19]. Focusing on the third generation, each monomial of the ANF can be evaluated as a serial multiplication of freshy encrypted ciphertexts, then the whole function is evaluated as a sum of multiplicative chains, giving an error-growth quasi-additive in n [MCJS19]. On the cryptographic point of view, the DSM can be obtained by recursively applying the direct sum construction, which is convenient to determine the resiliency and nonlinearity, but not to study the exact behavior of the algebraic properties such as the algebraic immunity and the dimension of annihilators (non null) of minimal degree.

First we recall some properties on direct sums (see *e.g.* [MJSC16]).

Lemma 1 (Direct Sum Properties ([MJSC16] Lemma 3)). Let F be the direct sum of f and g with n and m variables respectively. Then F has the following cryptographic properties:

1. Resiliency: $\mathsf{res}(F) = \mathsf{res}(f) + \mathsf{res}(g) + 1$.
2. Non Linearity: $\mathsf{NL}(F) = 2^m \mathsf{NL}(f) + 2^n \mathsf{NL}(g) - 2\mathsf{NL}(f)\mathsf{NL}(g)$.
3. Algebraic Immunity: $\max(\mathsf{AI}(f), \mathsf{AI}(g)) \leq \mathsf{AI}(F) \leq \mathsf{AI}(f) + \mathsf{AI}(g)$.
4. Fast Algebraic Immunity: $\mathsf{FAI}(F) \geq \max(\mathsf{FAI}(f), \mathsf{FAI}(g))$.

The previous lemma is sufficient to determine the resiliency and the nonlinearity of any direct sums of monomials.

Lemma 2 (Resiliency of Direct Sum of Monomials). *Let $f \in \mathbb{F}_2^n$ be a Boolean function obtained by direct sums of monomials with associated direct sum vector $= [m_1, \ldots, m_k]$, its resiliency is:*

$$\mathsf{res}(f) = m_1 - 1$$

Lemma 3 (Nonlinearity of Direct Sum of Monomials). *Let $f \in \mathbb{F}_2^n$ be a Boolean function obtained by direct sums of monomials with associated direct sum vector $\mathbf{m}_f = [m_1, \ldots, m_k]$, its nonlinearity is:*

$$\mathsf{NL}(f) = 2^{n-1} - \frac{1}{2} \left(2^{(n - \sum_{i=2}^k i m_i)} \prod_{i=2}^k \left(2^i - 2 \right)^{m_i} \right)$$

Now, we give the algebraic immunity, a lower bound on the FAI and an upper bound on th $\mathcal{D}$AN of a direct sum of monomials.

Theorem 1 (Algebraic Immunity of Direct Sums of Monomials). *Let $f \in \mathbb{F}_2^n$ be a Boolean function obtained by direct sums of monomials with associated direct sum vector $\mathbf{m}_f = [m_1, \ldots, m_k]$, its algebraic immunity is:*

$$\mathsf{AI}(f) = \min_{0 \leq d \leq k} \left(d + \sum_{i=d+1}^k m_i \right).$$

Lemma 4 (Fast Algebraic Immunity of Direct Sums of Monomials). *Let $f \in \mathbb{F}_2^n$ be a Boolean function obtained by the direct sum of monomials with associated direct sum vector $\mathbf{m}_f = [m_1, \ldots, m_k]$ such that $\mathsf{AI}(f) = \deg(f)$, and $\mathsf{AI}(f) > 1$, its fast algebraic immunity is:*

$$\mathsf{FAI}(f) = \begin{cases} \mathsf{AI}(f) + 1 & \text{if } m_k = 1, \\ \mathsf{AI}(f) + 2 & \text{otherwise.} \end{cases}$$

Note that this lemma does not consider the case $\mathsf{AI}(f) = 1$ (of linear functions or monomial functions), for this case the fast algebraic immunity is not very relevant as the algebraic attack already targets a linear system.

Theorem 2. *Let f be a DSM with associated direct sum vector $\mathbf{m}_f = [m_1, \ldots, m_k]$. Let us consider the set $\mathsf{S}_d(f)$ such that:*

$$\mathsf{S}_d(f) = \begin{cases} \{0 \leq d \leq k \mid d + \sum_{i>d} m_i = \mathsf{AI}(f)\} & \text{if } m_1 \neq 1, \\ \{0 < d \leq k \mid d + \sum_{i>d} m_i = \mathsf{AI}(f)\} & \text{if } m_1 = 1. \end{cases}$$

Then, we have the following relation:

$$\mathcal{D}\mathsf{AN}(f) \leq \sum_{d \in \mathsf{S}_d(f)} \prod_{i>d}^{k+1} i^{m_i}.$$

Note that when $\mathsf{dAN}(f) = \mathsf{AI}(f)$ the bound is reached.

This formula gives a tight upper bound on the dimension of the annihilators of a DSM.

4   Parameters of Threshold, and **XOR-THR** Functions

In this section we exhibit the parameters of threshold functions and xor-threshold functions. The threshold functions are a generalization of majority functions, which are known to reach the optimal algebraic immunity. Despite a very dense ANF, other representations can lead to low noise ciphertexts, as shown in [MCJS19] inspired by branching programs. Using multiplexers, the noise for third generation schemes is quasi additive in the number n of variables of the threshold function. Combining it in direct sum with a xor function allows to compensate the resiliency, and also to connect it with the predicates considered secure to instantiate local PRGs [Gol00, App13]. Therefore, the xor-threshold functions are appealing as filtering functions, and the threshold functions are an intermediate step to exhibit the cryptographic parameters of xor-threshold functions. Regarding homomorphic evaluation, the noise mostly depends on the threshold part, therefore for the third generation the noise is quasi additive in the number of variables. Focusing on the second generation of FHE, secure instantiations from xor-threshold functions can be chosen with a multiplicative depth between 3 and 7 [MCJS19].

On the cryptographic point of view, the parameters of majority functions are well investigated, but lesser is known on the properties of threshold functions in general, such as their nonlinearity. The direct sum with a xor function allows to derive the resiliency and nonlinearity from the parameters of the threshold function. Nevertheless, the exact immunity and dimension of annihilators (non null) of minimal degree require more than the general results of direct sum constructions.

4.1 Threshold Functions

Threshold functions are symmetric functions, which have been much studied relatively to cryptographic significant criteria (*e.g.* [CV05]). The existence of optimal symmetric functions relatively to a specific criterion has been widely investigated, but the class of symmetric functions is too wide for their parameters to be studied globally; here we focus on the exact parameters of the subfamily of threshold function.

We first give the resiliency and nonlinearity of such functions.

Lemma 5 (Resiliency of Threshold Functions). *Let f be the threshold function $\mathsf{T}_{d,n}$,*

$$\mathsf{res}(\mathsf{T}_{d,n}) = \begin{cases} 0 & \text{if } n = 2d - 1, \\ -1 & \text{otherwise.} \end{cases}$$

Theorem 3 (Nonlinearity of Threshold Functions). *Let n be a non null positive integer, the threshold function $\mathsf{T}_{d,n}$ has the following nonlinearity:*

$$NL(\mathsf{T}_{d,n}) = \begin{cases} 2^{n-1} - \binom{n-1}{(n-1)/2} & \text{if } d = \frac{n+1}{2}, \\ \displaystyle\sum_{k=d}^{n} \binom{n}{k} = \mathsf{w_H}(\mathsf{T}_{d,n}) & \text{if } d > \frac{n+1}{2}, \\ \displaystyle\sum_{k=0}^{d-1} \binom{n}{k} = 2^n - \mathsf{w_H}(\mathsf{T}_{d,n}) & \text{if } d < \frac{n+1}{2}. \end{cases}$$

We then investigate the algebraic immunity of threshold functions. As Boolean functions used for cryptography, the majority functions have been introduced as functions reaching the optimal algebraic immunity (case of $\mathsf{T}_{(n+1)/2,n}$ and $\mathsf{T}_{n/2+1,n}$ as proven in [BP05, DMS06]), but their nonlinearity is not good. As far as we know, the exact algebraic immunity have not been investigated for all threshold functions, but it can be determined in various ways as for the majority functions.

Lemma 6 (Algebraic Immunity of Threshold Functions). *Let n be a non null positive integer, the threshold function $\mathsf{T}_{d,n}$ has the following algebraic immunity:*

$$\mathsf{AI}(\mathsf{T}_{d,n}) = \min(d, n - d + 1).$$

Lemma 7. ($\mathcal{DAN}$ of Threshold Functions). *Let n be a non null positive integer, the threshold function $\mathsf{T}_{d,n}$ has the following $\mathcal{DAN}$:*

$$\mathcal{DAN}(\mathsf{T}_{d,n}) = \begin{cases} 0 & \text{if } d < \frac{n+1}{2}, \\ \binom{n}{d-1} & \text{if } d \geq \frac{n+1}{2}. \end{cases}$$

Corollary 1 (Lower Bound on the Fast Algebraic immunity of Threshold Functions). *Let n be a non null positive integer, the fast algebraic immunity of the threshold function $\mathsf{T}_{d,n}$ follows the following bound:*

$$\mathsf{FAI}(\mathsf{T}_{d,n}) \geq \begin{cases} \min(2d, n - d + 2) & \text{if } d \leq \frac{n+1}{2}, \\ \min(2(n - d + 1), d + 1) & \text{if } d > \frac{n+1}{2}. \end{cases}$$

Remark 2. Note that this bound can be reached, as proven in [TLD16] for the majority functions $T_{2^{m-1},2^m}$ and $T_{2^{m-1}+1,2^m+1}$ for all integers $m \geq 2$.

4.2   Parameters of **XOR-THR** Functions

The particular structure of XOR-THR functions make the resiliency and nonlinearity parameters easy to determine from the ones of these two components.

Lemma 8. (Resiliency of XOR-THR Functions). *Let f be the XOR-THR function* $\text{XOR}_k + T_{d,n}$, *then:*

$$\text{res}(\text{XOR}_k + T_{d,n}) = \begin{cases} k & \text{if } n = 2d - 1, \\ k - 1 & \text{otherwise.} \end{cases}$$

Lemma 9. (Nonlinearity of XOR-THR Functions). *Let f be the XOR-THR function* $\text{XOR}_k + T_{d,n}$, *then:*

$$\text{NL}(\text{XOR}_k + T_{d,n}) = \begin{cases} 2^{n+k-1} - 2^k \binom{n-1}{(n-1)/2} & \text{if } d = \frac{n+1}{2}, \\ 2^k \sum_{i=d}^{n} \binom{n}{i} & \text{if } d > \frac{n+1}{2}, \\ 2^k \sum_{i=0}^{d-1} \binom{n}{i} & \text{if } d < \frac{n+1}{2}. \end{cases}$$

Then we focus on the exact algebraic immunity of these functions, a lower bound on the fast algebraic immunity, and the exact $\mathcal{DAN}$.

Lemma 10 (Algebraic Immunity of XOR Threshold Functions). *Let f be the* $\text{XOR}_k + T_{d,n}$ *function:*

$$\text{AI}(\text{XOR}_k + T_{d,n}) = \begin{cases} \frac{n+1}{2} & \text{if } d = \frac{n+1}{2}, \\ \min\{k,1\} + \min\{d, n-d+1\} & \text{otherwise.} \end{cases}$$

Lemma 11. (Fast Algebraic Immunity of a XOR-THR Function). *Let f be the* $\text{XOR}_k + T_{d,n}$ *function:*
if $k = 0$, then:

$$\text{FAI}(\text{XOR}_0 + T_{d,n}) \geq \min(2\min(d, n-d+1), 1 + \max(d, n-d+1)).$$

If $k > 0$

$$\text{FAI}(\text{XOR}_k + T_{d,n}) \geq \begin{cases} \frac{n+3}{2} & \text{if } d = \frac{n+1}{2}, \\ 2 + \min(d, n-d+1) & \text{otherwise.} \end{cases}$$

Lemma 12. ($\mathcal{DAN}$ of XOR-THR Functions). *Let $\text{XOR}_k + T_{d,n}$ be a XOR-THR function such that $k > 0$, $n \in \mathbb{N}$, and $1 \leq d \leq n$, then:*

$$
\mathcal{D}\mathrm{AN}(\mathrm{XOR}_k + \mathsf{T}_{d,n}) =
\begin{cases}
\binom{n}{d} & \text{if } d < \frac{n}{2}, \\
\binom{n+1}{d+1} & \text{if } d = \frac{n}{2}, \\
\binom{n}{d} & \text{if } d = \frac{n+1}{2}, \\
\binom{n+1}{d} & \text{if } d = \frac{n}{2} + 1, \\
\binom{n}{d-1} & \text{otherwise.}
\end{cases}
$$

Furthermore $\mathcal{D}\mathrm{AN}(\mathrm{XOR}_k + \mathsf{T}_{d,n}) = \mathcal{D}\mathrm{AN}(1 + \mathrm{XOR}_k + \mathsf{T}_{d,n})$.

References

[ACG+06] Armknecht, F., Carlet, C., Gaborit, P., Künzli, S., Meier, W., Ruatta, O.: Efficient computation of algebraic immunity for algebraic and fast algebraic attacks. In: Vaudenay, S. (ed.) EUROCRYPT 2006. LNCS, pp. 147–164. Springer, Heidelberg (2006). https://doi.org/10.1007/11761679_10

[AL16] Applebaum, B., Lovett, S.: Algebraic attacks against random local functions and their countermeasures. SIAM J. Comput. **47**(1), 52–79 (2016)

[App13] Applebaum, B.: Cryptographic hardness of random local functions-survey. In: TCC, p. 599 (2013)

[ARS+16] Albrecht, M.R., Rechberger, C., Schneider, T., Tiessen, T., Michael, Z.: Ciphers for MPC and FHE. IACR Cryptology ePrint Archive, p. 687 (2016)

[BGV12] Brakerski, Z., Gentry, C., Vaikuntanathan, V.: (Leveled) fully homomorphic encryption without bootstrapping. In: ITCS, pp. 309–325 (2012)

[BP05] Braeken, A., Preneel, B.: On the algebraic immunity of symmetric boolean functions. In: Maitra, S., Veni Madhavan, C.E., Venkatesan, R. (eds.) INDOCRYPT 2005. LNCS, vol. 3797, pp. 35–48. Springer, Heidelberg (2005). https://doi.org/10.1007/11596219_4

[BV11] Brakerski, Z., Vaikuntanathan, V.: Efficient fully homomorphic encryption from (standard) LWE. In: FOCS, pp. 97–106 (2011)

[Car04] Carlet, C.: On the degree, nonlinearity, algebraic thickness, and nonnormality of boolean functions, with developments on symmetric functions. IEEE Trans. Inf. Theory **50**, 2178–2185 (2004)

[Car10] Carlet, C.: Boolean Functions for Cryptography and Error-Correcting Codes. Encyclopedia of Mathematics and its Applications, pp. 257–397. Cambridge University Press, Cambridge (2010)

[CCF+16] Canteaut, A., et al.: Stream ciphers: a practical solution for efficient homomorphic-ciphertext compression. J. Cryptol. **31**, 885–916 (2016)

[CGGI16] Chillotti, I., Gama, N., Georgieva, M., Izabachène, M.: Faster fully homomorphic encryption: bootstrapping in less than 0.1 seconds. In: Cheon, J.H., Takagi, T. (eds.) ASIACRYPT 2016. LNCS, vol. 10031, pp. 3–33. Springer, Heidelberg (2016). https://doi.org/10.1007/978-3-662-53887-6_1

[CV05] Canteaut, A., Videau, M.: Symmetric boolean functions. IEEE Trans. Inf. Theory **51**, 2791–2811 (2005)

[DEG+18] Dobraunig, C., et al.: Rasta: a cipher with low ANDdepth and few ANDs per bit. In: Shacham, H., Boldyreva, A. (eds.) CRYPTO 2018. LNCS, vol. 10991, pp. 662–692. Springer, Cham (2018). https://doi.org/10.1007/978-3-319-96884-1_22

[DLR16] Duval, S., Lallemand, V., Rotella, Y.: Cryptanalysis of the FLIP family of stream ciphers. In: Robshaw, M., Katz, J. (eds.) CRYPTO 2016. LNCS, vol. 9814, pp. 457–475. Springer, Heidelberg (2016). https://doi.org/10.1007/978-3-662-53018-4_17

[DM15]   Ducas, L., Micciancio, D.: FHEW: bootstrapping homomorphic encryption in less than a second. In: Oswald, E., Fischlin, M. (eds.) EUROCRYPT 2015. LNCS, vol. 9056, pp. 617–640. Springer, Heidelberg (2015). https://doi.org/10.1007/978-3-662-46800-5_24

[DMS06] Dalai, D.K., Maitra, S., Sarkar, S.: Basic theory in construction of boolean functions with maximum possible annihilator immunity. Des. Codes Crypt. **40**, 41–58 (2006)

[FV12]   Fan, J., Vercauteren, F.: Somewhat practical fully homomorphic encryption. IACR Cryptology ePrint Archive, p. 144 (2012)

[Gen09] Gentry, C.: Fully homomorphic encryption using ideal lattices. In: STOC, pp. 169–178 (2009)

[Gol00] Goldreich, O.: Candidate one-way functions based on expander graphs. Electron. Colloquium Comput. Complexity (ECCC) **7**(90) 2000

[GSW13] Gentry, C., Sahai, A., Waters, B.: Homomorphic encryption from learning with errors: conceptually-simpler, asymptotically-faster, attribute-based. In: Canetti, R., Garay, J.A. (eds.) CRYPTO 2013. LNCS, vol. 8042, pp. 75–92. Springer, Heidelberg (2013). https://doi.org/10.1007/978-3-642-40041-4_5

[HS14]   Halevi, S., Shoup, V.: Algorithms in HElib. In: Garay, J.A., Gennaro, R. (eds.) CRYPTO 2014. LNCS, vol. 8616, pp. 554–571. Springer, Heidelberg (2014). https://doi.org/10.1007/978-3-662-44371-2_31

[KGV14] Khedr, A., Gulak, G., Vaikuntanathan, V.: SHIELD: scalable homomorphic implementation of encrypted data-classifiers. Cryptology ePrint Archive, Report 2014/838 (2014)

[MCJS19] Méaux, P., Carlet, C., Journault, A., Standaert, F.-X.: Improved filter permutators: combining symmetric encryption design, boolean functions, low complexity cryptography, and homomorphic encryption, for private delegation of computations. Cryptology ePrint Archive, Report 2019/483 (2019). https://eprint.iacr.org/2019/483

[MJSC16] Méaux, P., Journault, A., Standaert, F.-X., Carlet, C.: Towards stream ciphers for efficient FHE with low-noise ciphertexts. In: Fischlin, M., Coron, J.-S. (eds.) EUROCRYPT 2016. LNCS, vol. 9665, pp. 311–343. Springer, Heidelberg (2016). https://doi.org/10.1007/978-3-662-49890-3_13

[QFLW09] Qu, L., Feng, K., Liu, F., Wang, L.: Constructing symmetric boolean functions with maximum algebraic immunity. IEEE Trans. Inf. Theory **55**, 2406–2412 (2009)

[QLF07] Qu, L., Li, C., Feng, K.: A note on symmetric boolean functions with maximum algebraic immunity in odd number of variables. IEEE Trans. Inf. Theory **53**, 2908–2910 (2007)

[Reg05] Regev, O.: On lattices, learning with errors, random linear codes, and cryptography. In: STOC, pp. 84–93 (2005)

[SM07]   Sarkar, P., Maitra, S.: Balancedness and correlation immunity of symmetric boolean functions. Discrete Math. **307**, 2351–2358 (2007)

[TKP68] Lin, S., Kasami, T., Peterson, W.W.: Polynomial codes. IEEE Trans. Inf. Theory **14**, 807–814 (1968)

[TLD16] Tang, D., Luo, R., Du, X.: The exact fast algebraic immunity of two subclasses of the majority function. IEICE Trans. Fundam. Electron. Commun. Comput. Sci. **99**, 2084–2088 (2016)

Jacobian Versus Infrastructure in Split Hyperelliptic Curves

Monireh Rezai Rad[1], Michael J. Jacobson Jr.[2($\boxtimes$)], and Renate Scheidler[1]

[1] Department of Mathematics and Statistics, University of Calgary,
2500 University Drive NW, Calgary, AB T2N 1N4, Canada
{mrezaira,rscheidl}@ucalgary.ca
[2] Department of Computer Science, University of Calgary,
2500 University Drive NW, Calgary, AB T2N 1N4, Canada
jacobs@ucalgary.ca

Abstract. Split (also known as real) hyperelliptic curves admit two main algebraic structures: the Jacobian and the infrastructure. In this paper, we describe exactly how the infrastructure and the Jacobian are related. We show that computations in the infrastructure using a new modified notion of distance and computations in a particular subgroup of the Jacobian heuristically have exactly the same cost for curves defined over sufficiently large finite fields. We also present a novel set of explicit formulas for genus three split hyperelliptic curves that improves on the current state-of-the-art.

Keywords: Split hyperelliptic curve · Jacobian · Balanced divisor · Infrastructure · Explicit formulas

1 Introduction

The Jacobian of a hyperelliptic curve defined over a finite field is a finite abelian group at the center of a number of important open questions in algebraic geometry and number theory. Sutherland [23] surveyed some of these, including the computation of the associated L-functions and zeta functions used in his investigation of Sato-Tate distributions [15]. Many of these problems lend themselves to numerical investigation, and as emphasized by Sutherland, fast arithmetic in the Jacobian is crucial for their efficiency.

Hyperelliptic curves are categorized as ramified and split models according to their number of points at infinity. Ramified curves have one point at infinity, whereas split curves have two. There is also a third model with no infinite points called inert or unusual, but these are usually avoided in practice as they have cumbersome Jacobian arithmetic and can be transformed to a split model over a quadratic extension of the base field. Jacobian arithmetic differs on ramified and split models. The split scenario is more complicated and is most efficiently

M. J. Jacobson Jr. and R. Scheidler—Supported in part by NSERC of Canada.

C. T. Gueye et al. (Eds.): A2C 2019, CCIS 1133, pp. 183–203, 2019.
https://doi.org/10.1007/978-3-030-36237-9_11

realized via a divisor arithmetic framework referred to as *balanced*. As a result, optimizing divisor arithmetic on split hyperelliptic curves has received less attention from the research community. However, split models have many interesting properties; most importantly, they exist for a much larger array of hyperelliptic curves compared to ramified models. Thus, exhaustive computations such as those in [15] conduct the bulk of their work on split models by necessity.

Split hyperelliptic curves support another algebraic structure, called the infrastructure, that can also be used for numerical investigations of the Jacobian. For example, Stein and Teske [22] used arithmetic in the infrastructure for computing the order of the Jacobian by first computing the regulator, i.e. the (usually very large) order of the class of the divisor D given by the difference of the two points at infinity. One attractive aspect of the infrastructure is that it supports an especially fast operation, called a *baby step*, that can be exploited to speed up various applications, including the Jacobian order computation algorithms used in [22] and [15].

Although both structures in split models can be employed for computational problems, it was originally not clear how they are connected or which one provides better performance in practice. Mireles Morales [17] related the infrastructure to the cyclic subgroup G of the Jacobian generated by the class of D. He showed that any computation in one structure can be reduced to an analogous computation in the other. Moreover, he asserted that performing computations in the Jacobian using balanced divisor arithmetic should always be more efficient than infrastructure arithmetic. However, he did not verify his claim via proof or implementation and also did not take into account the improved infrastructure arithmetic from [12] in his analysis. Therefore, the question of which setting provides better performance has to date not been decisively settled.

This paper offers two main contributions. First, we provide a definitive answer to the aforementioned performance question through rigorous methodology. We describe exactly how the infrastructure and the Jacobian are related. We investigate the assertion by Mireles Morales, considering state-of-the-art algorithms for arithmetic in both settings, including the results of [12]. An initial numerical investigation [11] suggested that the Jacobian offers better performance than the infrastructure, raising the question of whether this deficiency was an inherent property of the infrastructure or could be overcome through arithmetic improvements. We confirm the latter answer by introducing an alternative notion of distance on the infrastructure. Using this new distance, we prove that computations in the infrastructure and in G have exactly the same cost under the assumption of some reasonable, widely accepted heuristics.

Our second contribution is a suite of new explicit formulas for arithmetic in both the Jacobian and the infrastructure of a split genus 3 hyperelliptic curve. As pointed out earlier, these formulas represent a highly useful tool for computations involving such curves. Specifically, conforming to best practices, the operations in our formulas are described completely in terms of finite field operations as opposed to polynomial arithmetic. We present two sets of formulas, one for curves defined over finite fields of odd characteristic, and a second set for characteristic 2. Our formulas for odd characteristic offer improvements over those of [23] for baby steps and addition, but not doubling. For characteristic 2, our explicit formulas are the first to appear in the literature.

Following [23], our formulas use an affine model where each operation requires one field inversion. Much of the literature on efficient Jacobian arithmetic for low genus hyperelliptic curves assumes applications to cryptography, for which inversion-free (projective) formulas are preferred because inversion is computationally expensive for finite fields of the sizes required for cryptography. In the context of computational number theoretic applications like those in [15], affine formulas are superior because group order algorithms such as baby-step giant-step require frequent equality tests. As representations of group elements using projective coordinates are not unique, this implies a non-negligible computational cost per test, and precludes the use of more efficient searchable data structures for the baby steps such as hash tables. Furthermore, as described in [23], in these types of application the inversions can often be combined using a trick due to Montgomery, allowing a batch of inversions to be computed with only one field inversion and a small number of field multiplications. Thus, in this paper, as cryptographic applications are not our motivation, we chose to only present affine formulas. If projective formulas are required for some other application, our formulas can readily be converted to that setting; such formulas can be found in [18].

2   Hyperelliptic Curves

We refer the reader to [7,18,19] for more background on hyperelliptic curves. Throughout, let $\mathbb{F}_q$ be a finite field where q is a power of a prime, $\mathbb{F}_q[x]$ the univariate polynomial ring over $\mathbb{F}_q$, and $\mathbb{F}_q(x)$ the field of rational functions over $\mathbb{F}_q$. A *hyperelliptic curve* C *of genus* g is a smooth projective curve with an affine equation of the form $y^2 + h(x)y = f(x)$ where $h, f \in \mathbb{F}_q[x]$ satisfy certain properties. More specifically, C is said to be

- *ramified (or imaginary)* if $\deg(f) = 2g+1$, f is monic, $h = 0$ when $\mathrm{char}(\mathbb{F}_q) \neq 2$, and h is monic of degree at most g when $\mathrm{char}(\mathbb{F}_q) = 2$;
- *split (or real)* if $\deg(f) = 2g + 2$, $h = 0$ and f is monic when $\mathrm{char}(\mathbb{F}_q) \neq 2$, h is monic of degree $g + 1$ and the leading coefficient of f is of the form $e^2 + e$ for some $e \in \mathbb{F}_q^*$ when $\mathrm{char}(\mathbb{F}_q) = 2$;

The *coordinate ring* and *function field* of C are $\mathbb{F}_q[C] = \mathbb{F}_q[x, y]$ and $\mathbb{F}_q(C) = \mathbb{F}_q(x, y)$, respectively. Ramified curves C have a unique $\mathbb{F}_q$-rational point at infinity, denoted by ∞, whereas a split curve C has two such points, denoted by ∞^+ and ∞^-, respectively. For $\alpha \in \mathbb{F}_q(C)$, we put $\deg(\alpha) = -v_{\infty^+}(\alpha)$ and define $\lfloor \alpha \rfloor$ to be the unique polynomial $A \in \mathbb{F}_q[x]$ with $\deg(\alpha - A) < 0$. Then the polynomial $\lfloor y \rfloor \in \mathbb{F}_q[x]$ is well defined and has degree $g + 1$. It will be required in several algorithms.

2.1   Jacobian

Let $Cl^0(C) = Cl^0_{\mathbb{F}_q}(C)$ denote the *Jacobian* of C over $\mathbb{F}_q$; this is the group of classes of degree zero divisors defined over $\mathbb{F}_q$ under linear equivalence. The equivalence class of a degree zero divisor D is denoted $[D]$. The inverse of a class $[D]$ in $Cl^0(C)$ is $[\overline{D}]$ where $\overline{D}$ is the image of D under the *hyperelliptic involution*. This map sends a point $P = (a, b)$ on C to the point $\overline{P} = (a, -b - h(a))$ on C. On ramified models, it leaves the point at infinity fixed, whereas on split models, it sends ∞^+ to ∞^- and vice versa.

In the case of split hyperelliptic curves, the cyclic subgroup

$$G = \langle [\infty^+ - \infty^-] \rangle$$

of $Cl^0(C)$ will be of key importance later on. Its order $R = |G|$ is the *regulator* of C. Schmidt [20] showed that $|Cl^0(C)| = Rh'$, where h' is the ideal class number of $\mathbb{F}_q[C]$. For most split hyperelliptic curves, h' is small; frequently $h' = 1$, so $Cl^0(C) = G$. Since the Hasse-Weil bounds establish $(\sqrt{q} - 1)^{2g} \le |Cl^0(C)| \le (\sqrt{q} + 1)^{2g}$, the regulator R is generally of magnitude q^g for large q.

A divisor on C is *affine* if it contains no infinite points. An affine divisor is *reduced* if its degree is at most g and it is not supported simultaneously at P and $\overline{P}$ for any affine point P. A degree zero divisor is reduced if its affine part is reduced. Reduced divisors are said to be *generic* if their affine part has degree g and *degenerate* otherwise.

Every affine reduced divisor D is uniquely determined by its *Mumford basis*[1] which consists of two polynomials $u, v \in \mathbb{F}_q[x]$ such that u is monic, u divides $v^2 - hv - f$ and $\deg(u) = \deg(D) \le g$. Here, u is unique and v is unique modulo u, and we write $D = [u, v]$. In the *standard* or *adapted* Mumford basis of D, we choose v such that $\deg(v) < \deg(u)$; this is the typical choice for ramified models. For split models, we usually work with the *reduced* Mumford basis which satisfies $\deg(v - y - h) < \deg(u) < \deg(y + v)$. If $[u, v]$ is the standard basis of a reduced divisor, then the corresponding reduced basis is $[u, v']$ where $v' = \lfloor y \rfloor + h + v - ((\lfloor y \rfloor + h) \bmod u)$.

Every degree zero divisor class on a ramified model contains a unique reduced representative of the form $D - \deg(D)\infty$ with D affine. In contrast, on split models, since there are two points at infinity, a degree zero divisor class may contain many reduced divisors. In [7], Galbraith et al. introduced a unique representative for each degree zero divisor class which is "balanced at infinity". Put $D_\infty = \lceil g/2 \rceil \infty^+ + \lfloor g/2 \rfloor \infty^-$. Then every degree zero divisor can be written as $D = D_0 - D_\infty$ where $D_0 = D'_0 + n\infty^+ + m\infty^-$ with D'_0 affine and $n, m \in \mathbb{Z}$. Galbraith et al. proved that every element of $Cl^0(C)$ has a unique *balanced* representative, i.e. a reduced divisor D such that $0 \le n \le g - \deg(D'_0)$; since

---

[1] There is a discrepancy between [7] and [18, Definition 2.2.13] in the definition of the Mumford representation for split models. The signs of the polynomial v are opposite, which leads to slight differences in the descriptions of the algorithms for Jacobian arithmetic in the two sources. [7] is consistent with literature on ramified models, whereas [18] follows previous literature on split models such as [5, 11, 12, 14, 19, 21].

$m = g - \deg(D'_0) - n$, this implies $0 \leq m \leq g - \deg(D'_0)$. We write $D = (D'_0, n)$ or $D = ([u, v], n)$ where $[u, v]$ is the reduced or adapted Mumford basis of D'_0. A reduced divisor is generic if and only if $n = m = 0$; in this case, we simply write $D = [u, v]$. It is thus clear that every generic reduced divisor is balanced. Conversely, heuristically, almost all balanced divisors are generic. More specifically, we expect the probability that an element of G is represented by a degenerate balanced divisor to approach h'/q as $q \to \infty$; when $h' = 1$. i.e. $Cl^0(C) = G$, this was proved by Fontein (see, for example, [11]). Hence, heuristically, degenerate balanced divisors are extremely rare when q is of cryptographic size. This motivates the term "balanced", since generically, the infinite support of D is approximately equally balanced between the two points at infinity.

The group law on the Jacobian of a split hyperelliptic curve consists of performing the compound operation of formal addition of divisors (realized in practice as composition), divisor reduction and balancing, in this order. It is referred to as *addition* or, when the two inputs are identical, *doubling*. Adding divisor classes via divisor composition and reduction is akin to the standard arithmetic in the Jacobian of a ramified hyperelliptic curve as described, for example, by Cantor [2], and produces a reduced divisor $D = (D', n)$ that need not be balanced. A *balancing step*, denoted $D \to D_+$, decreases the value of n, whereas an *inverse balancing step*, denoted $D \to D_-$, increases this value; see Algorithm 3 of [7] and Algorithms 3 and 4 of [18]. Given any two balanced divisors D_1 and D_2 on C, the balanced representative of the class of $D_1 + D_2$ is denoted $D_1 \oplus D_2$. Thus, the group law on $Cl^0(C)$ can be expressed as $[D_1] + [D_2] = [D_1 \oplus D_2]$. Algorithm 5 of [18] and Algorithm 4 of [7] compute the divisor $D_1 \oplus D_2$. This algorithm can be considered as the main operation on the Jacobian of a split hyperelliptic curve. It is one of the fundamental operations required for many applications related to split hyperelliptic curves. See [7] and [18] for more details.

2.2   Infrastructure

Let C be a split hyperelliptic curve. In this section, we summarize the main properties of the infrastructure of C; details can be found in [18, 19, 21].

Every non-zero $\mathbb{F}_q[C]$-ideal $\mathfrak{a}$ is an $\mathbb{F}_q[x]$-module of rank 2 with a basis of the form $\{su, s(v + y)\}$ where u divides $v^2 - vh - f$; write $\mathfrak{a} = [su, s(v + y)]$. If we take s and u to be monic, then s and u are unique and v is unique modulo u. The ideal $\mathfrak{a}$ is *primitive* if $s = 1$ and *reduced* if additionally $\deg(u) \leq g$. The basis $[u, v]$ of a primitive ideal $\mathfrak{a} = [u, y + v]$ is *adapted* if $\deg(v) < \deg(u)$, and *reduced* if $\deg(v - h - y) < \deg(u) < \deg(y + v)$.

The *divisor associated to* $\mathfrak{a} = [u, v + y]$ is the affine divisor $D = \operatorname{div}(\mathfrak{a})$ whose Mumford representation is $[u, v]$. The *degree* of $\mathfrak{a}$ is $\deg(\mathfrak{a}) = \deg(\operatorname{div}(\mathfrak{a}))$, i.e. the degree of its associated divisor. A *generic* (resp. *degenerate*) reduced ideal is one whose associated divisor is generic (resp. degenerate), so $\mathfrak{a}$ is a generic ideal if and only if the balanced divisor $D = (\operatorname{div}(\mathfrak{a}), 0)$ is generic.

Definition 1. *The* infrastructure *of C is the (finite) collection $\mathscr{R}$ of reduced principal $\mathbb{F}_q[C]$-ideals. For an ideal $\mathfrak{a} \in \mathscr{R}$, the* distance *of $\mathfrak{a}$ is defined to be $\delta(\mathfrak{a}) = \deg(\alpha)$, where α is the unique generator of $\mathfrak{a}$ with $0 \leq \deg(\alpha) < R$.*

The fact that no two infrastructure ideals have the same distance imposes an ordering on $\mathscr{R}$ by distance. Hence, if we put $\mathfrak{a}_1 = (1)$ and $\delta_i = \delta(\mathfrak{a}_i)$, then we can write

$$\mathscr{R} = \{\mathfrak{a}_1, \mathfrak{a}_2, \dots, \mathfrak{a}_r\} \qquad 0 = \delta_1 < \delta_2 < \cdots < \delta_r < R .$$

Computing the distance of an infrastructure ideal given its $\mathbb{F}_q[x]$-basis is believed to be computationally infeasible; this is the *infrastructure discrete logarithm problem* on which the security of infrastructure-based cryptosystems is based [12,14,19]. However, the relative distance between two successive ideals can be efficiently computed. In [12] and [21], it was proved that if $\mathfrak{a}_i = [u_{i-1}, y + v_{i-1}]$ then

$$\delta_{i+1} = \delta_i + g + 1 - \deg(u_{i-1}) \quad \text{for } 1 \leq i \leq r - 1. \tag{1}$$

Thus, $\delta_{i+1} = \delta_i + 1$ when $\mathfrak{a}_i$ is generic. Hence, most integers between 0 and $R - 1$ occur as distance values, but this is not true for all of them: for example, $\delta_2 = g + 1$ shows that there are no infrastructure ideals with distance between 1 and g.

The infrastructure supports two main operations. A *baby step* computes $\mathfrak{a}_{i+1}$ from $\mathfrak{a}_i$, along with the relative distance $\delta_{i+1} - \delta_i$. Applying the same operation a sufficient number of times to a non-reduced principal $\mathbb{F}_q[C]$-ideal produces an infrastructure ideal; this process is referred to as ideal reduction. A *giant step* computes on input $\mathfrak{a}, \mathfrak{b} \in \mathscr{R}$ the first reduced ideal equivalent to the product $\mathfrak{a}\mathfrak{b}$ when applying reduction; this ideal is denoted $\mathfrak{a} \otimes \mathfrak{b}$ and is obtained after at most (and generically exactly) $\lceil g/2 \rceil$ reduction steps. The *conjugate* ideal of an ideal $\mathfrak{a} = [u, v] \in \mathscr{R}$ is the ideal $\bar{\mathfrak{a}} = [u, v - h - y] \in \mathscr{R}$; it satisfies $\mathrm{div}(\bar{\mathfrak{a}}) = \overline{\mathrm{div}(\mathfrak{a})}$ and has distance $\delta(\bar{\mathfrak{a}}) = R + \deg(u) - \delta(\mathfrak{a})$. With these notions, $\mathscr{R}$ is "almost" an abelian group under $\otimes$, where the identity is $\mathfrak{a}_1$, the "inverse" of an infrastructure ideal is its conjugate ideal, and $\mathscr{R}$ fails associativity only barely; specifically, if $\mathfrak{a}, \mathfrak{b} \in \mathscr{R}$, then

$$\delta(\mathfrak{a} \otimes \mathfrak{b}) = \delta(\mathfrak{a}) + \delta(\mathfrak{b}) - d \quad \text{with } 0 \leq d \leq 2g.$$

The quantity d is very small and is expected to be equal to $\lceil g/2 \rceil$ almost always as explained in Sect. 2.3; see also [12,14,19] for further details. This "shortfall" in distance is the reason that $\mathscr{R}$ is not associative under giant steps. This can be rectified for almost all giant steps by shifting distances down by $\lceil g/2 \rceil$. To that end, we define a new distance on $\mathscr{R}$ as

$$\gamma(\mathfrak{a}) = \delta(\mathfrak{a}) - \lceil g/2 \rceil .$$

We refer to δ as the classic distance and to $\mathscr{R}$ under δ as the classic infrastructure. Note that γ preserves the ordering on $\mathscr{R}$ defined by δ, but the first distance value is now negative; specifically, $\gamma(\mathfrak{a}_1) = -\lceil g/2 \rceil < 0$ and $\gamma(\mathfrak{a}_2) = \lfloor g/2 \rfloor + 1 > 0$. Moreover,

$$\gamma(\mathfrak{a} \otimes \mathfrak{b}) = \gamma(\mathfrak{a}) + \gamma(\mathfrak{b}) + (\lceil g/2 \rceil - d) \quad \text{with} \ \ 0 \leq d \leq 2g. \tag{2}$$

Since we almost always expect $d = \lceil g/2 \rceil$, the new distance is additive for most inputs. The ordering on $\mathscr{R}$ determined by γ shows that for every integer N with $-\lfloor g/2 \rfloor \leq N < R - \lceil g/2 \rceil$, there exists a unique ideal $\mathfrak{a}_i \in \mathscr{R}$ such that $\gamma(\mathfrak{a}_i) \leq N < \gamma(\mathfrak{a}_{i+1})$. This ideal $\mathfrak{a}_i$, referred to as the infrastructure ideal *below* N (with respect to γ) and denoted $\mathfrak{a}[N]$, can be efficiently computed, along with the "error" $N - \gamma(\mathfrak{a}[N])$. We expect $\gamma(\mathfrak{a}[N]) = N$ most of the time.

2.3   Connection Between G and $\mathscr{R}$

Mireles Morales in [17] introduced an injective map from the infrastructure into the Jacobian of a split hyperelliptic curve C. In [11], a shift of $\lceil g/2 \rceil [\infty^+ - \infty^-]$ was applied to the images under this map, yielding the injection

$$\phi : \mathscr{R} \to Cl^0(C)$$
$$\phi(\mathfrak{a}) = [\operatorname{div}(\mathfrak{a}) + (g - \deg(\mathfrak{a}))\infty^- - D_\infty] = [(\operatorname{div}(\mathfrak{a}), 0)].$$

This modification of the Mireles Morales map was a precursor to the analogous shift of $\lceil g/2 \rceil$ used to define the new distance γ on $\mathscr{R}$ and ensures that the representatives of the elements in the image $\phi(\mathscr{R})$ are balanced. By Proposition 4.2 of [11], we have

$$\phi(\mathfrak{a}) = \gamma(\mathfrak{a}) [\infty^+ - \infty^-]. \tag{3}$$

This identity shows that the image of $\mathscr{R}$ under ϕ is the subset of G consisting of precisely those classes $m[\infty^+ - \infty^-] \in G$ for which $(m \bmod R)$ is the new distance of some infrastructure ideal.

Based on the presumed scarcity of degenerate divisors and ideals, [12] and [11] formulated two heuristic assumptions, denoted (H1) and (H2) in these sources. Reformulating these statements with the new distance, (H1) asserts that with heuristic probability $1 - O(q^{-1})$ as $q \to \infty$, $\mathfrak{a}_i$ is generic, or equivalently, $\gamma(\mathfrak{a}_{i+1}) = \gamma(\mathfrak{a}_i) + 1$ for $2 \leq i \leq r - 1$. Heuristic (H2) states that $d = \lceil g/2 \rceil$ in (2) under the same assumption, or equivalently, $\gamma(\mathfrak{a} \otimes \mathfrak{b}) = \gamma(\mathfrak{a}) + \gamma(\mathfrak{b})$ for all $\mathfrak{a}, \mathfrak{b} \in \mathscr{R} \backslash \{\mathfrak{a}_1\}$. Hence, infrastructure arithmetic using the new distance γ is more natural and less complicated than using the classic distance δ. The assertion that $d = \lceil g/2 \rceil$ in (2) also implies that the number of reduction steps required in a giant step is heuristically equal to $\lceil g/2 \rceil$; see [18, Remark 3.2.4] for a formal proof. Both heuristics also imply that $\gamma(\mathfrak{a}[N]) = N$ almost always.

Jacobson et al. in [12] obtained improvements to scalar multiplication in the infrastructure by taking advantage of the fact that these heuristics eliminate the need to keep track of all relative distances when performing infrastructure arithmetic. In cryptographic protocols such as infrastructure Diffie-Hellman, even though the respective computations of the two communicants are different, the two parties "skip over" the same degenerate ideals (i.e. exceptions to the heuristics) and are hence expected to obtain the same target ideal. Extensive numerical computations in [12] confirm this. The description of the map ϕ in (3) makes it possible to extend the improvements of [12] to computations in G by deriving analogous heuristics for divisors in $\phi(\mathscr{R}) \subset G$. These modified heuristics assert that as $q \to \infty$, the following properties hold with heuristic probability $1 - O(q^{-1})$:

(H1) $\phi(\mathfrak{a}_{i+1}) = \phi(\mathfrak{a}_i) + [\infty^+ - \infty^-]$ for $2 \leq i \leq r - 1$.
(H2) $\phi(\mathfrak{a} \otimes \mathfrak{b}) = \phi(\mathfrak{a}) + \phi(\mathfrak{b})$ for $\mathfrak{a}, \mathfrak{b} \in \mathscr{R} \setminus \{\mathfrak{a}_1\}$.

(H2) explicitly links the arithmetic of $\mathscr{R}$ with that of G. This implies, as already noted in [17], that the infrastructure discrete logarithm problem can be reduced to the discrete logarithm problem in G and vice versa. Note that (1) shows that the cardinalities of $\mathscr{R}$ and G are of the same magnitude when g is small.

For $1 \leq i \leq r$, let $D_i = (\mathrm{div}(\mathfrak{a}_i), 0)$ be the balanced divisor representing the class $\phi(\mathfrak{a}_i)$. Applying Algorithm 3 of [18] (also Algorithm 3 of [7]) to a generic reduced divisor D shows that applying a balancing step to D is arithmetically the same as adding the divisor class $[\infty^+ - \infty^-]$ to $[D]$; this was already remarked in [7]. On the other hand, heuristically, $[D_{i+1}] = [D_i] + [\infty^+ - \infty^-]$ for $2 \leq i \leq r - 1$ by (H1). As a result, we see that a balancing step in the Jacobian and a baby step in the infrastructure act identically; more exactly, the formulas for a balancing step applied to a Mumford basis $[u, v]$ of a reduced divisor are identical to those for a baby step applied to the infrastructure ideal $\mathfrak{a} = [u, v+y]$. An analogous argument applies to inverse balancing steps and backward baby steps. Moreover, if $D = (\mathrm{div}(\mathfrak{a}), 0)$ and $E = (\mathrm{div}(\mathfrak{b}), 0)$, then heuristically, the balanced representative of the class $[D] + [E] = [D + E]$ is $(\mathrm{div}(\mathfrak{a} \otimes \mathfrak{b}), 0)$. It is obtained by applying at most $\lceil g/2 \rceil$ reduction steps to $D+E$, and no subsequent balancing steps are needed. This shows that the operations $\otimes$ on $\mathscr{R}$ and $\oplus$ on G have identical cost, generically. More detailed formal proofs of these results can be found in Sect. 3.3 of [18]. Finally, if $[D] = \phi(\mathrm{div}(\mathfrak{a}))$, then heuristically, $[nD] = \phi(\mathrm{div}(\mathfrak{b}))$ where $\mathfrak{b}$ is the infrastructure ideal with $\gamma(\mathfrak{b}) = n\gamma(\mathfrak{a})$. In other words, heuristically, scalar multiplying a generic divisor in G by n results in multiplying the associated distance value on $\mathscr{R}$ by n. Therefore, using the new distance, exponentiation on $\mathscr{R}$ and scalar multiplication on G are arithmetically identical.

We conclude this section with a remark on generic inversion in G, an operation that is required for some applications. Recall that $D_\infty = \lceil g/2 \rceil \infty^+ + \lfloor g/2 \rfloor \infty^-$, so $\overline{D}_\infty = D_\infty$ when g is even, and $\overline{D}_\infty = D_\infty - (\infty^+ - \infty^-)$ when g is odd. As a result, if $D = (D', 0)$ is a generic balanced divisor, then assuming (H1), the balanced representative of $[\overline{D}]$ is $(\overline{D'}, 0)$ if g is even and $(E', 0)$ if g is odd, where E' is obtained by applying a balancing step (i.e. an addition of $[\infty^+ - \infty^-]$ by our earlier remarks) to $\overline{D'}$.

3 Explicit Formulas

All algorithms in the Jacobian and infrastructure operate on the Mumford polynomials of a divisor or its corresponding ideal. Explicit formulas describe these operations in terms of the coefficients of the polynomials, allowing better optimization. We developed novel explicit formulas for genus 3 split hyperelliptic curves over both odd characteristic and characteristic 2 fields in affine representation (requiring one field inversion). Our input divisors are all reduced and given by their reduced Mumford basis.

The first step, as is standard in the literature, is to apply curve isomorphisms to cause as many coefficients of f and h as possible to vanish. For a genus 3 split curve C with equation $y^2 + h(x)y = f(x)$, we write $h(x) = \sum_{i=0}^{4} h_i x^i$ and $f(x) = \sum_{i=0}^{8} f_i x^i$. In odd characteristic, $h(x) = 0$ and the transformation $x \to x - f_7/8$ eliminates the x^7 term of $f(x)$. In characteristic 2, when $h_3 = 0$ the substitutions $x \to x$ and

$$y \to y + f_7 x^3 + (f_6 + f_7^2)x^2 + (f_7 h_2 + f_5)x + (h_2 f_6 + h_2 f_7^2 + f_6^2 + f_7^2 + h_1 f_7 + f_4)$$

cause the x^7, x^6, x^5, x^4 terms in $f(x)$ to vanish, as can easily be verified by direct substitution and simplification. A similar transformation exists when $h_3 \neq 0$, but we restrict to the case $h_3 = 0$ because having the x^3 coefficient of y being zero results in more efficient formulas. It is straightforward to compute the coefficients of $\lfloor y \rfloor = \sum_{i=0}^{4} y_i x^i$ by equating symbolically the coefficients of $y^2 + h(x)y$ with those of $f(x)$.

When implementing addition in the Jacobian and infrastructure, following Cantor's algorithm [2] literally is not the best strategy. It is instead preferable to use the method suggested by Gaudry and Harley in [9], in which the computations are broken into sub-expressions that can be re-used as much as possible. As in [9], we also specialize to the frequently occurring cases where the input divisors $D_1 = [u_1, v_1]$ and $D_2 = [u_2, v_2]$ for addition and $D = [u, v]$ for doubling are generic (i.e. $\deg(u_1) = \deg(u_2) = \deg(u) = 3$) and satisfy $\gcd(u_1, u_2) = \gcd(u, h + 2v) = 1$. In the rare cases where this does not hold, an implementation can resort to Cantor's algorithm. Specializations of Harley's addition and doubling algorithms for split hyperelliptic curves of genus 3 are given in Appendix A as Algorithms A.3 and A.4, respectively. Furthermore, balancing and inverse balancing steps in $Cl^0(C)$, or equivalently, baby steps and backward baby steps in the infrastructure, are given as Algorithms A.1 and A.2.

We convert these algorithms to explicit formulas using standard methods from the literature as follows; see, for example, [6, 8, 16, 24].

1. Both addition and doubling require the computation of a resultant, with both inputs of degree 3 for addition and inputs of degree 3 and 4 for doubling. We employ the method presented in [1] based on Cramer's rule to compute the quantity *inv* in step 1 of Algorithms A.3 and A.4 directly, yielding the smallest number of finite field operations required for this computation.
2. We use Karatsuba's method for polynomial multiplication and modular reduction whenever possible.
3. All four algorithms require exact polynomial division. In this context, we need not compute all the coefficients of the inputs since, as explained in [8] for example, the division of two polynomials of respective degrees d_1 and d_2 with $d_1 > d_2$ depends only on the $d_1 - d_2 + 1$ highest coefficients of the dividend.
4. As field inversions are significantly more costly than other operations (eg. estimates suggest that one field inversion in odd characteristic costs approximately the same as 100 multiplications), we seek to eliminate as many of these as possible. We use Montgomery's trick [3, Algorithm 10.3.4] to invert the product of all elements that must be inverted and recover the individual inverses at the cost of a few extra multiplications. Specifically, putting $I = (ab)^{-1}$, we have $a^{-1} = bI$ and $b^{-1} = aI$. This technique produces formulas that require only a single field inversion.

Our explicit formulas using affine coordinates for a balancing step, inverse balancing step, addition and doubling are presented in Appendix B. They can also be found in [18, Sections 4.6.1 and 4.6.2], along with further details and a more comprehensive exposition of the techniques employed.

To test our formulas, they were initially implemented in Sage version 5.2 using the small finite fields $\mathbb{F}_{7919}$ and $\mathbb{F}_{2^{13}}$. The testing was done in three stages. First, we implemented Cantor's algorithm with no modifications, then Harley's algorithm, and finally, our explicit formulas. We then compared all three outputs. We used the computer algebra library NTL to generate split hyperelliptic curves over the desired finite fields and reduced divisors on these curves. Our testing was done on 20 different examples. To test our explicit formulas for large finite fields, we implemented the Diffie-Hellman key exchange protocol in C++. The correctness of our formulas was confirmed by the fact that both parties always obtained the same shared key, after hundreds of thousands of iterations of the protocol over numerous randomly generated curves and base fields.

Operation counts for all our formulas are presented in Table 1. For each operation, we give the number of field squarings (S), multiplications (M), and inversions (I). We did not count multiplications or squarings involving coefficients of $\lfloor y \rfloor$, h, and f exclusively (such as $h_2 y_4$ for example), since such quantities can be precomputed for a given curve and can be optimized as multiplications by constants.

In addition to our new formulas, we compare operation counts for the best available formulas in the literature. Counts for affine formulas for genus 3 ram-

ified curves in odd characteristic are taken from [6] We considered the explicit
formulas presented in Tables B.1 and C.1 of [10] for genus 3 ramified curves in
characteristic 2. The closest analogue to a balancing step on a ramified model is
the addition of a generic divisor class represented by a divisor of the form $P - \infty$
where P is an $\mathbb{F}_q$-rational point on the curve. We used Tables VIII and XIII
in [6] for the baby/balancing step equivalent in affine coordinates on ramified
curves.

For genus 2 split hyperelliptic curves, we used the explicit formulas from [5].
For genus 3 split hyperelliptic curves in odd characteristic, we also include the
operation counts from [23], for which the input divisors are represented by their
adapted basis.

Table 1. Explicit Formula Operation Counts for Jacobian Arithmetic

	Operation	Split $g = 2$	Split $g = 3$	Split $g = 3$ from [23]	Ramified $g = 3$
$\mathbb{F}_q$	Baby/Bal Step	1I+4M+4S	1I+9M+1S	1I+14M	1I+21M
	Inv Baby/Bal Step	1I+4M+2S	1I+9M+1S	-	1I+21M
	Addition	1I+26M+2S	1I+74M+1S	1I+79M	1I+67M
	Doubling	1I+28M+4S	1I+84M+2S	1I+82M	1I+68M
$\mathbb{F}_{2^n}$	Baby/Bal Step	1I+5M+1S	1I+9M+1S	-	1I+20M
	Inv Baby/Bal Step	1I+5M+1S	1I+9M+1S	-	1I+20M
	Addition	1I+27M+1S	1I+81M	-	1I+64M+4S
	Doubling	1I+29M+2S	1I+89M+1S	-	1I+64M+5S

As expected, the number of operations for split models exceeds those for
ramified models. The main reason is that $\deg(f) = 7$ for ramified models, but
8 for split models. Our addition and baby/balancing step formulas are more
efficient than those given in [23], but our doublings are more costly. Note that
for use in the baby-step giant-step algorithm, faster addition is in fact favorable,
as addition is the main operation required. Note also that [23] only contains
formulas for curves in odd characteristic.

4   Conclusion

Our investigations show that, computationally, there is no advantage to using
Jacobian arithmetic over infrastructure arithmetic for number theoretic computations on split hyperelliptic curves. Jacobian arithmetic on balanced divisors
realizes arithmetic in a group (as opposed to an ordered set) and is hence perhaps
more natural and conceptually simpler than infrastructure arithmetic. Moreover,
it covers the entire Jacobian rather than just a fixed (albeit large) subgroup.
Hence, we recommend using Jacobian arithmetic in general. The computational
advantages offered by the inexpensive infrastructure baby step operation, as
recognized by Stein and Teske [22] for example, can be gained in general by
computing in the subgroup G as opposed to the infrastructure.

Our new explicit formulas for genus 3 split hyperelliptic curves will be useful for researchers doing large scale computations in the Jacobian, especially for applications that rely heavily on addition and baby steps, as they require fewer field operations as compared to those of Sutherland [23]. We are currently investigating whether further improvements may be obtained by using other approaches to arithmetic such as the geometric methods described by Costello and Lauter [4] and the Shanks' NUCOMP algorithm as adapted to hyperelliptic curves by Jacobson, Scheidler and Stein [13].

A   Basic Algorithms

All input divisors are assumed to be given by a reduced Mumford basis $\{u, v\}$ with $u = x^3 + u_2 x^2 + u_1 x + u_0$ and $v = v_4 x^4 + v_3 x^3 + v_2 x^2 + v_1 x + v_0$, with v_4 and v_3 are determined by f and h and are never computed. We use the notation $D = [u_2, u_1, u_0, v_2, v_1, v_0]$. In Algorithms A.3 and A.4, s_n and s'_n denote the leading coefficient of the polynomials s and s', respectively.

Algorithm A.1. Balancing Step

Input: Generic reduced divisor $D_0 = [u_0, v_0]$.
Output: $D_+ = [u_1, v_1]$.
1: $v_1 = \lfloor y \rfloor + h - (v_0 + \lfloor y \rfloor) \pmod{u_0}$, $u_1 = \frac{v'^2 - hv' - f}{u_0}$ made monic
2: **return** $[u_1, v_1]$

Algorithm A.2. Inverse Balancing Step

Input: Generic reduced divisor $D_0 = [u_0, v_0]$.
Output: $D_- = [u_1, v_1]$.
1: $u_1 = \frac{v_0^2 - hv_0 - f}{u_0}$ made monic, $v_1 = \lfloor y \rfloor + h - (v_0 + \lfloor y \rfloor) \pmod{u_1}$
2: **return** $[u_1, v_1]$

Algorithm A.3. Harley's Algorithm for Divisor Class Addition (genus 3)

Input: Generic reduced divisors $D_1 = [u_1, v_1]$, $D_2 = [u_2, v_2]$ with $\gcd(u_1, u_2) = 1$.
Output: $D' = D_1 \oplus D_2$.
1: Compute the resultant r of u_1 and u_2; $inv = r(u_2^{-1}) \pmod{u_1}$
2: $s' = (v_1 - v_2)inv \pmod{u_1}$, $s = s'/r$ and $s_{monic} = s/s_n$
3: $l = s_{monic}u_2$, $k = \frac{f + hv_2 - v_2^2}{u_2}$
4: $u_t = \frac{s_{monic}(s_{monic}u_2 + 2v_2 w - hw) - kw^2}{u_2}$ where $w = r/s'_n$, $v_t = (su_2 + v_2) \pmod{u_t}$
5: $v'_t = \lfloor y \rfloor + h - (v_t + \lfloor y \rfloor) \pmod{u_t}$
6: $u' = \frac{f + hv'_t - (v'_t)^2}{u_t}$ made monic, $v' = h + \lfloor y \rfloor - (\lfloor y \rfloor + v'_t) \pmod{u'}$
7: **return** $[u', v']$

Algorithm A.4. Harley's Algorithm for Divisor Class Doubling (genus 3)

Input: Generic reduced divisor $D_1 = [u_1, v_1]$ with $\gcd(u_1, h + 2v_1) = 1$.
Output: $D' = D_1 \oplus D_1$.
1: Compute the resultant r of $2v_1 - h$ and u_1; $inv = r(h + 2v_1)^{-1} \pmod{u}$
2: $k = \frac{f - hv_1 - v_1^2}{u_1}$, $k' = k \pmod{u_1}$
3: $s' = k'inv \pmod{u_1}$, $s = s'/r$ and $s_{monic} = s/s_n$
4: $l = s_{monic}u_1$
5: $u_t = s_{monic}^2 + \frac{w s_{monic}(2v_1 - h) - w^2 k}{u_1}$ where $w = r/s'_n$, $v_t = (v_1 + su_1) \pmod{u_t}$
6: $v'_t = \lfloor y \rfloor + h - (v_t + \lfloor y \rfloor) \pmod{u_t}$
7: $u' = \frac{f + hv'_t - (v'_t)^2}{u_t}$ made monic, $v' = h + \lfloor y \rfloor - (\lfloor y \rfloor + v'_t) \pmod{u'}$
8: **return** $[u', v']$

B   Genus 3 Explicit Formulas

Balancing and Inverse Balancing

Balancing Step, Odd Characteristic		
Input	$D = [u_2, u_1, u_0, v_2, v_1, v_0]$	
Output	$D_+ = [u_2', u_1', u_0', v_2', v_1', v_0']$	
Step	Expression	Operations
Computing $v' = \lfloor y \rfloor + h - [(\lfloor y \rfloor + v) \pmod u)]$		
1	$v_2' = -v_2 + 2u_1 - 2u_2^2, \quad v_1' = -v_1 + 2u_0 - 2u_1u_2, \quad v_0' = -v_0 - 2u_0u_2$	2M, 1S
Precomputation for Step 2		
	$w_0 = 2y_0 - 2v_0', \quad w_1 = 2y_1 - 2v_1', \quad w_2 = 2y_2 - 2v_2'$	
Computing $u' = \mathrm{Monic}((f + hv' - v'^2)/u)$		
General case: $w_2 \neq 0$ ($\deg(u') = 3$)		
2	$I = w_2^{-1}, \quad u_2' = Iw_1 - u_2, \quad u_1' = Iw_0 + (y_2 + v_2')/2 - u_1 - u_2u_2'$	
	$u_0' = I(f_3 - f_6v_1') + v_1' - u_0 - u_1u_2' - u_2u_1'$	1I, 7M
Special case 1: $w_2 = 0$ and $w_1 \neq 0$ ($\deg(u') = 2$)		
2	$I = w_1^{-1}, \quad u_1' = Iw_0 - u_2, \quad u_0' = I(f_3 - f_6y_1) + y_2 - u_1 - u_2u_1'$	1I, 3M
Special case 2: $w_2 = w_1 = 0$ and $w_0 \neq 0$ ($\deg(u') = 1$)		
2	$I = w_0^{-1}, \quad u_0' = I(f_3 - f_6y_1) - u_2$	1I, 1M
Special case 3: $v_2' = y_2, \quad v_1' = y_1$ and $v_0' = y_0$ ($\deg(u') = 0$)		
2	$u' = 1, \quad v' = y$	
Total	General Case	1I, 9M, 1S
	Special Case 1	1I, 5M, 1S
	Special Case 2	1I, 3M, 1S
	Special Case 3	2M, 1S

Balancing Step, Characteristic 2		
Input	$D = [u_2, u_1, u_0, v_2, v_1, v_0]$	
Output	$D_+ = [u_2', u_1', u_0', v_2', v_1', v_0']$	
Step	Expression	Operations
Computing $v' = \lfloor y \rfloor + h - [(\lfloor y \rfloor + v) \pmod u)]$		
1	$v_2' = h_2 + v_2 + u_1 + u_2^2, \quad v_1' = h_1 + v_1 + u_0 + u_1u_2, \quad v_0' = h_0 + v_0 + u_0u_2$	2M, 1S
Precomputation for Step 2		
	$w_0 = y_0 + h_0 + v_0', \quad w_1 = y_1 + h_1 + v_1', \quad w_2 = y_2 + h_2 + v_2'$	
Computing $u' = \mathrm{Monic}((f + hv' - v'^2)/u)$		
General case: $w_2 \neq 0$ ($\deg(u') = 3$)		
2	$I = w_2^{-1}, \quad u_2' = Iw_1 + u_2, \quad u_1' = Iw_0 + w_2 + h_2 + u_1 + u_2u_2'$	
	$u_0' = I(f_3 + h_2w_1) + h_1 + u_0 + u_1u_2' + u_2u_1'$	1I, 7M
Special case 1: $w_2 = 0$ and $w_1 \neq 0$ ($\deg(u') = 2$)		
2	$I = w_1^{-1}, \quad u_1' = Iw_0 + u_2, \quad u_0' = If_3 + h_2 + u_1 + u_2u_1'$	1I, 3M
Special case 2: $w_2 = w_1 = 0$ and $w_0 \neq 0$ ($\deg(u') = 1$)		
2	$I = w_0^{-1}, \quad u_0' = If_3 + u_2$	1I, 1M
Special case 3: $v_2' = y_2, v_1' = y_1$ and $v_0' = y_0$ ($\deg(u') = 0$)		
2	$u' = 1, \quad v' = y$	
Total	General Case	1I, 9M, 1S
	Special Case 1	1I, 5M, 1S
	Special Case 2	1I, 3M, 1S
	Special Case 3	2M, 1S

Inverse Balancing Step, Odd Characteristic,		
Input	$D = [u_2, u_1, u_0, v_2, v_1, v_0]$	
Output	$D_- = [u_2', u_1', u_0', v_2', v_1', v_0']$	
Step	Expression	Operations
Precomputation		
	$w_0 = 2y_0 - 2v_0, \quad w_1 = 2y_1 - 2v_1, \quad w_2 = 2y_2 - 2v_2$	
General case: $v_2 \neq y$ ($\deg(u') = 3$)		
Computing $u' = \mathrm{Monic}((f + hv - v^2)/u)$		
1	$u' = x^3 + u_2'x^2 + u_1'x + u_0'$	1I, 7M
	$I = w_2^{-1}, \quad u_2' = Iw_1 - u_2, \quad u_1' = Iw_0 + (y_2 + v_2)/2 - u_1 - u_2u_2'$	
	$u_0' = I(f_3 - f_6v_1) + v_1 - u_0 - u_2u_1' - u_1u_2'$	
Computing $v' = \lfloor y \rfloor + h - [(\lfloor y \rfloor + v) \pmod{u'}]$		
2	$v' = x^4 + v_2'x^2 + v_1'x + v_0'$	2M, 1S
	$v_2' = -v_2 + 2u_1' - 2(u_2')^2, \quad v_1' = -v_1 + 2u_0' - 2u_1'u_2', \quad v_0' = -v_0 - 2u_0'u_2'$	
Special case 1: $w_2 = 0$ and $w_1 \neq 0$ ($\deg(u') = 2$)		
Computing $u' = \mathrm{Monic}((f + hv - v^2)/u)$		
1	$u' = x^2 + u_1'x + u_0'$	1I, 3M
	$I = w_1^{-1}, \quad u_1' = Iw_0 - u_2, \quad u_0' = I(f_3 - f_6y_1) + y_2 - u_1 - u_2u_1'$	
Computing $v' = \lfloor y \rfloor + h - [(\lfloor y \rfloor + v) \pmod{u'}]$		
2	$v' = x^4 + y_2x^2 + v_1'x + v_0'$	2M, 1S
	$t_1 = y_2 - u_0' + (u_1')^2, \quad v_1' = -v_1 + 2(t_1 - u_0')u_1', \quad v_0' = -v_0 + 2t_1u_0'$	
Special case 2: $w_2 = w_1 = 0$ and $w_0 \neq 0$ ($\deg(u') = 1$)		
Computing $u' = \mathrm{Monic}((f + hv - v^2)/u)$		
1	$u' = x + u_0'$	1I, 1M
	$I = w_0^{-1}, \quad u_0' = I(f_3 - f_6y_1) - u_2$	
Computing $v' = \lfloor y \rfloor + h - [(\lfloor y \rfloor + v) \pmod{u'}]$		
2	$v' = x^4 + y_2x^2 + y_1x + v_0'$	2M, 1S
	$t_1 = (u_0')^2, \quad v_0' = -v_0 - 2(y_2 + t_1)t_1 + 2y_1u_0'$	
Special case 3: $w_2 = w_1 = w_0 = 0$ ($\deg(u') = 0$)		
	$u' = 1, \quad v' = y + h$	
Total	General Case	1I, 9M, 1S
	Special Case 1	1I, 5M, 1S
	Special Case 2	1I, 3M, 1S

Inverse Balancing Step, Characteristic 2,		
Input	$D = [u_2, u_1, u_0, v_2, v_1, v_0]$	
Output	$D_- = [u_2', u_1', u_0', v_2', v_1', v_0']$	
Step	Expression	Operations
Precomputation		
1	$w_0 = y_0 + h_0 + v_0, \quad w_1 = y_1 + h_1 + v_1, \quad w_2 = y_2 + h_2 + v_2$	
General case: $w_2 \neq 0$ $(\deg(u') = 3)$		
Computing $u' = \mathrm{Monic}((f + hv - v^2)/u)$		
2	$u' = x^3 + u_2' x^2 + u_1' x + u_0'$	1I, 7M
	$I = w_2^{-1}, \quad u_2' = Iw_1 + u_2, \quad u_1' = Iw_0 + h_2 + w_2 + u_1 + u_2 u_2'$	
	$u_0' = I(f_3 + h_2 w_1) + h_1 + u_0 + u_1 u_2' + u_2 u_1'$	
Computing $v' = \lfloor y \rfloor + h - [(\lfloor y \rfloor + v) \pmod{u'}]$		
3	$v' = (y_4 + 1)x^4 + (h_3 + y_3)x^3 + v_2' x^2 + v_1' x + v_0'$	2M, 1S
	$v_2' = h_2 + v_2 + u_1' + (u_2')^2, \quad v_1' = h_1 + v_1 + u_0' + u_1' u_2', \quad v_0' = h_0 + v_0 + u_0' u_2'$	
Special case 1: $w_2 = 0$ and $w_1 \neq 0$ $(\deg(u') = 2)$		
Computing $u' = \mathrm{Monic}((f + hv - v^2)/u)$		
2	$u' = x^2 + u_1' x + u_0'$	1I, 3M
	$I = w_1^{-1}, \quad u_1' = Iw_0 + u_2, \quad u_0' = If_3 + h_2 + u_1 + u_2 u_1'$	
Computing $v' = \lfloor y \rfloor + h - [(\lfloor y \rfloor + v) \pmod{u'}]$		
3	$v' = (y_4 + 1)x^4 + (h_3 + y_3)x^3 + (h_2 + y_2)x^2 + v_1' x + v_0'$	2M, 1S
	$t_1 = (u_1')^2 + h_2, \quad v_1' = h_1 + v_1 + t_1 u_1', \quad v_0' = h_0 + v_0 + (t_1 + u_0')u_0'$	
Special case 2: $w_2 = w_1 = 0$ and $w_0 \neq 0$ $(\deg(u') = 1)$		
Computing $u' = \mathrm{Monic}((f + hv - v^2)/u)$		
2	$u' = x + u_0'$	1I, 1M
	$I = w_0^{-1}, \quad u_0' = If_3 + u_2$	
Computing $v' = \lfloor y \rfloor + h - [(\lfloor y \rfloor + v) \pmod{u'}]$		
3	$v' = (y_4 + 1)x^4 + (h_3 + y_3)x^3 + (h_2 + y_2)x^2 + (h_1 + y_1)x + v_0'$	2M, 1S
	$t_1 = (u_0')^2, \quad v_0' = h_0 + v_0 + h_1 u_0' + (h_2 + t_1)t_1$	
Special case 3: $w_2 = w_1 = w_0 = 0$ $(\deg(u') = 0)$		
2	$u' = 1, \quad v' = y + h$	
Total	General Case	1I, 9M, 1S
	Special Case 1	1I, 5M, 1S
	Special Case 2	1I, 3M, 1S

Addition and Doubling

Addition, Odd Characteristic	
Input $D_1 = [u_{12}, u_{11}, u_{10}, v_{12}, v_{11}, v_{10}]$, $D_2 = [u_{22}, u_{21}, u_{20}, v_{22}, v_{21}, v_{20}]$	
Output $D_1 \oplus D_2 = [u_2', u_1', u_0', v_2', v_1', v_0']$	
Step **Expression**	**Operations**
Computing the resultant $r = \text{resultant}(u_1, u_2)$ and $inv = r(u_2^{-1})$ (mod u_1)	
1 $r = u_{10}(i_2 t_3 + i_1 t_2) - i_0 t_0$, $inv = i_2 x^2 + i_1 x + i_0$	15M
$t_0 = u_{10} - u_{20}$, $t_1 = u_{11} - u_{21}$, $t_2 = u_{12} - u_{22}$, $t_3 = t_1 - t_2 u_{12}$	
$t_4 = t_0 - t_2 u_{11}$, $t_5 = t_4 - t_3 u_{12}$, $t_6 = t_2 u_{10} + t_3 u_{11}$	
$i_0 = t_4 t_5 + t_3 t_6$, $i_1 = -(t_2 t_6 + t_1 t_5)$, $i_2 = t_1 t_3 - t_2 t_4$	
Computing $s' = (v_1 - v_2)inv$ (mod u_1)	
2 $s' = s_2' x^2 + s_1' x + s_0'$	10M
$w_0 = v_{10} - v_{20}$, $w_1 = v_{11} - v_{21}$, $w_2 = v_{12} - v_{22}$	
$t_0 = w_0 i_0$, $t_1 = w_1 i_1$, $t_2 = w_2 i_2$	
$t_3 = u_{12} t_2$, $t_4 = (w_1 + w_2)(i_1 + i_2) - t_1 - t_2 - t_3$	
$t_5 = u_{10} + u_{12}$, $t_6 = (t_5 + u_{11})(t_2 + t_4)$, $t_7 = (t_5 - u_{11})(t_2 - t_4)$	
$s_0' = t_0 - t_4 u_{10}$, $s_1' = (w_0 + w_1)(i_0 + i_1) - (t_7 + t_6)/2 - t_0 - t_1 + t_3$	
$s_2' = (w_0 + w_2)(i_0 + i_2) + (t_7 - t_6)/2 + t_1 - s_0' - t_2$	
Call Cantor algorithm if $s_2' = 0$	
Computing Required Inverses	
3 $\hat{u}_3 = s_2'(s_2' u_{11} - s_0') - s_1'(s_2' u_{12} - s_1')$	1I, 13M
$t_1 = s_2' r$, $w' = t_1 \hat{u}_3$, $I = w'^{-1}$, $t_2 = I t_1$, $t_3 = I \hat{u}_3$	
$I_r = t_3 s_2'$ (the inverse of r), $I_s = t_3 r$ (the inverse of s_2')	
$I_{u'} = t_1 t_2$ (the inverse of the leading coefficient of u' in step 9), $w = I_s r$, $s_2 = I_r s_2'$	
Computing $s = s'/s_2' = x^2 + s_1 x + s_0$	
4 $s_1 = I_s s_1'$, $s_0 = I_s s_0'$	2M
Computing $l = s u_2$	
5 $l = x^5 + l_4 x^4 + l_3 x^3 + l_2 x^2 + l_1 x + l_0$	4M
$t_1 = u_{20} + u_{22}$, $t_2 = (s_0 + s_1)(t_1 + u_{21})$, $t_3 = (s_0 - s_1)(t_1 - u_{21})$, $t_4 = s_1 u_{22}$	
$l_0 = s_0 u_{20}$, $l_1 = (t_2 - t_3)/2 - t_4$, $l_2 = (t_2 + t_3)/2 - l_0 + u_{20}$	
$l_3 = t_4 + s_0 + u_{21}$, $l_4 = s_1 + u_{22}$	
Computing $k = (v_2^2 - h v_2 - f)/u_2$	
6 $k_3 = 2(y_2 - v_{22})$	
Computing $u_t = (s(l + wh + 2wv_2) - kw^2)/u_1$	
7 $u_t = x^4 + u_{t3} x^3 + u_{t2} x^2 + u_{t1} x + u_{t0}$	12M
$t_0 = l_4 + 2w$, $t_1 = t_0 s_1$, $t_2 = s_0 l_3$, $t_3 = 2 v_{22} w$	
$u_{t3} = t_0 + s_1 - u_{12}$, $t_4 = u_{12} u_{t3}$, $u_{t2} = t_1 + l_3 + s_0 - u_{11} - t_4$, $t_5 = u_{11} u_{t2}$	
$u_{t1} = (s_0 + s_1)(l_3 + t_0) - (u_{11} + u_{12})(u_{t2} + u_{t3}) - t_1 - t_2 + t_3 + l_2 + t_4 + t_5 - u_{10}$	
$u_{t0} = s_1(t_3 + l_2) + w(2 v_{21} - k_3 w) + t_2 + l_1 - t_5 - u_{12} u_{t1} - u_{10} u_{t3}$	
Computing $v_t = (v_2 + l s_2)$ (mod u_t)	
8 $v_t = v_{t3} x^3 + v_{t2} x^2 + v_{t1} x + v_{t0}$	9M
$t_1 = s_2(u_{t3} - l_4) - 1$	
$v_{t3} = t_1 u_{t3} - s_2(u_{t2} - l_3)$, $v_{t2} = t_1 u_{t2} - s_2(u_{t1} - l_2) + v_{22}$	
$v_{t1} = t_1 u_{t1} - s_2(u_{t0} - l_1) + v_{21}$, $v_{t0} = t_1 u_{t0} + s_2 l_0 + v_{20}$	
Computing $u' = (f + h v_t' - (v_t')^2)/u_t$ where $v_t' = h + \lfloor y \rfloor - (v_t + \lfloor y \rfloor)$ (mod u_t)	
9 $u' = x^3 + u_2' x^2 + u_1' x + u_0'$	6M, 1S
$w_3 = v_{t3} - u_{t3}$, $w_2 = v_{t2} - u_{t2}$, $w_1 = v_{t1} - u_{t1}$, $w_0 = v_{t0} - u_{t0} + y_0$	
$u_2' = I_{u'}(w_2 + y_2) - w_3/2 - u_{t3}$, $u_1' = I_{u'}(w_1 + y_1) - w_2 - u_2' u_{t3} - u_{t2}$	
$u_0' = I_{u'}((y_2^2 - w_2^2)/2 + w_0) - w_1 - u_1' u_{t3} - u_2' u_{t2} - u_{t1}$	
Computing $v' = h + \lfloor y \rfloor - (\lfloor y \rfloor + v_t')$ (mod u')	
10 $v' = x^4 + v_2' x^2 + v_1' x + v_0'$	3M
$t_1 = w_3 + 2 u_2'$	
$v_2' = -t_1 u_2' + 2 u_1' + w_2$, $v_1' = -t_1 u_1' + 2 u_0' + w_1$, $v_0' = -t_1 u_0' - y_0 + w_0$	
Total	1I, 74M, 1S

Addition, Characteristic 2		
Input	$D_1 = [u_{12}, u_{11}, u_{10}, v_{12}, v_{11}, v_{10}], \quad D_2 = [u_{22}, u_{21}, u_{20}, v_{22}, v_{21}, v_{20}]$	
Output	$D_1 \oplus D_2 = [u_2', u_1', u_0', v_2', v_1', v_0']$	
Step	Expression	Operations
Computing the resultant $r =$ resultant(u_1, u_2) and $inv = r(u_2^{-1})$ (mod u_1)		
1	$r = i_0 t_0 + u_{10}(i_2 t_3 + i_1 t_2), \quad inv = i_2 x^2 + i_1 x + i_0$ $t_0 = u_{20} + u_{10}, \quad t_1 = u_{21} + u_{11}, \quad t_2 = u_{22} + u_{12}, \quad t_3 = t_1 + t_2 u_{12}$ $t_4 = t_0 + t_2 u_{11}, \quad t_5 = t_4 + t_3 u_{12}, \quad t_6 = t_2 u_{10} + t_3 u_{11}$ $i_0 = t_4 t_5 + t_3 t_6, \quad i_1 = t_2 t_6 + t_1 t_5, \quad i_2 = t_2 t_4 + t_1 t_3$	15M
Computing $s' = (v_1 + v_2)inv$ (mod u_1)		
2	$s' = s_2' x^2 + s_1' x + s_0'$ $w_0 = v_{10} + v_{20}, \quad w_1 = v_{11} + v_{21}, \quad w_2 = v_{12} + v_{22}, \quad t_0 = w_0 i_0, \quad t_1 = w_1 i_1,$ $t_2 = w_2 i_2$ $t_3 = (w_1 + w_2)(i_1 + i_2) + t_1 + t_2 + t_2 u_{12}, \quad t_4 = t_2 u_{11}, \quad t_5 = t_3 u_{10}$ $s_2' = (w_0 + w_2)(i_0 + i_2) + t_0 + t_1 + t_2 + t_4 + t_3 u_{12}$ $s_1' = (w_0 + w_1)(i_0 + i_1) + (t_3 + t_2)(u_{10} + u_{11}) + t_0 + t_1 + t_4 + t_5, \quad s_0' = t_0 + t_5$ Call Cantor algorithm if $s_2' = 0$	11M
Computing Required Inverses		
3	$\hat{u}_3 = s_2'(s_2' u_{11} + s_0') + s_1'(s_2' u_{12} + s_1')$ $t_1 = s_2' r, \quad w' = t_1 \hat{u}_3, \quad I = w'^{-1}, \quad t_2 = I t_1, \quad t_3 = I \hat{u}_3$ $I_r = t_3 s_2'$ (the inverse of r), $\quad I_s = t_3 r$ (the inverse of s_2') $I_{u'} = t_1 t_2$ (the inverse of the leading coefficient of u' in step 9), $\quad w = I_s r,$ $s_2 = I_r s_2'$	1I, 13M
Computing $s = s'/s_2' = x^2 + s_1 x + s_0$		
4	$s_1 = I_s s_1', \quad s_0 = I_s s_0'$	2M
Computing $l = s u_2$		
5	$l = x^5 + l_4 x^4 + l_3 x^3 + l_2 x^2 + l_1 x + l_0$ $\tilde{w}_0 = s_0 u_{21}, \quad \tilde{w}_1 = s_1 u_{22}, \quad l_4 = s_1 + u_{22}, \quad l_3 = \tilde{w}_1 + s_0 + u_{21}$ $l_2 = (s_0 + s_1)(u_{21} + u_{22}) + \tilde{w}_1 + \tilde{w}_0 + u_{20}, \quad l_1 = s_1 u_{20} + \tilde{w}_0, \quad l_0 = s_0 u_{20}$	5M
Computing $k = (v_2^2 + h v_2 + f)/u_2$		
6	$k_3 = h_2 + y_2 + v_{22}$	
Computing $u_t = s(l + wh + 2w v_2) + k w^2)/u_1$		
7	$u_t = x^4 + u_{t3} x^3 + u_{t2} x^2 + u_{t1} x + u_{t0}$ $t_0 = w h_2, \quad a_1 = s_1(w + l_4), \quad a_0 = s_0 l_3$ $u_{t3} = w + l_4 + s_1 + u_{12}, \quad t_1 = u_{12} u_{t3}, \quad u_{t2} = a_1 + t_1 + l_3 + s_0 + u_{11},$ $t_2 = u_{11} u_{t2}$ $u_{t1} = (s_0 + s_1)(w + l_3 + l_4) + (u_{11} + u_{12})(u_{t2} + u_{t3}) + t_0 + a_0 + a_1 + l_2 + t_1 + t_2 + u_{10}$ $u_{t0} = w(w k_3 + h_1) + s_1(t_0 + l_2) + a_0 + l_1 + t_2 + u_{12} u_{t1} + u_{10} u_{t3}$	12M
Computing $v_t = (v_2 + l s_2)$ (mod u_t)		
8	$v_t = v_{t3} x^3 + v_{t2} x^2 + v_{t1} x + v_{t0}$ $t_1 = s_2(u_{t3} + l_4) + y_4 + 1$ $v_{t3} = t_1 u_{t3} + s_2(u_{t2} + l_3), \quad v_{t2} = t_1 u_{t2} + s_2(u_{t1} + l_2) + v_{22}$ $v_{t1} = t_1 u_{t1} + s_2(u_{t0} + l_1) + v_{21}, \quad v_{t0} = t_1 u_{t0} + s_2 l_0 + v_{20}$	9M
Computing $u' = \frac{f + h v_t' + (v_t')^2}{u_t}$ where $v_t' = h + \lfloor y \rfloor - (v_t + \lfloor y \rfloor)$ (mod u_t)		
9	$u' = x^3 + u_2' x^2 + u_1' x + u_0'$ $w_3 = y_4 u_{t3} + v_{t3}, \quad w_2 = y_4 u_{t2} + v_{t2} + y_2, \quad w_1 = y_4 u_{t1} + v_{t1} + y_1,$ $w_0 = y_4 u_{t0} + v_{t0} + y_0$ $u_2' = I_u w_2 + w_3 + u_{t3}, \quad u_1' = I_u w_1 + h_2 + u_{t2} + u_{t3} u_2'$ $u_0' = I_u(w_2(w_2 + h_2) + w_0) + h_1 + u_{t1} + u_{t3} u_1' + u_{t2} u_2'$	11M
Computing $v' = h + \lfloor y \rfloor + (\lfloor y \rfloor + v_t')$ (mod u')		
10	$v' = (h_4 + y_4) x^4 + (h_3 + y_3) x^3 + v_2' x^2 + v_1' x + v_0'$ $t_1 = w_3 + u_2'$ $v_2' = t_1 u_2' + w_2 + y_2 + u_1', \quad v_1' = t_1 u_1' + w_1 + y_1 + u_0', \quad v_0' = t_1 u_0' + w_0 + y_0$	3M
Total		1I, 81M

Doubling, Odd Characteristic		
Input	$D = [u_2, u_1, u_0, v_2, v_1, v_0]$	
Output	$D \oplus D = [u_2', u_1', u_0', v_2', v_1', v_0']$	
Step	Expression	Operations
Computing the resultant $r =$ resultant$(2v - h, u)$ and $inv = r(2v - h)^{-1} \pmod{u}$		
1	$r = \hat{w}_0 i_0 - u_0(t_1 i_2 + \hat{w}_2 i_1), \quad inv = i_2 x^2 + i_1 x + i_0$	17M, 1S
	$\hat{w}_0 = 2(v_0 + u_0 u_2), \quad \hat{w}_1 = 2(v_1 - u_0 + u_1 u_2), \quad \hat{w}_2 = 2(v_2 - u_1 + u_2^2)$	
	$t_1 = \hat{w}_1 - u_2 \hat{w}_2, \quad t_2 = \hat{w}_0 - u_1 \hat{w}_2, \quad t_3 = t_2 - u_2 t_1, \quad t_4 = u_0 \hat{w}_2 + u_1 t_1$	
	$i_0 = t_2 t_3 + t_1 t_4, \quad i_1 = -(\hat{w}_1 t_3 + \hat{w}_2 t_4), \quad i_2 = \hat{w}_1 t_1 - \hat{w}_2 t_2$	
Computing $k' = (f + hv - v^2)/u, \quad k = k' \pmod{u}$		
2	$k = k_2 x^2 + k_1 x + k_0$	7M
	$t_1 = y_2 - v_2, \quad t_2 = u_2 t_1, \quad k_3' = 2t_1, \quad k_2 = -4t_1 + 2y_1 - 2v_1$	
	$k_1 = 2u_2(t_2 - y_1 + v_1) + t_2(y_2 + v_2 - 4u_1) + 2y_0 - 2v_0$	
	$k_0 = f_3 - u_2 k_1 - 4u_0 t_1 - 2u_1(y_1 - v_1) - 2v_1 v_2$	
Computing $s' = k.inv \pmod{u}$		
3	$s' = s_2 x^2 + s_1 x + s_0$	10M
	$t_0 = k_0 i_0, \quad t_1 = k_1 i_1, \quad t_2 = k_2 i_2, \quad t_3 = u_2 t_2$	
	$t_4 = (k_1 + k_2)(i_1 + i_2) - t_1 - t_2 - t_3, \quad t_5 = t_4 u_0, \quad t_6 = u_0 + u_2, \quad t_7 = t_6 + u_1$	
	$t_8 = t_6 - u_1, \quad t_9 = t_7(t_4 + t_2), \quad t_{10} = t_8(t_4 - t_2)$	
	$s_2' = (k_0 + k_2)(i_0 + i_2) - (t_{10} + t_9)/2 - t_0 - t_2 + t_1 + t_5$	
	$s_1' = (k_0 + k_1)(i_0 + i_1) + (t_{10} - t_9)/2 - t_0 - t_1 + t_3, \quad s_0' = t_0 - t_5$	
	If $s_2' = 0$ then call Cantor Algorithm	
Computing Required Inverses		
4	$\hat{u}_3 = s_2'(s_2' u_2 - s_0') - s_1'(s_1' - s_2' u_2)$	1I, 13M
	$t_1 = s_2' r, \quad w' = t_1 \hat{u}_3, \quad I = w'^{-1}, \quad t_2 = I t_1, \quad t_3 = I \hat{u}_3$	
	$I_r = t_3 s_2'$ (the inverse of r), $\quad I_s = t_3 r$ (the inverse of s_2')	
	$I_{u'} = t_1 t_2$ (the inverse of the leading coefficient of u' in step 9), $\quad w = I_s r$,	
	$s_2 = I_r s_2'$	
Computing $s = s'/s_2' = x^2 + s_1 x + s_0$		
5	$s_1 = I_s s_2', \quad s_0 = I_s s_0'$	2M
Computing $l = su$		
6	$l = x^5 + l_4 x^4 + l_3 x^3 + l_2 x^2 + l_1 x + l_0$	4M
	$t_1 = u_0 + u_2, \quad t_2 = (s_0 + s_1)(t_1 + u_1), \quad t_3 = (s_0 - s_1)(t_1 - u_1), \quad \tilde{w}_1 = s_1 u_2$	
	$l_0 = s_0 u_0$	
	$l_1 = (t_2 - t_3)/2 - \tilde{w}_1, \quad l_2 = (t_2 + t_3)/2 - l_0 + u_0, \quad l_3 = \tilde{w}_1 + s_0 + u_{21},$	
	$l_4 = s_1 + u_{22}$	
Computing $u_t = s^2 + (sw(h + 2v) - w^2 k')/u$		
7	$u_t = x^4 + u_{t3} x^3 + u_{t2} x^2 + u_{t1} x + u_{t0}$	13M
	$u_{t3} = 2(w + s_1), \quad t_1 = u_2 u_{t3}$	
	$u_{t2} = s_1(2w + s_1) + 2\tilde{w}_1 - t_1 + 2s_0, \quad t_2 = u_1 u_{t2}$	
	$u_{t1} = 2w(s_0 + v_2) + s_1(\tilde{w}_1 + 2s_0) - (u_{t2} + u_{t3})(u_1 + u_2) + 2l_2 - 2u_0 + t_1 + t_2$	
	$u_{t0} = w(2s_1 v_2 - 2wk_3' + 2v_1) + s_1(l_2 + u_0) + s_0(s_0 + \tilde{w}_1 + 2u_1) - t_2 - u_2 u_{t1} - u_0 u_{t3}$	
Computing $v_t = (v + s_2 l) \pmod{u_t}$		
8	$v_t = v_{t3} x^3 + v_{t2} x^2 + v_{t1} x + v_{t0}$	9M
	$t_1 = s_2(u_{t3} - l_4) - 1$	
	$v_{t3} = u_{t3} t_1 - s_2(u_{t2} - l_3), \quad v_{t2} = u_{t2} t_1 - s_2(u_{t1} - l_2) + v_2$	
	$v_{t1} = u_{t1} t_1 - s_2(u_{t0} + l_1) + v_1, \quad v_{t0} = u_{t0} t_1 + s_2 l_0 + v_0$	
Computing $u' = (f + hv_t' - (v_t')^2)/u_t$ where $v_t' = h + \lfloor y \rfloor - (v_t + \lfloor y \rfloor) \pmod{u_t}$		
9	$u' = u_3' x^3 + u_2' x^2 + u_1' x + u_0'$	6M, 1S
	$w_3 = v_{t3} - u_{t3}, \quad w_2 = v_{t2} - u_{t2}, \quad w_1 = v_{t1} - u_{t1}, \quad w_0 = v_{t0} - u_{t0} + y_0$	
	$u_2' = I_{u'}(w_2 + y_2) - w_3/2 - u_{t3}, \quad u_1' = I_{u'}(w_1 + y_1) - w_2 - u_{t3} u_2' - u_{t2}$	
	$u_0' = I_{u'}((y_2^2 - w_2^2)/2 + w_0) - w_1 - u_{t3} u_1' - u_{t2} u_2' - u_{t1}$	
Computing $v' = h + \lfloor y \rfloor - (\lfloor y \rfloor + v_t') \pmod{u}$		
10	$v' = (h_4 + y_4)x^4 + (h_3 + y_3)x^3 + v_2' x^2 + v_1' x + v_0'$	3M
	$t_1 = w_3 + 2u_2'$	
	$v_2' = -t_1 u_2' + 2u_1' + w_2, \quad v_1' = -t_1 u_1' + 2u_0' + w_1, \quad v_0' = -t_1 u_0' + w_0 - y_0$	
Total		1I, 84M, 2S

Doubling, Characteristic 2		
Input	$D = [u_2, u_1, u_0, v_2, v_1, v_0]$	
Output	$D \oplus D = [u'_2, u'_1, u'_0, v'_2, v'_1, v'_0]$	
Step	Expression	Operations
Precomputations		
Computing the resultant r =resultant$(2v + h, u)$ and $inv = r(2v + h)^{-1} \pmod u$		
1	$r = \hat{w}_0 i_0 + u_0(t_1 i_2 + \hat{w}_2 i_1), \quad inv = i_2 x^2 + i_1 x + i_0$ $\hat{w}_0 = h_0 + u_0 u_2, \quad \hat{w}_1 = h_1 - u_0 + u_1 u_2, \quad \hat{w}_2 = h_2 - u_1 + u_2^2$ $t_1 = \hat{w}_1 + u_2 \hat{w}_2, \quad t_2 = \hat{w}_0 + u_1 \hat{w}_2, \quad t_3 = t_2 + u_2 t_1, \quad t_4 = u_0 \hat{w}_2 + u_1 t_1$ $i_0 = t_2 t_3 + t_1 t_4, \quad i_1 = \hat{w}_1 t_3 + \hat{w}_2 t_4, \quad i_2 = \hat{w}_1 t_1 + \hat{w}_2 t_2$	17M, 1S
Computing $k' = (f + hv + v^2)/u, \quad k = k' \pmod u$		
2	$k = k_2 x^2 + k_1 x + k_0$ $k'_3 = h_2 + y_2 + v_2, \quad k_2 = h_1 + y_1 + v_1$ $k_1 = u_2(k'_3 u_2 + k_2) + v_2(h_2 + v_2) + h_0 y_4 + h_0 + v_0$ $k_0 = f_3 + k_2(h_2 + u_1) + k'_3 h_1 + k_1 u_2$	6M
Computing $s' = k.inv \pmod u$		
3	$s' = s'_2 x^2 + s'_1 x + s'_0$ $t_0 = k_0 i_0, \quad t_1 = k_1 i_1, \quad t_2 = k_2 i_2, \quad t_3 = (k_1 + k_2)(i_1 + i_2) + t_1 + t_2 + t_2 u_2,$ $t_4 = t_2 u_1, \quad t_5 = t_3 u_0$ $s'_2 = (k_0 + k_2)(i_0 + i_2) + t_0 + t_1 + t_2 + t_4 + t_3 u_2$ $s'_1 = (k_0 + k_1)(i_0 + i_1) + (t_2 + t_3)(u_0 + u_1) + t_0 + t_1 + t_4 + t_5, \quad s'_0 = t_0 + t_5$ If $s'_2 = 0$ then call Cantor Algorithm	11M
Computing Required Inverses		
4	$\hat{u}_3 = s'_1(s'_1 + s'_2 u_2) + s'_2(s'_0 + s'_2 u_1), \quad t_1 = s'_2 r, \quad w' = t_1 \hat{u}_3, \quad I = w'^{-1},$ $t_2 = It_1, \quad t_3 = I\hat{u}_3$ $I_r = t_3 s'_2$ (the inverse of r, $I_s = t_3 r$ (the inverse of s'_2) $I_{u'} = t_1 t_2$ (the inverse of the leading coefficient of u' in step 9), $w = I_s r,$ $s_2 = I_r s'_2$	1I, 13M
Computing $s = s'/s'_2 = x^2 + s_1 x + s_0$		
5	$s_1 = I_s s'_1, \quad s_0 = I_s s'_0$	2M
Computing $l = su$		
6	$l = x^5 + l_4 x^4 + l_3 x^3 + l_2 x^2 + l_1 x + l_0$ $\tilde{w}_1 = s_1 u_2, \quad \tilde{w}_0 = s_0 u_1, \quad l_4 = s_1 + u_2, \quad l_3 = \tilde{w}_1 + s_0 + u_1$ $l_2 = (s_0 + s_1)(u_1 + u_2) + \tilde{w}_0 + \tilde{w}_1 + u_0, \quad l_1 = \tilde{w}_0 + s_1 u_0, \quad l_0 = s_0 u_0$	5M
Computing $u_t = s^2 + (ws(h + 2v) + w^2 k')/u$		
7	$u_t = x^4 + u_{t3} x^3 + u_{t2} x^2 + u_{t1} x + u_{t0}$ $t_1 = wu_2, \quad u_{t3} = w, \quad u_{t2} = s_1(w + s_1) + t_1, \quad t_2 = u_1 u_{t2}$ $u_{t1} = w(s_0 + h_2) + s_1 \tilde{w}_1 + (w + u_{t2})(u_1 + u_2) + t_1 + t_2$ $u_{t0} = w(s_1 h_2 + wk'_3 + h_1 + u_0) + s_1(l_2 + u_0) + s_0(s_0 + w_1) + u_2 u_{t1} + t_2$	12M
Computing $v_t = (v + s_2 l) \pmod{u_t}$		
8	$v_t = v_{t3} x^3 + v_{t2} x^2 + v_{t1} x + v_{t0}$ $t_1 = s_2(w + l_4) + y_4 + 1$ $v_{t3} = wt_1 + s_2(u_{t2} + l_3), \quad v_{t2} = u_{t2} t_1 + s_2(u_{t1} + l_2) + v_2$ $v_{t1} = u_{t1} t_1 + s_2(u_{t0} + l_1) + v_1, \quad v_{t0} = u_{t0} t_1 + s_2 l_0 + v_0$	9M
Computing $u' = (f + hv'_t - (v'_t)^2)/u_t$ where $v'_t = h + \lfloor y \rfloor - (v_t + \lfloor y \rfloor) \pmod{u_t}$		
9	$u' = u'_3 x^3 + u'_2 x^2 + u'_1 x + u'_0$ $w_3 = v_{t3} + y_4 w, \quad w_2 = y_2 + v_{t2} + y_4 u_{t2}, \quad w_1 = y_1 + v_{t1} + y_4 u_{t1},$ $w_0 = y_0 + v_{t0} + y_4 u_{t0}$ $u'_2 = I_{u'} w_2 + w_3 + w, \quad u'_1 = I_{u'} w_1 + h_2 + wu'_2 + u_{t2}$ $u'_0 = I_{u'}(w_2(w_2 + h_2) + w_0) + h_1 + wu'_1 + u_{t2} u'_2 + u_{t1}$	11M
Computing $v' = h + \lfloor y \rfloor + (\lfloor y \rfloor + v'_t) \pmod{u'}$		
10	$v' = (h_4 + y_4)x^4 + (h_3 + y_3)x^3 + v'_2 x^2 + v'_1 x + v'_0$ $t_1 = u'_2 + w_3$ $v'_2 = u'_2 t_1 + u'_1 + w_2 + y_2, \quad v'_1 = u'_1 t_1 + u'_0 + w_1 + y_1, \quad v'_0 = u'_0 t_1 + w_0 + y_0$	3M
Total		1I, 89M, 1S

References

1. Avanzi, R., Thériault, N., Wang, Z.: Rethinking low genus hyperelliptic Jacobian arithmetic over binary fields: interplay of field arithmetic and explicit formuae. J. Math. Cryptol. **2**(3), 227–255 (2008)
2. Cantor, D.G.: Computing in the Jacobian of a hyperelliptic curve. Math. Comp. **48**(177), 95–101 (1987)
3. Cohen, H.: A Course in Computational Algebraic Number Theory. Graduate Texts in Mathematics, vol. 138. Springer, Berlin (1993)
4. Costello, C., Lauter, K.: Group law computations on Jacobians of hyperelliptic curves. In: Miri, A., Vaudenay, S. (eds.) SAC 2011. LNCS, vol. 7118, pp. 92–117. Springer, Heidelberg (2012). https://doi.org/10.1007/978-3-642-28496-0_6
5. Erickson, S., Jacobson Jr., M.J., Stein, A.: Explicit formulas for real hyperelliptic curves of genus 2 in affine representation. Adv. Math. Commun. **5**(4), 623–666 (2011)
6. Fan, X., Wollinger, T.J., Gong, G.: Efficient explicit formulae for genus 3 hyperelliptic curve cryptosystems over binary fields. IET Inf. Secur. **1**(2), 65–81 (2007)
7. Galbraith, S.D., Harrison, M., Mireles Morales, D.J.: Efficient hyperelliptic arithmetic using balanced representation for divisors. In: van der Poorten, A.J., Stein, A. (eds.) ANTS 2008. LNCS, vol. 5011, pp. 342–356. Springer, Heidelberg (2008). https://doi.org/10.1007/978-3-540-79456-1_23
8. Gathen, J.V.Z., Gerhard, J.: Modern Computer Algebra, 3rd edn. Cambridge University Press, Cambridge (2013)
9. Gaudry, P., Harley, R.: Counting points on hyperelliptic curves over finite fields. In: Bosma, W. (ed.) ANTS 2000. LNCS, vol. 1838, pp. 313–332. Springer, Heidelberg (2000). https://doi.org/10.1007/10722028_18
10. Guyot, C., Kaveh, K., Patankar, V.M.: Explicit algorithm for the arithmetic on the hyperelliptic Jacobians of genus 3. J. Ramanujan Math. Soc. **19**, 75–115 (2004)
11. Jacobson Jr., M.J., Rezai Rad, M., Scheidler, R.: Comparison of scalar multiplication on real hyperelliptic curves. Adv. Math. Commun. **8**(4), 389–406 (2014)
12. Jacobson Jr., M.J., Scheidler, R., Stein, A.: Cryptographic protocols on real hyperelliptic curves. Adv. Math. Commun. **1**(2), 197–221 (2007)
13. Jacobson Jr., M.J., Scheidler, R., Stein, A.: Fast arithmetic on hyperelliptic curves via continued fraction expansions. Advances in Coding Theory and Cryptography. Coding Theory Cryptology, vol. 3, pp. 200–243. World Scientific Publishing, Hackensack (2007)
14. Jacobson Jr., M.J., Scheidler, R., Stein, A.: Cryptographic aspects of real hyperelliptic curves. Tatra Mountains Math. Publ. **47**(1), 31–65 (2010)
15. Kedlaya, K.S., Sutherland, A.V.: Computing L-series of hyperelliptic curves. In: van der Poorten, A.J., Stein, A. (eds.) ANTS 2008. LNCS, vol. 5011, pp. 312–326. Springer, Heidelberg (2008). https://doi.org/10.1007/978-3-540-79456-1_21
16. Lange, T.: Formula for arithmetic on genus 2 hyperelliptic curves. Appl. Algebra Eng. Commun. Comput. **15**(5), 295–328 (2005)
17. Mireles Morales, D.J.: An analysis of the infrastructure in real function fields. Cryptology ePrint Archive no. 2008/299 (2008)
18. Rezai Rad, M.: A complete evaluation of arithmetic in real hyperelliptic curves. Ph.D. thesis, University of Calgary (2016). https://prism.ucalgary.ca/bitstream/handle/11023/3293/ucalgary_2016_rezairad_monireh.pdf
19. Scheidler, R., Stein, A., Williams, H.C.: Key-exchange in real quadratic congruence function fields. Des. Codes Crypt. **7**(1), 153–174 (1996). Special issue dedicated to Gustavus J. Simmons

20. Schmidt, F.K.: Analytische Zahlentheorie in Körpern der Charakteristik p. Math. Z. **33**(1), 1–32 (1931)
21. Stein, A.: Explicit infrastructure for real quadratic function fields and real hyperelliptic curves. Glasnik Matematički. Serija III **44**(1), 89–126 (2009)
22. Stein, A., Teske, E.: Optimized baby step-giant step methods. J. Ramanujan Math. Soc. **20**(1), 27–58 (2005)
23. Sutherland, A.V.: Fast Jacobian arithmetic for hyperelliptic curves of genus 3. In: ANTS XIII–Proceedings of the Thirteenth Algorithmic Number Theory Symposium. Open Book Series, vol. 2, pp. 425–442. Mathematical Sciences Publishers, Berkeley (2019)
24. Wollinger, T.J., Pelzl, J., Paar, C.: Cantor versus Harley: optimization and analysis of explicit formulae for hyperelliptic curve cryptosystems. IEEE Trans. Comput. **54**(7), 861–872 (2005)

Proposition of a Model for Securing the Neighbor Discovery Protocol (NDP) in IPv6 Environment

Bôa Djavph Yoganguina$^{(\boxtimes)}$, Khadidiatou Wane ep Keîta, Idy Diop,
Khaly Tall, and Sidi Mouhamed Farssi

Polytechnic Institute (ESP), Universite Cheikh Anta Diop, Dakar, Senegal
yoganguina5@gmail.com

Abstract. This article proposes a model for securing the Neighbor Discovery Protocol, to enable a secure exchange of IPv6 mobiles for insertion into another network. As part of the Neighbor Discovery Protocol, we have listed all the features and demonstrated that they can all be attacked though our particular focus is on appraising the existing one. The article demonstrates that it is possible to secure the most critical points in the Neighbor Discovery Protocol features, including the IP address and prefix. The security model using the IPsec AH protocol combination, and the CGA Protocol.

Keywords: Security · Neighbor Discovery Protocol · IPv4/IPv6 · IPsec

1 Introduction

IPv6 was designed to address the weaknesses of IPv4. It offers a larger address space. It integrates security and allows mobility. It also allows for auto-configuration making it possible for IPv6 mobile nodes to acquire a valid IPv6 address in host networks. This auto-configuration is made possible by the Neighbor Discovery Protocol. Neighbor discovery includes router discovery and redirection. Router discovery is used by an IPv6 host to detect the presence of routers and network settings. This enables the host to configure the elements that are needed. However, NDP's security is ensured by the IPsec protocol, to initiate the use of IKE, IPsec cannot be used manually with a manual security association, which makes this approach impractical [1]. In the absence of security, this Protocol is vulnerable and is subject to network attacks in all its variants. In order to address this issue, a security mechanism based on the number of hops that is determined at 255 bits has been proposed [2]. But this security mechanism does not allow effective security between nodes in a communication. A solution involving the use of AH protocol with MAC address within the Neighbor Discovery Protocol for securing this Protocol [1], but this mechanism is inefficient and unoptimized. Securing neighbor discovery messages using

© Springer Nature Switzerland AG 2019
C. T. Gueye et al. (Eds.): A2C 2019, CCIS 1133, pp. 204–215, 2019.
https://doi.org/10.1007/978-3-030-36237-9_12

Secure Neighbor Discovery (SeND) [3] and IPsec [4] outlined by the NDP standard does not specify how to use both in IPv6 local communication including router discoveries. This article is part of a contribution to securing the neighbor discovery while minimizing the major concern of usurpation. Our approach is to provide an authentication model for packets exchanged by the Neighbor Discovery Protocol while providing proof of address ownership by IPv6 nodes, thus securing the most critical points. Especially addresses and discovery of prefixes. The document is structured as follows: a section deals with the Neighbor Discovery mechanism and its limitation. The following section reviews the literature on the security of the Neighbor Discovery Protocol. Section 4 describes the proposed model. Section 5 deals with the validation of the proposed model. Section 6 analyzes and tests the performance of CGA SeND and CGA AH with the various hashing algorithms. Then a conclusion.

2   Background

2.1   Neighbor Discovery Mechanism

The Neighbor Discovery Protocol, in lieu of the ARP request in IPv4, allows the IPv6 mobile to acquire a valid IPv6 address for release into a host network through DHCPv6. In the mobility process, the IPv6 mobile emits signals to detect its closest neighbor (default router). The Neighbor Discovery Protocol uses five (5) message types: the solicitation message sent by a mobile node to solicit the default router (RS), the message in response to the mobile solicitation message by the router (RA), the message from Neighbor Discovery on the link, sent by the mobile or router (NS), the redirect message (RM). And the message in response to the neighbor discovery message on the link (NA). Initially the mobile node sends an RS solicitation message, which contains its MAC address to the router. The router in possession of the MAC address, responds to the mobile by sending the prefix. The mobile uses the prefix to construct its IPv6 address. The mobile again sends an NS message to its neighbors to inform them of its status. The neighbors respond, by an acknowledgment of receipt, with an NA message. Figure 1 shows the unsecured swap between the mobile and its default router.

Fig. 1. Neighborhood discovery process

Figure 2 shows the neighbor discovery message in detail.

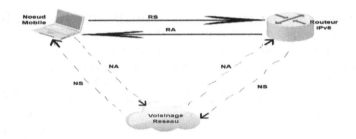

Fig. 2. NDP message interchange

However, this process does not happen in a secure manner.

2.2 Authentication Header (IPSec AH)

Packets from IPsec sender are authenticated by the IPsec receiver. The authentication is ensured by the data integrity service, thus making it possible to identify the sent message actually coming from the source. The data integrity service is made possible by the Authentication Header (AH) extension. Thus, the sent message is not altered during transmission [3]. It can optionally be used to detect replay attacks. Authentication headers provide integrity, source authentication, and provides protection against replayed packets. The IPSec AH protocol is used in UDP mode (non-connection-oriented mode). Figure 3 shows packet authentication with IPsec AH.

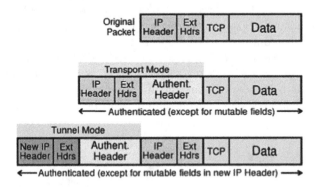

Fig. 3. Packet authentication with IPsec AH

2.3 Cryptographic Address Generation (CGA SeND)

CGA or cryptographically generated addresses, is a protocol for generating addresses in a cryptographic way thus ensuring the authenticity of nodes during exchanges [3]. An address generated with CGA has a security setting (sec) that allows it to cope with brute-force attacks. The security setting is a three-bit unsigned integer, and it is encoded in the three leftmost bits (that is, bits 0 through 2) of the identifier interface. CGA's are generated from a cryptographic hash function of a public key and the auxiliary settings. Each CGA is associated with a CGA data structure parameter (RFC 3972).

3   State of the Art

Neighbor Discovery is the process by which a node in a network determines the total number and identity of the other nodes in its vicinity. This is a fundamental element of many protocols, including localization [5], routing [6], leader election [7], and group management [8], time-based communications and many media access control mechanisms. Agencies [9] rely on accurate neighborhood information. Neighbor Discovery is especially important for the smooth operation of wireless networks. Fifteen years ago, an alternative was developed to identify several security vulnerabilities in Neighbor Discovery Protocol and to propose a security strategy based on the secure Neighbor Discovery Protocol, among these are the security function developed and incorporated in the CGA module and the SeND module ensuring the securing of information transiting on the network and addressing paralysis [10]. However, it should be noted that the SEND IETF working group has demonstrated that the SEND protection does not guarantee total reliability. Even though it protects the node, it does not allow functions that typically require a third-party node to modify or send NDP messages and disable any form of address sharing. The use of the multi-key cryptography (MCGA) security mechanism is made to secure Neighbor Discovery (ND) proxy for mobility and a reconnaissance test on SeND [11]. However, it should be noted that the signature can be generated using the private key of any node, but verification requires the public key of all other nodes making the system less efficient. NDP covers host initialization and automatic address configuration. The IPsec protocol, which is supposed to ensure the security of the link-local does not provide this security, this is due to IKE initiation issue [12], IPsec cannot simply be used because at the beginning of the connection the IPv6 nodes do not know the source and destination IPv6 addresses for the construction of the security tunnel. This makes this approach impractical [13]. In the absence of security this Protocol is very vulnerable and is exposed to any type of attack. SeND is used to ensure the security of local links but its deployment is too complex [12]. The Neighbor Discovery Protocol (NDP) standard for IPv6, RFC 4861, details how to secure neighbor discovery messages using Secure Neighbor Discovery (SeND) [12] and IPsec [6]. However, it does not show how to use both security protocols on local IPv6 link communication, including router discovery. The authors of [1] explain how SeND prevents the neighbor discovery threat and vulnerability listed in [7]. However, many studies on the implementation of SeND on RD security, such as [9], have also encountered some issues on SeND such as calculation exhaustion, bandwidth consumption and lack of confidentiality. These studies show that the implementation of SeND is very limited.

Neighbor Discovery (ND) packets have an initial value of the hop limit set to 255, and a value of negative 1 when passing a router. Upon receipt of the packet, the receiver checks the IPv6 hop limit value when receiving an ND packet. If the value equals 255, which means that the packet derives from the legitimate nodes in the home link, the receiver would accept; otherwise the packet is deleted [10]. In addition, NDP security authentication through number of hops validation in ND packets is not sufficient, preventing only network threats launched by adjacent nodes in foreign links and avoiding security attacks from malicious nodes. As a result, security threats between nodes in the communication of domestic links cannot be effectively guaranteed. Authors of [16] present an authentication scheme based on unsigned encryption, and the use of the encryption scheme ensures confidentiality, reliability during the message interaction process. The authors of [17] proposed a secure access authentication scheme for the localized mobility management protocol, for the Next Generation Mobile Network (PMIPv6). The authentication between the mobile node and the mobile access gateway is performed by the intra-domain and inter-domain scenarios using the signature based on the identity and service level contract [17].

However, the schema does not consider confidentiality when authenticating PMIPv6 access based on the proposed schema. After a study of the existing, two of their work have caught our attention. The first model is based on the CGA SeND model, the use of SeND and CGA addresses provides security for the Neighbor Protocol. CGA addresses can be used to generate addresses in a cryptographic way, thus ensuring the legitimacy of nodes during interchanges. The SeND Protocol is used to ensure packet authentication during messages exchanged between nodes. However, SeND can be difficult to deploy and may still be prone to certain types of denial-of-service attacks. The remaining vulnerabilities and current deployment issues were also addressed [10]. The second model is a method using the MAC address combination and the IPsec AH protocol to improve the security mechanism for the Neighbor Discovery Protocol. This model is exposed to several attacks related to the MAC address, including arpspoofing, and the Man in the Middle attack. Our model, unlike the existing model, makes it possible to ensure the legitimacy of the nodes, and the authentication of the packets during the mobility episode (auto-configuration).

4 Proposed Model

The use of CGA and SeND addresses is unreliable in practice because SeND is not able to provide assurance about the identity of the actual node and cannot guarantee that the CGA address is used by the appropriate node [10]. SeND can also be vulnerable to denial-of-service attacks. The implementation of SeND is still rare. Most modern operating systems support NDP, but do not fully support SeND [18]. To address the first, we found that combining CGA protocol with the IPSEC AH protocol is reliable to implement our security model for the protection of the Neighbor Discovery Protocol. In the second model the combination of IPSec AH and the MAC address is unreliable because the Address Resolution Protocol (ARP) is a protocol that resolves the IP address to a MAC address. If the MAC address is not present in the ARP table, the node will broadcast the ARP request. All nodes will compare their IP address with this. Only one of the nodes identifies and responds to it. ARP does not provide any means of verifying authenticity. The attacker can send any IP or MAC address. The victim's ARP table stores this malicious content [7]. ARP poisoning threats include: 1. Packet sniffing, 2. Session hijacking, 3. Data manipulation, 4. Attack of the man in the middle, 5. Misuse of the connection, 6. Resetting of the connection, 7. Passwords, and denial of service (DOS) attacks [19]. The IPSec AH protocol is integrated into our model because of its effectiveness. Our contribution in this section is to provide a security model using the AH protocol for packet authentication and CGA for IP address generation in an encrypted manner, see Fig. 4.

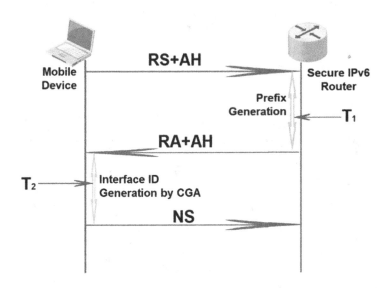

Fig. 4. Proposed model for NDP security

According to Fig. 4, the L latency of the NDP procedure via the proposed CGA + AH method is as follows.

$$L = L(RS_AH) + T_1 + L(RA_AH) + T_2 + L_N S + L_D DD \tag{1}$$

Formula 1: Formula for latency averages. Where: is the transmission time of the RS packet, and the processing time of the AH header. The time between the reception of the RS packet and the time when the RA + AH packet is sent. Is the transmission time of the RA packet and the processing time of the AH header.

Is the time between receiving the RA packet and the time the ID interface is generated by CGA. Is the transmission time of the NS packet. DAD latency (duplicate address detection), default 1 s. The IPv6 mobile before sending the solicitation message (RS) signs the packet that contains the MAC address by the AH protocol (Fig. 5).

```
Demarage du client IPv6
Generation d'une Association de Securite (SA) en mode AH
Envoie d'un paquet Router Solicitation
Securisation via AH
###[ IPv6 ]###
    version  = 6
    tc       = 0
    fl       = 0
    plen     = 48
    nh       = AH Header
    hlim     = 255
    src      = fe80::6661:ea58:9385:bde8
    dst      = ff02::2
###[ AH ]###
       nh         = ipv6_icmp
       payloadlen = 4
       reserved   = None
       spi        = 0xdeadbeef
       seq        = 1
       icv        = 327d08cf27bc9cfb45284869
       padding    = None
###[ ICMPv6 Neighbor Discovery - Router Solicitation ]###
          type     = Router Solicitation
          code     = 0
          cksum    = 0x6056
          res      = 0
###[ ICMPv6 Neighbor Discovery Option - Source Link-Layer Address ]###
             type     = 1
             len      = 1
             lladdr   = 00:0c:29:ad:45:ed
###[ ICMPv6 Neighbor Discovery Option - MTU ]###
             type     = 5
             len      = 1
             res      = 0x0
             mtu      = 1280

Envoie du paquet du paquet Router Sollicitation en mode transport AH
Begin emission:
..Finished sending 1 packets.
............................*
```

Fig. 5. Generating RS security packages.

Once the packet is received by the Secure IPv6 router, it is decrypted. The router retrieves the Mac address. It encapsulates in an RA packet, then it responds to the router with a RA message containing the prefix. And in the same way as mobile, the router also executes the signature of the RA message.

When the packet is received, the mobile decrypts and comes into possession of the prefix. Based on the prefix, it automatically generates its IPV6 address cryptographically. This is made possible by the CGA protocol, whose role is to generate IP addresses in a secure way.

5   CGA SEND and CGA AH Performance Analysis and Test

In this section we have chosen to analyze CGA SeND and CGA AH, as CGA SeND is an improvement in the Neighbor Discovery process. So if we position ourselves in relation to CGA SeND, we do so implicitly in the relative processes used. We will then make a comparative analysis of the performance of the proposed CGA SEND and CGA AH model. Depending on the encryption keys used, the size of the RS, RA, and NS packets will vary. Table 1 shows the size of the packets based on the type of encryption and the type of key used. The goal is to show the increase in latencies, because of the facts highlighted in Fig. 8. Depending on this latency we can do a performance analysis. For experiment, we used SCAPY [23] to generate the packets. It should be noted that SCAPY has in addition a module that takes into account the IPSEC stack. SCAP [24], enables packet captures. The CGA module [25] allows us to generate the ID interface. For different types of packets we have different packet sizes. This information is listed in the following table (Fig. 7).

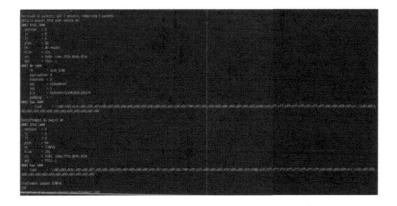

Fig. 6. Receiving the RA message from the client.

Fig. 7. Generating the CGA address by mobile.

Table 1. Summary of different packet types and their sizes.

Authentication algorithm	RS	RA	NS
CGA + SEND	78	118	86
CGA + AH (HMAC-SHA1-96)	102	142	86
CGA + AH (SHA2-256-128)	110	150	86
CGA + AH (SHA2-384-192)	118	158	86
CGA + AH (SHA2-512-256)	126	166	86
CGA + AH (SHA2-MD5-96)	102	142	86
CGA + AH (AES-CMAC-96)	102	142	86

Table 2. Summary of latency averages (CGA AH and CGA SEND).

Authentication algorithm	Averages of latencies
CGA + SEND	1,6799 s
CGA + AH (HMAC-SHA1-96)	1,6832 s
CGA + AH (SHA2-256-128)	1,7072
CGA + AH (SHA2-384-192)	1,7088
CGA + AH (SHA2-512-256)	1,7274
CGA + AH (SHA2-MD5-96)	1,7433
CGA + AH (AES-CMAC-96)	1,6731

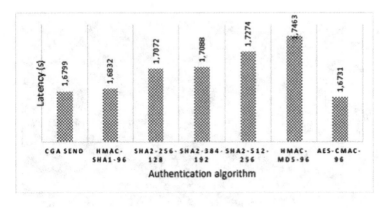

Fig. 8. Histogram of latency averages (in seconds).

Table 2 and Fig. 6 show the latency of the NDP procedure through the CGA SeND methods and the proposed CGA AH method. Overall, latency increases based on the size and complexity of the authentication algorithm used. However, this increase is low. Much of the latency is caused by the AH authentication mechanism and the generation of the ID interface that use cryptographic operations. In addition, it is noted that the AES-CMAC-96 authentication algorithm has the lowest latency. This can be explained by the lightness of the AES algorithm [43]. In sum, the performance results show that the proposed CGA + AH method can be used to secure the NDP procedure. However, for best performance, the AES-CMAC-96 authentication algorithm could be used. The table below shows the result of the safe analysis between the proposed CGA SeND and CGA AH. Table 3: summary of comparisons.

Table 3. Confirms the performance that could be associated with CGA HA.

	CGA SEND	CGA AH
Reliability	Unreliable	Reliable
Router identity spoofing	High risk of spoofing	Low risk of spoofing
Mobile node spoofing	Risk of spoofing the weak node	Risk of spoofing the weak node
Latency	Substantially the same	Substantially the same

CGA SeND and CGA AH both provide node legitimacy through the CGA protocol. The CGA AH reliably authenticates packets using the AH protocol of the IPsec stack. Moreover, it should be noted that the authentication at the level of CGA SeND is not reliable, this is due to the hash function used in SeND.

6   Conclusion

This article presents the mechanism of the Neighbor Discovery Protocol (NDP), with particular emphasis on the vulnerabilities and attacks to which it is exposed and proposes a model for its security. To date, several proposals for securing the Neighbor Discovery Protocol have been made. Most of this work has limitations listed in Sect. 3. Based on the Neighbor Discovery mechanism and on existing models, we have proposed a method for securing the most critical points of the Neighbor Discovery Protocol (the node and packets) using both the IPSec AH protocol for packet authentication and the CGA Protocol for cryptographically generating addresses. We then tested the model with the various types of cryptographic keys. Performance results show that the proposed solution has latencies similar to that of the CGA SEND method despite the increase in packet size. Moreover, securing via AES-CMAC-96 has the lowest latency. Therefore, this authentication algorithm could be handy for improving the security of the Neighbor Discovery Protocol mechanism.

References

1. Author, F.: Article title. Journal **2**(5), 99–110 (2016)
2. Author, F., Author, S.: Title of a proceedings paper. In: Editor, F., Editor, S. (eds.) CONFERENCE 2016, LNCS, vol. 9999, pp. 1–13. Springer, Heidelberg (2016). https://doi.org/10.10007/1234567890

3. Author, F., Author, S., Author, T.: Book title, 2nd edn. Publisher, Location (1999)
4. Author, A.-B.: Contribution title. In: 9th International Proceedings on Proceedings, pp. 1–2. Publisher, Location (2010)
5. LNCS Homepage. http://www.springer.com/lncs. Accessed 4 Oct 2017
6. Xiaorong, F., Jun, L., Shizhun, J.: security analysis for IPv6 neighbor discovery protocol. In: IMSNA, pp. 303–307 (2013)
7. Cunjiang, Y., Li, J.: Authentication algorithms of Internet Protocol security in an IPV6-based environment. In: Proceedings of 2010 4th International Conference on Intelligent Information Technology Application, vol. 4 (2010)
8. Arkko, J., (Ed.) et al.: Secure Neighbor Discovery (SEND), in request for comments 3971, Internet Engineering Task Force (2005)
9. Kent, S., Seo, K.: Security architecture for the Internet Protocol, in request for comments: 4301, Internet Engineering Task Force (2005)
10. Ramachandran, V., Nandi, S.: Detecting ARP spoofing: an active technique. In: Jajodia, S., Mazumdar, C. (eds.) ICISS 2005. LNCS, vol. 3803, pp. 239–250. Springer, Heidelberg (2005). https://doi.org/10.1007/11593980_18
11. Lootah, W., Enck, W., McDaniel, P.: Tarp: ticket-based address resolution protocol. Comput. Netw. **51**(15), 4322–4337 (2007)
12. Tzang, Y.-J., Chang, H.-Y., Tzang, C.-H.: Enhancing the performance and security against media-access-control table overflow vulnerability attacks. Secur. Commun. Netw. **8**, 1780–1793 (2015)
13. Xiaorong, F., Jun, L., Shizhun, J.: Security analysis for IPv6 neighbor discovery protocol. In: 2013 2nd International Symposium on Instrumentation and Measurement, Sensor Network and Automation (IMSNA) (2013)
14. Arkko, M., Kempf, J., Sommerfeld, B., Zill, B., Nikander, P., (ed.): SEcure Neighbor Discovery (SEND), RFC 3971, March 2005
15. Ahmed, A.S., Hassan, R., Othman, N.E.: Secure neighbor discovery (SeND): attacks and challenges. In: 2017 6th International Conference on Electrical Engineering and Informatics (ICEEI) (2017). https://doi.org/10.1109/iceei.2017.8312422
16. Sumathi, P., Patel, S., Prabhakaran.: Secure neighbor discovery (SEND) protocol challenges and approaches. In: 2016 10th International Conference on Intelligent Systems and Control (ISCO) (2016). https://doi.org/10.1109/ISCO.2016.7726976
17. Xiaorong, F., Jun, L., Shizhun, J.: Security analysis for IPv6 neighbor discovery protocol. In: 2013 2nd International Symposium on Instrumentation and Measurement, Sensor Network and Automation (IMSNA) (2013). https://doi.org/10.1109/imsna.2013.6743275
18. Nikander, P., Kempf, J., Nordmark, E., (ed.): IPv6 Neighbor Discovery (ND) trust models and threats, in request for comments 3756, Internet Engineering Task Force (2004)
19. Zhang, L.J., Tian-Qinga, O., Zhao, L.Y.: Authentication scheme based on certificateless signcryption in proxy mobile IPv6 network. Appl. Res. Comput. **29**(2), 640–643 (2012)
20. Zhang, Z., Wuhan.: Secure access authentication scheme in mobile IPv6 networks. Comput. Sci. **36**(12), 26–31 (2009)
21. Gracia, D.F.: Performance evaluation of advanced encryption standard algorithm. In: 2015 Second International Conference on Mathematics and Computers in Sciences and in Industry (MCSI) (2015). https://doi.org/10.1109/MCSI.2015.61
22. Alsadeh, A., Rafiee, H., Meinel, C.: Cryptographically generated addresses (CGAs): possible attacks and proposed mitigation approaches. In: 2012 IEEE 12th International Conference on Computer and Information Technology (CIT). IEEE (2012)

23. Anu, P., Vimala, S.: A survey on sniffing attacks on computer networks. In: 2017 International Conference on Intelligent Computing and Control (I2C2) (2017). https://doi.org/10.1109/I2C2.2017.8321914
24. Verma, A., Singh, A.: An approach to detect packets using packet sniffing. Int. J. Comput. Sci. Eng. Survey (IJCSES) 4(3), 21 (2013)
25. https://github.com/secdev/scapy
26. https://github.com/helpsystems/pcapy. Accessed Nov 2018
27. https://Github.com/bestrocker221/send-CGA-offloading. Accessed Nov 2018

BI-NTRU Encryption Schemes: Two New Secure Variants of NTRU

Michel Seck[(✉)] and Djiby Sow

Department of Mathematics and Computer Science, Cheikh Anta Diop University,
Dakar, Senegal
michel.seck@ucad.edu.sn, sowdjibab@yahoo.fr

Abstract. NTRU is one of the first public key cryptosystems not based on factorization or discrete logarithmic problems and is also considered secure even against quantum computer attacks. In 2011, Stehle and Steinfeld proposed a variant of the classical NTRU that is IND-CPA secure but for the key generation algorithm, they use Gaussian distribution with a large standard deviation to prove the uniformity of the public key by assuming the hardness of Ring Learning With Error (Ring-LWE) problem. In this paper, we present two variants of NTRUEncrypt called BI-NTRU-Product and BI-NTRU-LPR which are IND-CPA secure assuming the hardness of Ring-LWE problem. We also show how one can design an IND-CCA2 secure key encapsulation mechanism from our encryption schemes by using a variant of the Fujisaki-Okamoto Transformation (CRYPTO 1999 and Journal of Cryptology 2013).

Keywords: BI-NTRU-Product · BI-NTRU-LPR · NTRU · Ring-LWE · Lattice-based cryptography

1 Introduction

Most number-theoretic schemes, such as RSA cryptosystem and the Diffie-Hellman protocol, rely on the conjectured hardness of integer factorization or the discrete logarithm problem in certain groups. However, in 1994, Shor [20,21] discovers an efficient algorithm for quantum computers that can break all known classical cryptosystems whose hardness is related to the hardness of factoring or discrete logarithm problem. This quantum algorithm motivates several researchers to explore other theories such as lattice theory. Many problems in lattices have been proved NP-hard and no efficient quantum algorithm that can break those problems is known. The fundamental computational problems associated to a lattice are those of finding a shortest nonzero vector (SVP) in the lattice and of finding a vector in the lattice that is closest to a given nonlattice vector (CVP). One of the first cryptographic scheme whose security was expected to be tightly related to the shortest vector problem and the closest vector problem in a lattice is NTRU [9].

NTRU is a public key cryptosystem introduced by Hoffstein, Silverman and Pipher [9] in 1996. The Nth degree Truncated polynomial Ring Units (NTRU)

© Springer Nature Switzerland AG 2019
C. T. Gueye et al. (Eds.): A2C 2019, CCIS 1133, pp. 216–235, 2019.
https://doi.org/10.1007/978-3-030-36237-9_13

is a lattice-based alternative to RSA, El Gamal and ECC. It is a probabilistic cryptosystem and is based on algebraic operations on truncated polynomial rings and modular reduction. NTRU was first presented at the rump session of Crypto '96 and patented by the company NTRU Cryptosystems Inc. at the end of 1996. Coppersmith and Shamir [4] (1997) founded a lattice attack against NTRU before it formally published in 1998. The original NTRU doesn't have security proof; the NAEP variant have a security proof in the random oracle model [11]. Its strong points are moderate key-sizes, speed of encryption and decryption, and it is still expected to be secure even against quantum computer attacks. Since NTRU was published without security proof, many variants have been proposed in the two last decades. Many researchers have revised all algorithms in classical NTRU like key generation, encryption and decryption algorithm in order to get security proof.

In 2002, Banks and Shparlinski [1] introduced a new NTRU like cryptosystem. They have showed that the ciphertext and the plaintext are uncorrelated by using bounds for exponential sums. They avoided the potential problem of finding double invertible polynomials within very thin sets of polynomials, as in the original version of NTRU but the public key size and the encryption time of their scheme are roughly doubled.

In 2011, Stehle and Steinfeld [22] introduced a noise variant of NTRUEncrypt and proved that their NTRU is IND-CPA secure. They showed how to modify NTRUEncrypt to make it provably secure in the standard model, under the assumed quantum hardness of standard worst-case lattice problems, restricted to a family of lattices related to some cyclotomic fields. For their proof, the secret elements are chosen by Gaussian distribution. To prove that their public key is uniformly distributed, Stehle and Steinfeld use Gaussian distribution with large standard deviation. The use of a large standard deviation σ in their scheme increased the size of the keys and makes their cryptosystem inefficient. The security of their scheme follows from the already proven hardness of Ring-LWE problem [16, 19].

Using the Noise NTRU IND-CPA [22], Steinfeld et al. [23] design a new variant of NTRU called NTRUCCA, that is IND-CCA2 secure in the standard model assuming the worst-case quantum hardness of problems in ideal lattices. NTRUCCA scheme is built from an All-But-One (ABO) lossy trapdoor function by using a generalization of the generic Peikert-Waters construction [18] of IND-CCA2 encryption from ABO lossy trapdoor functions. This variant is neither simple nor efficient.

In December 2016, the National Institute of Standards and Technology (NIST) published a call for proposals for post-quantum public-key (PQ-PK) cryptographic algorithms [17]. For this call Bernstein et al. [3] have proposed NTRU Prime which are two Key Encapsulation Mechanism (KEM). They replace the ring in classical NTRU and in NTRU IND-CPA of Damien Stehle and Ron Steinfeld by a field. Their scheme is IND-CCA2 secure in the random oracle model by using the result of [6]. Other variants of the classical NTRU are also proposed namely NTRU-HRSS-KEM and SS-NTRU [12].

In this work, we propose two new variants of NTRUEncrypt: BI-NTRU-Product and BI-NTRU-LPR which are IND-CPA secure over $R_q = \mathbb{Z}_q[x]/(\Phi(x))$ where $\Phi(x)$ is square-free (cyclotomic) polynomial of degree n. We derive a KEM (BI-NTRU-KEM) from BI-NTRU-LPR by using a variant of FO transformation [7].

1. **BI-NTRU-Product:** The public key of this scheme is uniformly distributed by using a framework of Banks and Shparlinski [1]. We avoided using Gaussian distribution with a large standard deviation (which has reduced the efficiency of NTRU IND-CPA [22]) in key generation algorithm. Furthermore, as NTRU IND-CPA (Damien Stehle and Ron Steinfeld), this scheme is IND-CPA secure relative to the hardness of the decision version of Ring-LWE over $R_q = \mathbb{Z}_q[x]/(\mathbf{x^n + 1})$.
2. **BI-NTRU-LPR:** The key generation of this scheme is very similar to the Lyubashevsky, Peikert, and Regev (LPR) scheme [16]. The uniformity of the public key distribution and the IND-CPA security are due to the Ring-LWE problem.

This paper is organized as follows. Section 1 is devoted to the introduction. In Sect. 2, we recall some definitions and properties in lattices. We also give an overview on NTRU-like cryptosystems. We describe, in Sect. 3, our encryption schemes BI-NTRU-Product and BI-NTRU-LPR which are IND-CPA secure. We also discuss how to transform the BI-NTRU-LPR to an IND-CCA secure key encapsulation mechanism. In Sect. 4 we give a conclusion. And finally, in the Appendix A, we discuss about the uniformity of the public key of BI-NTRU-Product.

2 Preliminaries

We recall in the following subsections some definitions on lattices and give an overview on NTRU-like cryptosystems.

2.1 Basic Definitions in Lattices

Definition 1 (lattice). *Let m and n be two positive integers. Let $\mathcal{L}$ be a non-empty subset of $\mathbb{R}^m$. We say that $\mathcal{L}$ is a lattice if there exist linearly independent vectors $b_1, \ldots, b_n$ of $\mathbb{R}^m$ such that $\mathcal{L} = \sum_{i=1}^{n} \mathbb{Z}b_i = \{\sum_{i=1}^{n} x_i b_i, x_i \in \mathbb{Z}\}$. The integer n is the dimension of the lattice, and $(b_1, ..., b_n)$ is a basis of this lattice. If $n = m$, we have a full-rank lattice.*

Definition 2 (ideal lattice). *Let $\Phi(X) \in \mathbb{Z}[X]$ a monic polynomial of degree n. An ideal lattice $\mathcal{L}$ for $\Phi(X)$ is a sublattice of $\mathbb{Z}^n$ that corresponds to a non-zero ideal $\mathbf{I} \subseteq \mathbb{Z}[X]/(\Phi(X))$.*

Definition 3 (distance of a vector with respect to a lattice). *Let $\mathcal{L}$ be an n-dimensional lattice and v a vector of $\mathbb{R}^m$. The distance of v to $\mathcal{L}$ is defined by $\mathrm{dist}(v, \mathcal{L}) = \min_{x \in \mathcal{L}} \|x - v\|$*

Definition 4 (successive minima). *Let $\mathcal{L}$ be an n-dimension lattice of $\mathbb{R}^m$. For $i \in \{1, 2, ..., n\}$, we defined the ith successive minima as the smaller ball of radius centered at 0 containing i linearly independent vectors; i.e $\lambda_i(\mathcal{L}) = \inf\{r \mid \dim(\mathrm{span}(\mathcal{L} \cap \mathbf{B}(\mathbf{0}, \mathbf{r}))) \geq i\}$ where $\mathbf{B}(\mathbf{0}, \mathbf{r}) = \{x \in \mathbb{R}^m : \|x\| \leq r\}$.*

Definition 5 (The Shortest Vector Problem)

1. ***Approximate Shortest Vector Problem (SVP$_\gamma$):*** *Given a basis B of an n-dimensional lattice $\mathcal{L} = \mathcal{L}(B)$, find a nonzero vector $v \in \mathcal{L}$ for which $\|v\| \leq \gamma(n)\lambda_1(\mathcal{L})$.*
2. *If $\gamma(n) = 1$, we have the exact problem called **SVP**.*
3. ***Decisional Approximate SVP (GapSVP)$_\gamma$):*** *Given a basis B of an n-dimensional lattice $\mathcal{L} = \mathcal{L}(B)$ and an integer $d > 0$, decide if $\lambda_1(\mathcal{L}) < d$ or $\lambda_1(\mathcal{L}) \geq \gamma(n)d$.*

Definition 6 (The Closest Vector Problem)

1. ***Approximate Closest Vector Problem (CVP$_\gamma$):*** *Given a lattice $\mathcal{L}$ and a target vector v_0, find a nonzero vector $v \in \mathcal{L}$ which minimizes the norm $\gamma(n)(\|v - v_0\|)$.*
2. *If $\gamma(n) = 1$ we have the exact problem called **CVP**.*
3. ***Decisional Approximate CVP (GapCVP$_\gamma$):*** *Given an n-dimensional lattice $\mathcal{L}$, a target vector v_0 and a real $d > 0$, decide if $\mathrm{dist}(v_0, \mathcal{L}) < d$ or $\mathrm{dist}(v_0, \mathcal{L}) \geq \gamma(n)d$.*

Definition 7 (Gaussian function). *The n-dimensional Gaussian function $\rho_{s,c} : \mathbb{R}^n \to (0, 1]$ centered in c scaled by a factor of s is defined as $\rho_{s,c}(x) = \exp\left(\dfrac{-\pi\|x - c\|^2}{s^2}\right)$. When c is omitted, it is assumed to be 0.*

Definition 8 (Discrete Gaussian Distribution). *The discrete Gaussian distribution $D_{\mathcal{L}+c,s}$ with parameter s and c over a lattice coset $\mathcal{L} + c$ is the distribution that samples each element $x \in \mathcal{L} + c$ with probability $\dfrac{\rho_{s,c}(x)}{\rho_{s,c}(\mathcal{L} + c)}$, where $\rho_{s,c}(\mathcal{L} + c) = \sum_{y \in \mathcal{L}+c} \rho_{s,c}(y)$ is a normalization factor.*

Definition 9 (The Learning With Errors (LWE) Problem). *The LWE problem was introduced by Regev [19] and can be formulated as follows: Let $m = \mathrm{poly}(n)$ be arbitrary and let $s \in \mathbb{Z}_q^n$ be some vector. Given a random matrix $A \in \mathbb{Z}_q^{m \times n}$ and a vector $t = As + e \in \mathbb{Z}_q^m$ where each coordinate of the error vector $e \in \mathbb{Z}_q^m$ is chosen independently from a distribution χ on $\mathbb{Z}_q$, the goal is to recover s.*

The Ring Learning with Errors (Ring-LWE): Ring-LWE depends to the following parameters:

1. a ring R, which is represented generally as a polynomial quotient ring $R = \mathbb{Z}[X]/(f(X))$ for some irreducible $f(X)$, e.g., $f(X) = X^{2^k} + 1$ or another cyclotomic polynomial;

2. a positive integer modulus q defining the quotient ring $R_q = R/qR = \mathbb{Z}_q[X]/(f(X))$;
3. an error distribution χ over R;
4. a number of samples provided to the attacker.

Definition 10 (Ring-LWE Problem). *The Ring-LWE search problem is to find a uniformly random secret $s \in R_q$, given independent samples of the form*

$$(a_i, b_i = a_i \times s + e_i) \in R_q \times R_q$$

where each $a_i \in R_q$ is uniformly random and each e_i is drawn from the error distribution χ. The decision problem is to distinguish samples of the above form from uniformly random samples over $R_q \times R_q$.

Remark 1. Notice that Ring-LWE was introduced by Lyubashevsky et al. [15] and shown hard for specific error distributions χ.

Definition 11. *Variants of Ring-LWE*

- *When a_i are sampled from the set of invertible elements of R_q (denoted $R_q^\times$) instead of R_q, the corresponding modification of Ring-LWE is called Ring-LEW$^\times$. This variant remains hard [22].*
- *When the noises s_i are sampled from the error distribution χ, the corresponding modification of Ring-LWE is called Ring-LWE$_{HNF}$. This variant remains hard [22].*
- *When $s_i \in R_q$ is sampled from the error distribution χ and invertible modulo a small prime $p < q$. We call this variant Ring-LWE$_p$.*

Remark 2 (General Ring-LWE ([14], pages 23, 24)). Notice that one cannot distinguish samples of the form $\{(a_i, a_i s_j + e_{i,j})\}$ to samples $\{(a_i, u_{i,j})\}$ where $a_i, u_{i,j}$ are uniformly random in R_q and $s_j, e_{i,j}$ are drawn from the error distribution χ

Definition 12 (distribution $\bar{\Upsilon}_\alpha$ [22])

- *For $\sigma \in \mathbb{R}^n$ with positive coordinates, we define the ellipsoidal Gaussian ρ_σ as the row vector of independent Gaussians $(\rho_{\sigma_1}, \rho_{\sigma_2}, \ldots, \rho_{\sigma_n})$, where $\sigma_i = \sigma_{i+n/2}$ for $1 \leq i \leq n/2$.*
- *We define a sample from ρ'_σ as a sample from ρ_σ, multiplied first (from the right) by $\frac{1}{\sqrt{2}}\begin{pmatrix} 1 & 1 \\ i & -i \end{pmatrix} \otimes Id_{n/2} \in \mathbb{C}^{n \times n}$, and second by $V = \frac{1}{n}(\zeta^{-(2j+1)k})_{0 \leq j,k \leq n}$.*
- *We now define a sample from $\bar{\rho}'_\sigma$ as follows: compute a sample from ρ'_σ with absolute error $< 1/n^2$; if it is within distance $1/n^2$ of the middle of two consecutive integers, then restart; otherwise, round it to a closest integer and then reduce it modulo q.*

A distribution sampled from $\bar{\Upsilon}_\alpha$ for $\alpha \geq 0$ is defined as $\bar{\rho}'_\sigma$, where $\sigma_i = \sqrt{\alpha^2 q^2 + x_i}$ with the x_i's sampled independently from the exponential distribution $\text{Exp}(n\alpha^2 q^2)$.

2.2 NTRU-Like Cryptosystems

Classical NTRUEncrypt: The classical NTRUEncrypt [9] depends on three parameters (N, p, q) (where N, q are integers and p a polynomial or an integer) and four sets $\mathcal{L}_f, \mathcal{L}_g, \mathcal{L}_\phi, \mathcal{L}_m$ which are subsets of the polynomial ring $R = \mathbb{Z}_q[x]/(x^n - 1)$.

- **Key Generation:** The public key p_k is $h = g/f \pmod{q}$ with f selected randomly in $\mathcal{L}_f$ invertible modulo p and modulo q and g selected randomly in $\mathcal{L}_g$. The private key is $s_k = f$.
- **Encryption:** The ciphertext of a message $m \in \mathcal{L}_m$ is $C = p\phi.h + m \pmod{q}$ where ϕ is chosen at random from $\mathcal{L}_\phi$,
- **Decryption:** Denote by f_p the inverse of f modulo p. To decrypt C, we compute $a = f.C \pmod{q}$ and lift the coefficients of a in the interval from $-q/2$ to $q/2$. We then compute $f_p.a \pmod{p}$ to recover m.

NTRU with None Invertible Polynomials: In [1], Banks and Shparlinski made few modifications in key generation algorithm in the classical NTRU encryption scheme to avoid double invertible polynomials. The polynomial ring is $R = \mathbb{Z}_q[X]/(X^N - 1)$ as in the standard version of NTRU. This variant depends on four polynomial sets $\mathcal{L}_f, \mathcal{L}_g, \mathcal{L}_\phi, \mathcal{L}_m$ in R.

- **Key Generation:** Alice randomly selects polynomials $f \in \mathcal{L}_f, g \in \mathcal{L}_g$ and $G \in R$ such that G has an inverse modulo q (denoted G_q) and f has an inverse modulo p (denoted f_p). The public key of Alice will be (h, H) where $h = G_q.g \pmod{q}$ and $H = G_q.f \pmod{q}$ and the private key $s_k = (f, g, G)$.
- **Encryption:** To encrypt a message $m \in \mathcal{L}_m$, Bob first selects $\phi \in \mathcal{L}_\phi$ and then computes the ciphertext $C = p\phi.h + H.m \pmod{q}$ where (h, H) is Alice's public key.
- **Decryption:** To decrypt C, Alice first computes $a = G.C \pmod{q}$ and chooses the coefficients of a to lie in the interval from $-q/2$ to $q/2$. She finally computes $f_p.a \pmod{p}$ to recover m.

NTRU IND-CPA: In order to achieve CPA-security, Stehle and Steinfeld [22] make a few modifications to the original NTRU encryption scheme. The Polynomial ring used in this scheme is $R = \frac{\mathbb{Z}_q[x]}{(x^n + 1)}$ which is the ring of integers of a cyclotomic number field with q a prime integer such that $q = 1 \pmod{2n}$ and n is a power of 2.

- **Key Generation:** To generate a pair key (p_k, s_k). Alice samples f' and g from a discrete Gaussian distribution $D_{\mathbb{Z}^n, \sigma}$ such that $f = pf' + 1$ and g are invertible modulo q. Alice's key pair $(p_k = h = g/f \pmod{q}, s_k = f)$.
- **Encryption:** To encrypt a message m, Bob computes $C = hs + pe + m$ where e, s are sampled from the Ring-LWE error distribution and h is Alice's public key.
- **Decryption:** To decrypt C, Alice first computes $C' = f.C \in R_q$ and finally returns $C' \pmod{p}$.

NTRU Variants Proposed in NIST: Several post-quantum lattice-based cryptography like the classic NTRU cryptosystem and some of its variants are proposed for post-quantum schemes to NIST's call [17].

- **NTRUEncrypt:** Cong Chen, Jeffrey Hoffstein, William Whyte and Zhenfei Zhang proposed four variants of the classical NTRUEncrypt:
  - **ntru-pke:** this scheme is a public key encryption (PKE) scheme based on the classical NTRU encryption algorithm that uses NAEP transformation [10] in order to achieves CCA-2 security.
  - **ntru-kem:** this scheme is a key encapsulation mechanism (KEM) using the ntru-pke algorithms.
  - **ss-ntru-pke:** this scheme is a public key encryption scheme based on the NTRU IND-CPA of Stehle and Steinfeld [22] that achieves CCA-2 security via NAEP transformation [10].
  - **ss-ntru-kem:** this scheme is a key encapsulation mechanism (KEM) using the ss-ntru-pke.
- **NTRU-HRSS** and **NTRU-HRSS-KEM:** these schemes are proposed in NIST by Hülsing, Rijneveld, Schanck and Schwabe [12]. The first is a OWCPA-secure public key encryption scheme that is a direct parameterization of NTRUEncrypt. The second is a CCA2-secure KEM that uses a generic transformation from a OWCPA-secure public key encryption scheme and avoided the NAEP padding mechanism used in standard NTRUEncrypt.
- **NTRU LPRime** and **Streamline NTRU Prime:** Bernstein, Chuengsatiansup, Lange and van Vredendaal proposed NTRU Prime [3] which are two key-encapsulation mechanisms namely Streamline NTRU Prime and NTRU LPRime. Their cryptosystems all use a prime-degree number field with a large Galois group and an inert modulus $R/q = (\mathbb{Z}/q)[x]/(x^p - x - 1)$ and also share the same core design element, sending $m + hr \in R/q$ where m, r are small secrets and h is public.

3   BI-NTRU IND-CPA

In this section, we propose two variants of NTRUEncrypt called BI-NTRU-Product and BI-NTRU-LPR over $R = \mathbb{Z}[x]/(\Phi(x))$ where $\Phi(x)$ is a monic square free polynomial of degree n. Let q a prime integer, and p a prime integer small with respect to q ($p << q$). In the rest of this paper we take $\Phi(x) = x^n + 1$.

Notations

- $R_p = R \pmod{p}$ and $R_q = R \pmod{q}$
- For a polynomial $F = \sum F_i X^i \in R$, we denote by $\|F\|$ its Euclidean norm as a vector, $\|F\|_\infty$ its infinity norm and by $|F|_\infty$ the width of F, i.e $|F|_\infty = \max\{F_i\} - \min\{F_i\}$
- g_p the inverse of a polynomial $g \in R$ modulo p
- $\mathcal{L}_t = \{f \in R_q : f$ has coefficients lying between $-(t-1)/2$ and $(t-1)/2\}$ for an integer $t > 2$ and $\mathcal{L}_2 = \{f \in R_q : f$ has coefficients lying in $\{0,1\}\}$

- χ: a Ring-LWE error distribution
- $\chi(r)$: a Ring-LWE error distribution with seed r

Public Parameters: Our schemes use the following parameters.

1. n a power of 2 (the degree of $\Phi(x) = x^n + 1$);
2. q a prime integer such that $x^n + 1$ is square-free over $\mathbb{Z}_q[x]$. We know that such q exists since when $q = 1 \pmod{2n}$ the polynomial $x^n + 1$ has n distinct linear factors modulo q [22] and then square free in $\mathbb{Z}_q[x]$;
3. p a small prime number with respect to q ($p << q$);
4. The polynomial ring $\frac{\mathbb{Z}_q[x]}{(\mathbf{x^n + 1})}$ which is the ring of integers of a cyclotomic number field as in [22];
5. t an integer ≥ 2 which defines $\mathcal{L}_t$;
6. the Ring-LWE error distribution $\chi = \bar{\varUpsilon}_\alpha$;
7. $\mathcal{L}_m = \mathcal{L}_2$ the message space.

Notice that the security and the efficiency of our encryption schemes depend on the choice of the above parameters.

3.1 BI-NTRU-Product

The key generation algorithm of BI-NTRU-Product is described as follows.

Algorithm 1. Key Generation Algorithm for BI-NTRU-Product

Output: A public key **pk**
Output: A private key **sk**
1 Choose uniformly at random $B \in R_q$;
2 Choose uniformly at random a polynomial $r \in \mathcal{L}_t$;
3 Choose uniformly at random $g \in \mathcal{L}_t$ (if g is not invertible modulo p, resample);
4 Compute $G_1 = -gB \pmod{q}$, $G_2 = rB \pmod{q}$;
5 The public key **pk** is (G_1, G_2) and the private key **sk** $= (r, g)$.

Remark 3. Notice that in practice one can choose uniformly at random $f \in \mathcal{L}_t$ and set $g = 1 + pf$. In this case the inverse of g modulo p is trivial.

Lemma 1. *For almost all $B \in R_q$, the set $\{G = \phi . B : \phi \in \mathcal{L}_t\}$ is uniformly distributed and then the public key of BI-NTRU-Product scheme is uniformly distributed.*

Proof: It follows from the framework of Banks and Shparlinski [1] (for more details see the Appendix A). □

Algorithm 2. Encryption algorithm for BI-NTRU-Product

Input: A message $\mathbf{m} \in \mathcal{L}_m$, the public key **pk** $= (G_1, G_2)$
Output: the ciphertext $C = (C_1, C_2)$ of m
1 Choose randomly three "small" polynomials $u, v, e \in R_q$ from χ;
2 Compute $C_1 = p(uG_1 + v) \in R_q$ and $C_2 = p(uG_2 + e) + m \in R_q$;
3 **return** $C = (C_1, C_2) \in R_q \times R_q$

The decryption algorithm for BI-NTRU-Product is described as follows.

Algorithm 3. Decryption algorithm for BI-NTRU-Product

Input: The ciphertext $C = (C_1, C_2)$, the private key $\mathbf{sk} = (r, g)$
Output: the decrytion of C
1 Compute $D_1 = gC_2 + rC_1 \pmod{q}$;
2 Compute $D_2 = D_1 \pmod{p}$;
3 **return** $D = g_p D_2 \pmod{p}$

Lemma 2. *If $|gm + p(ge + rv)|_\infty < q$, then the decryption algorithm for BI-NTRU-Product output m. i.e $D = m$.*

Proof: Suppose $|gm + p(ge + rv)|_\infty < q$; We have

$$D_1 = gC_2 + rC_1 \pmod{q}$$
$$= (gm + gpuG_2 + pge) + (rpuG_1 + rpv) \pmod{q}$$
$$= gm + purgB + pge - purgB + rpv \pmod{q}$$
$$= gm + p(ge + rv) \pmod{q}$$

Since $|gm+p(ge+rv)|_\infty < q$ then $D_1 = gm+p(ge+rv) \pmod{q} = gm+p(ge+rv)$. Hence $D_2 = D_1 \pmod{p} = gm + p(ge + rv) \pmod{p} = gm \pmod{p}$. Finally $D = g_p D_2 \pmod{p} = m$. $\qquad\square$

Lemma 3 (adapted from Lemma 6 [22]). *Let $y, r \in R$, with r fixed and y sampled from $\tilde{\Upsilon}_\alpha$ such that $\alpha q \geq n^{1/4}$. Then*

$$Pr\left[\|yr\|_\infty \geq \alpha q w(\log n).\|r\|\right] \leq n^{-w(1)}$$

We show in the following Lemma 4 the probability that the decryption fails is negligible for appropriate parameters.

Lemma 4. *If $t = 3$, $\mathcal{L}_m \subset \mathcal{L}_t$, $\alpha q \geq n^{1/4}$ and $n^{3/2} + 2\alpha q n^{1/2} pw(\log n) < q/2$ then the decryption algorithm of BI-NTRU-Product recovers m with probability $1 - n^{-w(1)}$ over the choice of $u, v, e,$ and g.*

The following proof is similar to that one of [22], but for the sake of completeness let us give it.

Proof: In the decryption algorithm, we have $C = gm + p(ge + rv) \pmod{q} = gm + pge + prv \pmod{q}$. Let $C' = gm + pge + prv$ computed in R (not modulo q). If $\|C'\|_\infty < q/2$ (it means that $|C'|_\infty < q$) then we have $C = C'$ in R and hence the decryption algorithm succeeds. It thus suffices to give an upper bound on the probability that $\|C'\|_\infty \geq q/2$. We know that each polynomial in $\mathcal{L}_3$ have its coefficients in $\{-1, 0, 1\}$ and so $\|g\|_\infty = 1, \|r\|_\infty = 1$, and $\|m\|_\infty = 1$. We also have $\|r\| \leq \sqrt{n}, \|m\| \leq \sqrt{n}, \|g\| \leq \sqrt{n}$ and then $\|gm\|_\infty \leq \sqrt{n}\|g\|\|m\| \leq n^{3/2}$. Since, e and v are chosen from $\tilde{\Upsilon}_\alpha$, it follows from Lemma 3 that $\|pge\|_\infty \leq \alpha q w(\log n).p\|g\|$, and $\|prv\|_\infty \leq \alpha q w(\log n).p\|r\|$ with probability at least $1 - n^{-w(1)}$. Consequently, we have $\|pge\|_\infty \leq \alpha q \, n^{1/2} pw(\log n)$, and $\|prv\|_\infty \leq \alpha q n^{1/2} pw(\log n)$ with probability at least $1 - n^{-w(1)}$. This implies that $\|C'\|_\infty \leq n^{3/2} + 2\alpha q n^{1/2} pw(\log n)$ with probability at least $1 - n^{-w(1)}$. $\qquad\square$

Theorem 1. *Our BI-NTRU-Product is IND-CPA secure assuming that the Ring-LWE$_{HNF}$ is hard.*

Proof: We consider a sequence of games **Game0, Game1 and Game2**. Let $\mathcal{A}$ be an attacker against IND-CPA of our scheme. Our goal is to construct an algorithm $\mathcal{B}$ against the Ring-LWE$_{HNF}$.

1. **Game0:** In this game that is the real scheme, suppose that the public key is $[G_1 = -gB \pmod{q}, G_2 = rB \pmod{q}]$ where (r, g) is randomly selected in $\mathcal{L}_t \times \mathcal{L}_t$ and B in R_q. The algorithm $\mathcal{A}$ takes as input the public key (G_1, G_2) and output two challenge messages M_0 and M_1. Algorithm $\mathcal{B}$ selects b uniformly in $\{0, 1\}$, and computes the challenger ciphertext (C_{1b}, C_{2b}) where $C_{1b} = p(uG_1 + v) \pmod{q}$ and $C_{2b} = p(uG_2 + e) + M_b \pmod{q}$ with u, v, e selected from the error distribution $\tilde{\Upsilon}_\alpha$ and gives (C_{1b}, C_{2b}) to $\mathcal{A}$. The algorithm $\mathcal{A}$ taking as input (C_{1b}, C_{2b}) outputs b' the guess for b. $\mathcal{B}$ outputs 0 if $b' \neq b$ and 1 otherwise.
2. **Game1:** The same condition as in **Game0** but now (G_1, G_2) is randomly selected in $R_q \times R_q$. One can't distinguish **Game1** and **Game0** since the public key is uniformly distributed in R_q.
3. **Game2:** The same condition as in **Game1** but now $[C_{1b} = pC_1', C_{2b} = pC_2' + M_b]$ where (C_1', C_2') is uniformly selected in $R_q \times R_q$. Since u, v and e are sampled from $\tilde{\Upsilon}_\alpha$, then one can't distinguish **Game2** and **Game1** by Ring-LWE$_{HNF}$ (see Remark 2). Furthermore since (C_1', C_2') is uniformly selected in $R_q \times R_q$ and p is invertible in R_q, then pC_1' and pC_2' are uniformly random in R_q. Therefore $C_{2b} = pC_2' + M_b$ is a perfect one time pad. Hence it follows that algorithm $\mathcal{B}$ outputs 1 with probability $1/2$. □

BI-NTRU-Product and Other NTRU-Like Cryptosystems

- Most of variants of NTRU-like cryptosystems like NTRUEncrypt itself [9], NTRU Prime [3], and NTRU of Banks and Shparlinski [1] are not proven to achieve the IND-CPA security which is realized by our BI-NTRU-Product.
- To our knowledge, the first variant of NTRU that is IND-CPA secure is the one of Stehle and Steinfeld [22] but they use Gaussian distribution with large standard deviation in their key generation algorithm.

We show in the following subsection, how one can modify slightly BI-NTRU-Product to design a new IND-CPA secure variant of NTRUEncrypt whose security is entirely related to Ring-LWE or its variants.

3.2 BI-NTRU-LPR

We describe in this subsection the BI-NTRU-LPR encryption scheme which uses some properties of the LPR scheme [16]. The key generation algorithm is very similar to the one of BI-NTRU-Product scheme; we make a few modifications in order to get public key distribution uniform by the hardness of the decision version of Ring-LWE instead of the framework of Banks and Shparlinski [1].

The key generation algorithm of our BI-NTRU-LPR scheme is described as follows.

Algorithm 4. BI-NTRU-LPR.KeyGen()

Output: A public key **pk**
Output: A private key **sk**
1 Choose uniformly at random $B \in R_q$;
2 Choose randomly a "small" polynomial $r \in R_q$ from χ;
3 Choose randomly a "small" polynomial $g \in R_q$ from χ (if g is not invertible modulo p, resample);
4 Choose randomly two "small" polynomials $e_1, e_2 \in R_q$ from χ;
5 Compute $G_1 = -gB + e_1 \pmod{q}$, $G_2 = rB + e_2 \pmod{q}$;
6 The public key **pk** is (G_1, G_2) and the private key **sk** $= (r, g)$;

Remark 4. In practice, as in the key creation of BI-NTRU-Product, one can choose uniformly at random a "small" polynomial $f \in R_q$ from χ and set $g = 1 + pf$.

Lemma 5. *The public key of BI-NTRU-LPR scheme is uniformly distributed assuming that Ring-LWE$_{HNF}$ and its variant Ring-LWE$_p$ are hard.*

Proof: Suppose the Ring-LWE is hard. Since r and e_2 are sampled from the Ring-LWE error distribution χ and B is sampled uniformly at random in R_q, one cannot distinguish G_2 to an element uniformly selected in R_q by Ring-LWE$_{HNF}$. By using the same method, one can't also distinguish G_2 to a uniform element assuming the Ring-LWE$_p$ is hard.   □

The following encryption algorithm for BI-NTRU-LPR scheme is the same as encryption algorithm for BI-NTRU-Product.

Algorithm 5. BI-NTRU-LPR.Enc(p_k, m)

Input: A message $\mathbf{m} \in \mathcal{L}_m$, the public key **pk** $= (G_1, G_2)$
Output: the ciphertext $C = (C_1, C_2)$ of m
1 Choose randomly three "small" polynomials $u, v, e \in R_q$ from χ;
2 Compute $C_1 = p(uG_1 + v) \in R_q$ and $C_2 = p(uG_2 + e) + m \in R_q$;
3 **return** $C = (C_1, C_2) \in R_q \times R_q$

The decryption algorithm for BI-NTRU-LPR is described as follows.

Algorithm 6. BI-NTRU-LPR.Dec(s_k, C)

Input: The ciphertext $C = (C_1, C_2)$, the private key **sk** $= (r, g)$
Output: the decryption of C
1 Compute $D_1 = gC_2 + rC_1 \pmod{q}$;
2 Compute $D_2 = D_1 \pmod{p}$;
3 **return** $D = g_p D_2 \pmod{p}$

Lemma 6. *If* $|gm+p(ge+rv+rue_1+gue_2)|_\infty < q$*, then the decryption algorithm for BI-NTRU-LPR output* m*. i.e* $D = m$*.*

Proof: Suppose $|gm + p(ge + rv + rue_1 + gue_2|_\infty < q$; We have

$$D_1 = gC_2 + rC_1 \pmod{q}$$
$$= (gm + gupG_2 + gep) + (rupG_1 + rvp) \pmod{q}$$
$$= gm + gurpB + gue_2p + gep - gurpB + rue_1p + rvp \pmod{q}$$
$$= gm + p(ge + rv + rue_1 + gue_2) \pmod{q}$$

Since $|gm + p(ge + rv + rue_1 + gue_2)|_\infty < q$ then $D_1 = gm + p(ge + rv + rue_1 + gue_2) \pmod{q} = gm + p(ge + rv + rue_1 + gue_2)$. Hence $D_2 = D_1 \pmod{p} = gm + p(ge + rv + rue_1 + gue_2) \pmod{p} = gm \pmod{p}$. Finally $D = g_pD_2 \pmod{p} = (g_pg)m \pmod{p} = m$. □

Theorem 2. *Our BI-NTRU-LPR is IND-CPA secure assuming that the Ring-LWE is hard.*

Proof: It is similar to the proof of Theorem 1. □

BI-NTRU-Product vs. BI-NTRU-LPR: The main difference between these two encryption schemes is in the key generation algorithms. The uniformity of the public key of BI-NTRU-Product is not based on any assumption (proven hard problem) while the uniformity of the public key of NTRU-LPR is based on the decision version of Ring-LWE.

BI-NTRU Key Encapsulation Mechanism

We show here, how one can design an IND-CCA secure asymmetric scheme like in many post quantum schemes proposed at NIST [2,5]. We design BI-NTRU-KEM which is a key encapsulation mechanism consisting of the triplet of algorithms (BI-NTRU-KEM.keyGen, BI-NTRU-KEM.Encaps, BI-NTRU-KEM.Decaps). It is derived from the BI-NTRU-LPR scheme which is IND-CPA secure. We use the transformation of Hofheinz *et al.* [8] which is an alternative variant of the Fujisaki-Okamoto (FO) [7] and Dent [6] frameworks that allows a generic conversion from weak asymmetric encryption scheme to an asymmetric encryption scheme that is chosen-ciphertext secure in the random oracle model. The FO transformation require the underlying asymmetric encryption scheme to be perfectly correct, i.e., not having decryption failure while the ones of Hofheinz *et al.* [8] work for correct and $\delta-$correct asymmetric schemes. Let $\mathcal{G}$ and $\mathcal{H}$ two hash functions. First, we slightly modified the encryption of BI-NTRU-Product as follows.

Algorithm 7. BI-NTRU-LPR.Enc$'(p_k, m; r_s)$

---

1 (C_1, C_2):=BI-NTRU-LPR.Enc(p_k, m) ;   /* but we replace χ by
  $\chi(r_s)$ */
2 **return** (C_1, C_2)

The key generation algorithm and the encapsulation algorithm of BI-NTRU-KEM are described as follows.

Algorithm 8. BI-NTRU-KEM.KeyGen()

Output: A public key **pk**
Output: A private key **sk**
1 $(p'_k, s'_k) :=$ BI-NTRU-LPR.KeyGen();
2 Choose randomly $z \in \mathcal{L}_m$;
3 **return** $(\mathbf{pk} = p'_k, \mathbf{sk} = (s'_k, z))$

Algorithm 9. BI-NTRU-KEM.Encaps(p_k)

Input: $p_k = (G_1, G_2)$
Output: A ciphertext c and a key K
1 Choose randomly $m \in \mathcal{L}_m$;
2 $(\tilde{K}, r_s) := \mathcal{G}((G_1, G_2), m)$;
3 $c :=$ BI-NTRU-LPR.Encrypt$'(p_k, m; r_s)$; $// \ c = (C_1, C_2)$
4 $K := \mathcal{H}(\tilde{K}, c)$;
5 **return** (c, K)

The decapsulation algorithm for BI-NTRU-KEM is described as follows.

Algorithm 10. BI-NTRU-KEM.Decaps(s_k, c)

Input: the private key **sk**, the ciphertext $c = (C_1, C_2)$
Output: A key K
1 Parse $\mathbf{sk} = (s'_k, z)$;
2 $m' :=$ BI-NTRU-LPR.Decrypt(s'_k, c);
3 $(\tilde{K}', r'_s) := \mathcal{G}(p_k, m')$;
4 $(C'_1, C'_2) :=$ BI-NTRU-LPR.Encrypt(p_k, m');
5 **if** $(C'_1, C'_2) = (C_1, C_2)$ **then**
6 | $K = \mathcal{H}(\tilde{K}', c)$
7 **else**
8 | $K := \mathcal{H}(z, c)$
9 **end**
10 **return** K

---

4 Conclusion

In this paper, we have designed two encryption schemes BI-NTRU-Product and BI-NTRU-LPR which are IND-CPA secure. We have also discussed how to transform BI-NTRU-LPR to an IND-CCA secure key encapsulation mechanism by using the framework of Hofheinz *et al.* (15th Int. Conf., TCC 2017).

Acknowledgment. The authors would like to thank anonymous reviewers for their helpful comments and suggestions, and Igor E. Shparlinski for many online discussions.

A   Appendix

Characters on Finite Abelian Groups and Cauchy Inequality

Definition 13 (additive character [13]**).** *Let G be a finite (additive) abelian group. An additive character on G is a function $\chi : G \to \mathbb{C}$ such that*

$$\chi(g_1 + g_2) = \chi(g_1).\chi(g_2) \ and \ |\chi(g)| = 1$$

for any $g, g_1, g_2 \in G$. The character χ_0 with $\chi_0(g) = 1$ for all $g \in G$ is called the trivial character.

Remark 5. – One can define also in a similar way a multiplicative character with respect to a multiplicative character.
– The set of characters on G, together with the multiplication $(\chi_1 \chi_2)(g) = \chi_1(g)\chi_2(g)$ is an abelian group called character group of G, and denoted by $\bar{G}$.

Theorem 3. *Let G be a finite abelian group. Then there exists an isomorphism from G to $\bar{G}$. In particular, $|\bar{G}| = |G|$*

Theorem 4 (Orthogonality relations). *Let G be an abelian (additive) group of order n with character group $\bar{G}$ and identity element 0_G.*

1. for each $\chi \in \bar{G}$ we have $\sum_{a \in G} \chi(a) = \begin{cases} n & \text{if } \chi = \chi_0 \\ 0 & \text{if } \chi \neq \chi_0 \end{cases}$

2. for each $a \in G$ we have $\sum_{\chi \in \bar{G}} \chi(a) = \begin{cases} n & \text{if } a = 0_G \\ 0 & \text{if } a \neq 0_G \end{cases}$

Lemma 7 (Cauchy-Schwarz inequality [13]**).** *The inequality $\sum_{i=1}^{N} A_i B_i \leq \left(\sum_{i=1}^{N} A_i^\alpha \right)^{1/\alpha} \left(\sum_{i=1}^{N} B_i^\beta \right)^{1/\beta}$ holds for any two sequences of positive numbers A_i, B_i for $i = 1, 2, \ldots, N$ and any two positive numbers α, β with $1/\alpha + 1/\beta = 1$.*

Proof of the Uniformity of the Distribution of the Public Keys

Banks and Shparlinski in [1] show that for almost all $Q \in R_q^\times$ the set $\{Q.\phi : \phi \in \mathcal{L}\}$ is uniformly distributed where $R_q = \mathbb{Z}_q[X]/(\Phi(X))$ with Φ a square free polynomials of degree N and $\mathcal{L}$ is a subset of R_q. We will show that their result remains true for almost all $Q \in R_q$ i.e for almost all $B \in R_q$, the set $\{G = B.\phi : \phi \in \mathcal{L}_t\}$ is uniformly distributed.

Notice that Theorem 1 in [1] remains true if we replace $R_q^\times$ by R_q. For the sake of completeness, let us prove it:

Let $\Phi(X) = \Psi_1(X) \ldots \Psi_r(X)$ be the complete factorization of $\Phi(X)$ into square free polynomial in the ring $R_q = \mathbb{Z}_q[x]/(\Phi(X))$. Since $\Phi(X)$ is square-free in R_q, then all of these factors are pairwise distinct.

We recall that $\mathbb{F}_q[X]/\Psi(X) \cong F_{q^m}$ for any irreducible polynomials $\Psi(X) \in \mathbb{F}_q[X]$ with $\deg\Psi = m$ For each $j = 1, \ldots, r$, we fix a root α_j of $\Psi_j(X)$ and denote

$$\mathbb{K}_j = \mathbb{F}_{q^{n_j}} = \mathbb{F}_q(\alpha_j) \cong \mathbb{F}_q[X]/\Psi_j(X)$$

where $n_j = \deg\Psi_j$. For each j, let $Tr_j(z) = \sum_{k=0}^{n_j-1} z^{q^{n_j}}$ be the trace of $z \in \mathbb{K}_j$ to $\mathbb{F}_q$. We denote by A the direct product of fields, we recall that $A = \mathbb{K}_1 \times \mathbb{K}_2 \times \ldots \times \mathbb{K}_r$. Consider the map $G_\alpha : R_q \longrightarrow A : f \mapsto a_f = (f(\alpha_1), f(\alpha_2), \ldots, f(\alpha_r))$

One can show easily that the map G_α is an isomorphism. Which implies that $R_q \cong A$. For every vector $a = (a_1, \ldots, a_r) \in A$, let χ_a be the character of R_q given by

$$\chi_a(f) = \prod_{j=1}^{r} e(Tr_j(a_j f(\alpha_j))), \qquad f \in R_q \tag{1}$$

where $e(z) = \exp(2i\pi z/q)$. It is easy to show that $\{\chi_a, a \in A\}$ is the complete set of additive characters of R_q.

We have the following lemma

Lemma 8. *Let* $a = (a_1, \ldots, a_r) \in A$ *and let* $\mathcal{J} = \{1, \ldots, r\}$ *be the set of* j *with* $a_j \neq 0$. *Define* $W_a(\mathcal{L}_t) = \sum_{B \in R_q} \left| \sum_{\phi \in \mathcal{L}_t} \chi_a(B.\phi) \right|$ *for* $a \in A$. *Then we have* $W_a(\mathcal{L}_t) \leq q^N |\mathcal{L}_t|^{1/2} \prod_{j \notin \mathcal{J}} q^{n_j/2}$

Proof: Using the Cauchy-Schwarz inequality Lemma 7 (with $A_i = 1$ and $B_i = \left| \sum_{\phi \in \mathcal{L}_t} \chi_a(B.\phi) \right|$), we derive

$$(W_a(\mathcal{L}_t))^2 = \left(\sum_{B \in R_q} \left| \sum_{\phi \in \mathcal{L}_t} \chi_a(B.\phi) \right| \right)^2$$

$$\leq |R_q| \sum_{B \in R_q} \left| \sum_{\phi \in \mathcal{L}_t} \chi_a(B.\phi) \right|^2$$

$$= |R_q| \sum_{B \in R_q} \sum_{\phi_1, \phi_2 \in \mathcal{L}_t} \chi_a(B.(\phi_1 - \phi_2))$$

$$\leq |R_q| \sum_{\phi_1, \phi_2 \in \mathcal{L}_t} \sum_{B \in R_q} \prod_{j=1}^{r} e(Tr_j(a_j B(\alpha_j)(\phi_1(\alpha_j) - \phi_2(\alpha_j))))$$

$$\leq |R_q| \sum_{\phi_1, \phi_2 \in \mathcal{L}_t} \prod_{j=1}^{r} \sum_{x_j \in \mathbb{K}_j} e(Tr_j(a_j x_j(\phi_1(\alpha_j) - \phi_2(\alpha_j))))$$

$$= |R_q| \prod_{j \notin \mathcal{J}} q^{n_j} \sum_{\phi_1, \phi_2 \in \mathcal{L}_t} \prod_{j \in \mathcal{J}} \sum_{x_j \in \mathbb{K}_j} e(Tr_j(a_j x_j(\phi_1(\alpha_j) - \phi_2(\alpha_j))))$$

We have the following:

- if $\phi_1(\alpha_j) \neq \phi_2(\alpha_j)$ for some $j \in \mathcal{J}$, the product vanishes
- otherwise $\prod_{j \in \mathcal{J}} \sum_{x_j \in \mathbb{K}_j} e(Tr_j(a_j x_j(\phi_1(\alpha_j) - \phi_2(\alpha_j)))) = \prod_{j \in \mathcal{J}} q^{n_j}$

Hence,

$$(W_a(\mathcal{L}_t))^2 \leq |R_q| \prod_{j \notin \mathcal{J}} q^{n_j} \sum_{\substack{\phi_1, \phi_2 \in \mathcal{L}_t \\ \phi_1(\alpha_j) = \phi_2(\alpha_j) \forall j}} \prod_{j \in \mathcal{J}} q^{n_j}$$

$$= q^N \prod_{j \notin \mathcal{J}} q^{n_j} \prod_{j \in \mathcal{J}} q^{n_j} \sum_{\substack{\phi_1, \phi_2 \in \mathcal{L}_t \\ \phi_1(\alpha_j) = \phi_2(\alpha_j) \forall j}} 1$$

$$= q^{2N} \sum_{\substack{\phi_1, \phi_2 \in \mathcal{L}_t \\ \phi_1(\alpha_j) = \phi_2(\alpha_j) \forall j}} 1$$

Since $\{\Psi_j | j = 1, \ldots, r\}$ are irreducible polynomials, the condition $\phi_1(\alpha_j) = \phi_2(\alpha_j))$ is equivalent to $\Psi_j | (\phi_1 - \phi_2)$. Hence $(W_a(\mathcal{L}_t))^2 \leq q^{2N} M(\mathcal{J})$ where $M(\mathcal{J}) = \sum_{\substack{\phi_1, \phi_2 \in \mathcal{L}_t \\ \phi_1(\alpha_j) = \phi_2(\alpha_j) \forall j}} 1$ is the number of pairs $\phi_1, \phi_2 \in \mathcal{L}_t$ with $\phi_1 \equiv \phi_2 \pmod{\prod_{j \in \mathcal{J}} \Psi_j}$. For each $\phi_1 \in \mathcal{L}_t$ there are at most $q^N \prod_{j \in \mathcal{J}} q^{-n_j} = \prod_{j \notin \mathcal{J}} q^{n_j}$ such values for ϕ_2. Consequently $M(\mathcal{J}) \leq |\mathcal{L}_t| \prod_{j \notin \mathcal{J}} q^{n_j}$ and the lemma follows. □

Theorem 5. *Given polynomials $S \in R_q$ and $B \in R_q$, a set $\mathcal{L}_t \subset R_q$, and an integer d. We denote by $N_d(S, B, \mathcal{L}_t)$ the number of polynomials $\phi \in \mathcal{L}_t$ such that the inequality $\deg(S - B\phi) < d$ holds. Then for $q > 4$, the following bound holds.*

$$\frac{1}{|R_q|} \sum_{B \in R_q} \left| N_d(S, B, \mathcal{L}_t) - \frac{|\mathcal{L}_t|}{q^{N-d}} \right| \leq 3^{Nq^{-1/2}} |\mathcal{L}_t|^{1/2}$$

Proof: We know that $N_d(S, B, \mathcal{L}_t) = q^{-d} T_d(S, B, \mathcal{L}_t)$, where $T_d(S, B, \mathcal{L}_t)$ is the number of representations $B.\phi = S + \psi_1 - \psi_2$ with $\phi \in \mathcal{L}_t$ and polynomials $\psi_1, \psi_2 \in R_q$ of degree at most $d - 1$. We have

$$T_d(S, B, \mathcal{L}_t) = \frac{1}{q^N} \sum_{\phi \in \mathcal{L}_t} \sum_{\substack{\psi_1, \psi_2 \in R_q \\ \deg(\psi_1), \deg(\psi_1) \leq d-1}} \sum_{a \in A} \chi_a(B.\phi - S - \psi_1 + \psi_2)$$

$$= \frac{1}{q^N} \sum_{\phi \in \mathcal{L}_t} \sum_{\substack{\psi_1, \psi_2 \in R_q \\ \deg(\psi_1), \deg(\psi_1) \leq d-1}} \sum_{a \in A} \chi_a(B.\phi) \chi_a(-S) \chi_a(-\psi_1 + \psi_2))$$

$$= \frac{1}{q^N} \sum_{a \in A} \chi_a(-S) \sum_{\phi \in \mathcal{L}_t} \chi_a(B.\phi) \sum_{\substack{\psi_1, \psi_2 \in R_q \\ \deg(\psi_1), \deg(\psi_1) \leq d-1}} \chi_a(\psi_2 - \psi_1)$$

$$= \frac{1}{q^N} \sum_{a \in A} \chi_a(-S) \sum_{\phi \in \mathcal{L}_t} \chi_a(B.\phi) \left| \sum_{\substack{\psi \in R_q \\ \deg(\psi) \leq d-1}} \chi_a(\psi) \right|^2$$

$$= \frac{1}{q^N} \sum_{\substack{a \in A \\ a \neq 0}} \chi_a(-S) \sum_{\phi \in \mathcal{L}_t} \chi_a(B.\phi) \left| \sum_{\substack{\psi \in R_q \\ \deg(\psi) \leq d-1}} \chi_a(\psi) \right|^2 + q^{2d-N} |\mathcal{L}_t|$$

For any nonempty set $\mathcal{J} \subseteq \{1, \dots, r\}$, let $A_{\mathcal{J}}$ be the subset of A consisting of all $a = (a_1, \dots, a_r)$ such that $a_j = 0$ whenever $j \notin \mathcal{J}$. Then we obtain

$$\left| T_d(S, B, \mathcal{L}_t) - \frac{|\mathcal{L}_t|}{q^{N-2d}} \right| \leq \frac{1}{q^N} \sum_{\substack{\mathcal{J} \neq \emptyset \\ \mathcal{J} \subseteq \{1,\dots,r\}}} \sum_{\substack{a \neq 0 \\ a \in A_{\mathcal{J}}}} \left| \sum_{\phi \in \mathcal{L}_t} \chi_a(B.\phi) \right| \left| \sum_{\substack{\psi \in R_q \\ \deg(\psi) \leq d-1}} \chi_a(\psi) \right|^2$$

Applying Lemma 8, it follows that

$$\sum_{Q \in R_q} \left| T_d(S, B, \mathcal{L}_t) - \frac{|\mathcal{L}_t|}{q^{N-2d}} \right| \leq |\mathcal{L}_t|^{1/2} \sum_{\substack{\mathcal{J} \neq \emptyset \\ \mathcal{J} \subseteq \{1,\dots,r\}}} \prod_{j \notin \mathcal{J}} q^{n_j/2} \sum_{\substack{a \neq 0 \\ a \in A_{\mathcal{J}}}} \left| \sum_{\substack{\psi \in R_q \\ \deg(\psi) \leq d-1}} \chi_a(\psi) \right|^2$$

We have

$$\sum_{\substack{a \neq 0 \\ a \in A_{\mathcal{J}}}} \left| \sum_{\substack{\psi \in R_q \\ \deg(\psi) \leq d-1}} \chi_a(\psi) \right|^2 = -q^{2d} + \sum_{a \in A_{\mathcal{J}}} \left| \sum_{\substack{\psi \in R_q \\ \deg(\psi) \leq d-1}} \chi_a(\psi) \right|^2$$

$$= -q^{2d} + \sum_{a \in A_{\mathcal{J}}} \sum_{\substack{\phi, \psi \in R_q \\ \deg(\phi), \deg(\psi) \leq d-1}} \chi_a(\phi - \psi)$$

$$= -q^{2d} + \sum_{\substack{\phi, \psi \in R_q \\ \deg(\phi), \deg(\psi) \leq d-1}} \sum_{a \in A_{\mathcal{J}}} \chi_a(\phi - \psi)$$

$$= -q^{2d} + \sum_{\substack{\phi, \psi \in R_q \\ \deg(\phi), \deg(\psi) \leq d-1}} \sum_{a \in A_{\mathcal{J}}} \prod_{j \in \mathcal{J}} e(Tr_j(a_j(\phi(\alpha_j) - \psi(\alpha_j))))$$

$$= -q^{2d} + U \prod_{j \in \mathcal{J}} q^{n_j}$$

where U is the number of pairs of $\phi, \psi \in R_q$ with $\deg(\phi), \deg(\psi) \leq d-1$ and such that $\phi(\alpha_j) = \psi(\alpha_j)$ for all $j \in \mathcal{J}$. Since this condition is equivalent to the polynomial congruence $\phi(X) \equiv \psi(X) \pmod{\prod_{j \in \mathcal{J}} \Psi_j(X)}$ we derive that

$$U = \begin{cases} q^{2d} \prod_{j \in \mathcal{J}} q^{-n_j}, & \text{if } d \geq \sum_{j \in \mathcal{J}} n_j \\ q^d, & \text{otherwise} \end{cases}$$

Hence, in either case $0 \leq -q^{2d} + U \prod_{j \in \mathcal{J}} q^{n_j} \leq q^d \prod_{j \in \mathcal{J}} q^{n_j}$ and consequently

$$\sum_{\substack{a \neq 0 \\ a \in A_{\mathcal{J}}}} \left| \sum_{\substack{\psi \in R_q \\ \deg(\psi) \leq d-1}} \chi_a(\psi) \right|^2 \leq q^d \prod_{j \in \mathcal{J}} q^{n_j}$$

Therefore, we have

$$\frac{1}{|R_q|} \sum_{Q \in R_q} \left| T_d(S, B, \mathcal{L}_t) - \frac{|\mathcal{L}_t|}{q^{N-2d}} \right| \leq |R_q|^{-1} |\mathcal{L}_t|^{1/2} q^d \sum_{\substack{\mathcal{J} \neq \emptyset \\ \mathcal{J} \subseteq \{1,\ldots,r\}}} \prod_{j \notin \mathcal{J}} q^{n_j/2} \prod_{j \in \mathcal{J}} q^{n_j}$$

$$\leq |\mathcal{L}_t|^{1/2} q^{d-N/2} \sum_{\substack{\mathcal{J} \neq \emptyset \\ \mathcal{J} \subseteq \{1,\ldots,r\}}} \prod_{j \in \mathcal{J}} q^{n_j/2}$$

$$\leq |\mathcal{L}_t|^{1/2} q^{d-N/2} \left(\prod_{j=1}^{r} (1 + q^{n_j/2}) - 1 \right)$$

$$< |\mathcal{L}_t|^{1/2} q^{d-N/2} \prod_{j=1}^{r} (1 + q^{n_j/2})$$

$$= |\mathcal{L}_t|^{1/2} q^d \prod_{j=1}^{r} (1 + q^{-n_j/2})$$

Since $(1 + x) \leq 3^x$ for every $0 \leq x \leq 1$, and each term $q^{n_j/2} > 1$, we have

$$\prod_{j=1}^{r} (1 + q^{-n_j/2}) \leq \prod_{j=1}^{r} 3^{q^{-n_j/2}} \leq \prod_{j=1}^{r} 3^{q^{-1/2}} \leq 3^{Nq^{-1/2}}$$

Consequently

$$\frac{1}{|R_q|} \sum_{Q \in R_q} \left| T_d(S, B, \mathcal{L}_t) - \frac{|\mathcal{L}_t|}{q^{N-2d}} \right| \leq 3^{Nq^{-1/2}} |\mathcal{L}_r|^{1/2} q^d$$

Since $T_d(S, B, \mathcal{L}_t) = q^d N_d(S, B, \mathcal{L}_t)$, we have

$$\frac{1}{|R_q|} \sum_{B \in R_q} \left| N_d(S, B, \mathcal{L}_t) - \frac{|\mathcal{L}_t|}{q^{N-d}} \right| \leq 3^{Nq^{-1/2}} |\mathcal{L}_t|^{1/2}$$

$\square$

References

1. Banks, W.D., Shparlinski, I.E.: A Variant of NTRU with non-invertible polynomials. In: Menezes, A., Sarkar, P. (eds.) INDOCRYPT 2002. LNCS, vol. 2551, pp. 62–70. Springer, Heidelberg (2002). https://doi.org/10.1007/3-540-36231-2_6
2. Bos, J.W., et al.: CRYSTALS - Kyber: a CCA-secure module-lattice-based KEM. In: EuroS&P 2018, pp. 353–367 (2018)
3. Bernstein, D.J., Chuengsatiansup, C., Lange, T., van Vredendaal, C.: NTRU prime. Cryptology ePrint Archive, Report 2016/461 (2016)
4. Coppersmith, D., Shamir, A.: Lattice attacks on NTRU. In: Fumy, W. (ed.) EUROCRYPT 1997. LNCS, vol. 1233, pp. 52–61. Springer, Heidelberg (1997). https://doi.org/10.1007/3-540-69053-0_5

5. D'Anvers, J.-P., Karmakar, A., Sinha Roy, S., Vercauteren, F.: Saber: module-LWR based key exchange, CPA-secure encryption and CCA-secure KEM. In: Joux, A., Nitaj, A., Rachidi, T. (eds.) AFRICACRYPT 2018. LNCS, vol. 10831, pp. 282–305. Springer, Cham (2018). https://doi.org/10.1007/978-3-319-89339-6_16

6. Dent, A.W.: A designer's guide to KEMs. In: Paterson, K.G. (ed.) Cryptography and Coding 2003. LNCS, vol. 2898, pp. 133–151. Springer, Heidelberg (2003). https://doi.org/10.1007/978-3-540-40974-8_12

7. Fujisaki, E., Okamoto, T.: Secure integration of asymmetric and symmetric encryption schemes. In: Wiener, M. (ed.) CRYPTO 1999. LNCS, vol. 1666, pp. 537–554. Springer, Heidelberg (1999). https://doi.org/10.1007/3-540-48405-1_34

8. Hofheinz, D., Hövelmanns, K., Kiltz, E.: A modular analysis of the Fujisaki-Okamoto transformation. In: Kalai, Y., Reyzin, L. (eds.) TCC 2017, Part I. LNCS, vol. 10677, pp. 341–371. Springer, Cham (2017). https://doi.org/10.1007/978-3-319-70500-2_12

9. Hoffstein, J., Pipher, J., Silverman, J.H.: NTRU: a ring-based public key cryptosystem. In: Buhler, J.P. (ed.) ANTS 1998. LNCS, vol. 1423, pp. 267–288. Springer, Heidelberg (1998). https://doi.org/10.1007/BFb0054868

10. Howgrave-Graham, N., Silverman, J.H., Singer, A., Whyte, W.: NAEP: provable security in the presence of decryption failures. IACR Cryptology ePrint Archive, 2003:172 (2003)

11. Howgrave-Graham, N., Silverman, J.H., Whyte, W.: Choosing parameter sets for NTRUEncrypt with NAEP and SVES-3. IACR Cryptology ePrint Archive (2005). https://eprint.iacr.org/2005/045. ANTS-III, Springer LNCS vol. 1423, pp. 267–288 (1998)

12. Hülsing, A., Rijneveld, J., Schanck, J., Schwabe, P.: High-speed key encapsulation from NTRU. In: Fischer, W., Homma, N. (eds.) CHES 2017. LNCS, vol. 10529, pp. 232–252. Springer, Cham (2017). https://doi.org/10.1007/978-3-319-66787-4_12. http://cryptojedi.org/papers/#ntrukem

13. Konyagin, S., Shparlinski, I.: Character Sums with Exponential Functions and Their Applications. Cambridge University Press, Cambridge (1994)

14. López-Alt, A., Tromer, E., Vaikuntanathan, V.: On-the-fly multiparty computation on the cloud via multikey fully homomorphic encryption. In: Proceedings of the Forty-Fourth Annual ACM Symposium on Theory of Computing (STOC 2012), pp. 1219–1234. ACM, New York (2012). https://doi.org/10.1145/2213977.2214086

15. Lyubashevsky, V., Peikert, C., Regev, O.: On ideal lattices and learning with errors over rings. In: Gilbert, H. (ed.) EUROCRYPT 2010. LNCS, vol. 6110, pp. 1–23. Springer, Heidelberg (2010). https://doi.org/10.1007/978-3-642-13190-5_1

16. Lyubashevsky, V., Peikert, C., Regev, O.: On ideal lattices and learning with errors over rings. J. ACM **60**(6), 43:1–43:35 (2013). Preliminary version in EUROCRYPT'10

17. National Institute of Standards and Technology: Announcing request for nominations for public-key post-quantum cryptographic algorithms (2016). https://csrc.nist.gov/news/2016/public-key-post-quantum-cryptographic-algorithms

18. Peikert, C., Waters, B.: Lossy trapdoor functions and their applications. In: STOC 2008, pp. 187–196 (2008)

19. Regev, O.: On lattices, learning with errors, random linear codes, and cryptography. J. ACM **56**(6), 1–40 (2005). Preliminary version in STOC

20. Shor, P.: Algorithms for quantum computation: discrete logarithms and factoring. In: Proceedings of 35th Annual Symposium on Foundations of Computer Science, pp. 124–134. IEEE (1994)

21. Shor, P.-W.: Polynomial-time algorithms for prime factorization and discrete logarithms on a quantum computer. SIAM J. Comput. **26**(5), 1484–1509 (1997). Extended abstract in FOCS-94

22. Stehlé, D., Steinfeld, R.: Making NTRU as secure as worst-case problems over ideal lattices. In: Paterson, K.G. (ed.) EUROCRYPT 2011. LNCS, vol. 6632, pp. 27–47. Springer, Heidelberg (2011). https://doi.org/10.1007/978-3-642-20465-4_4

23. Steinfeld, R., Ling, S., Pieprzyk, J., Tartary, C., Wang, H.: NTRUCCA: how to strengthen NTRUEncrypt to chosen-ciphertext security in the standard model. In: Fischlin, M., Buchmann, J., Manulis, M. (eds.) PKC 2012. LNCS, vol. 7293, pp. 353–371. Springer, Heidelberg (2012). https://doi.org/10.1007/978-3-642-30057-8_21

Author Index

Printed in the United States
By Bookmasters